Lecture Notes in Networks and Systems

Volume 148

The series "Lecture Notes in Networks and Systems" publishes the latest developments in Networks and Systems—quickly, informally and with high quality. Original research reported in proceedings and post-proceedings represents the core of LNNS.

Volumes published in LNNS embrace all aspects and subfields of, as well as new challenges in, Networks and Systems.

The series contains proceedings and edited volumes in systems and networks, spanning the areas of Cyber-Physical Systems, Autonomous Systems, Sensor Networks, Control Systems, Energy Systems, Automotive Systems, Biological Systems, Vehicular Networking and Connected Vehicles, Aerospace Systems, Automation, Manufacturing, Smart Grids, Nonlinear Systems, Power Systems, Robotics, Social Systems, Economic Systems and other. Of particular value to both the contributors and the readership are the short publication timeframe and the world-wide distribution and exposure which enable both a wide and rapid dissemination of research output.

The series covers the theory, applications, and perspectives on the state of the art and future developments relevant to systems and networks, decision making, control, complex processes and related areas, as embedded in the fields of interdisciplinary and applied sciences, engineering, computer science, physics, economics, social, and life sciences, as well as the paradigms and methodologies behind them.

**** Indexing: The books of this series are submitted to ISI Proceedings, SCOPUS, Google Scholar and Springerlink ****

More information about this series at http://www.springer.com/series/15179

Kanad Ray · Krishna Chandra Roy ·
Sandeep Kumar Toshniwal ·
Harish Sharma · Anirban Bandyopadhyay
Editors

Proceedings of International Conference on Data Science and Applications

ICDSA 2019

Springer

Editors
Kanad Ray
Amity School of Applied Sciences
Amity University Rajasthan
Jaipur, Rajasthan, India

Krishna Chandra Roy
Kautilya Institute of Technology
and Engineering
Jaipur, Rajasthan, India

Sandeep Kumar Toshniwal
Kautilya Institute of Technology
and Engineering
Jaipur, Rajasthan, India

Harish Sharma
Department of Computer Science
and Engineering
Rajasthan Technical University
Kota, Rajasthan, India

Anirban Bandyopadhyay
Surface Characterization Group
Nano Characterization Unit
Advanced Key Technologies Division
National Institute for Materials Science
Tsukuba, Japan

ISSN 2367-3370 ISSN 2367-3389 (electronic)
Lecture Notes in Networks and Systems
ISBN 978-981-15-7560-0 ISBN 978-981-15-7561-7 (eBook)
https://doi.org/10.1007/978-981-15-7561-7

This Springer imprint is published by the registered company Springer Nature Singapore Pte Ltd.
The registered company address is: 152 Beach Road, #21-01/04 Gateway East, Singapore 189721, Singapore

Preface

The International Conference on Data Science and Application (ICDSA-2019) was organized at Kautilya Institute of Technology and Engineering, Jaipur, during December 2–3, 2019. The conference was sponsored by the TEQIP III and Rajasthan Technical University, Kota.

The objective of ICDSA-2019 was to provide a common platform to researcher, academicians, scientists and industrialists working in the area of data science.

This book that we wish to bring forth with great pleasure is an encapsulation of research papers presented during the two-day international conference. We hope that the efforts would be found informative and interesting to those who are keen to learn the technology that addresses the challenges of the exponentially growing information in the core and allied fields of data science.

We are thankful to the authors of the research papers for their valuable contribution to the conference and for bringing forth significant research and literature across the field of data science. The editors also express their sincere gratitude to ICDSA-2019 patrons, plenary speakers, keynote speakers, program committee members, international advisory committee and local organizing committee, sponsors and student volunteers without whose support the quality of the conference would not have been the same. We are thankful to Prof. Dhirendra Mathur, Er. C. K. Bafna, Er. Vimal Goleccha, Dr. Nilanjan Dey and Dr. Satyajit Sahu.

We would like to express our sincere gratitude to Prof. R. A. Gupta, VC, Rajasthan Technical University, for being the chief patron and taking out time for a plenary talk. We would also like to accord our special thanks to Prof. S. L. Dhingra and Prof. Ved Vyas Dwivedi for plenary deliberations.

We express our heartfelt indebtedness to Prof. Preecha Yupapin,Vietnam, Prof. Ram Prasad Khatiwada, Nepal, Prof. Anirban Bandyopadhyay, Japan, and Prof. M. Shamim Kaiser, Bangladesh, for their gracious presence during the conference and delivering their plenary talks.

We express special thanks to Springer and its team for valuable support in the publication of proceedings. With great fervor, we wish to bring together researchers and practitioners in the field of data science year after year to explore new avenues in the field.

Jaipur, India	Prof. Kanad Ray
Jaipur, India	Prof. Krishna Chandra Roy
Jaipur, India	Er. Sandeep Kumar Toshniwal
Kota, India	Dr. Harish Sharma
Tsukuba, Japan	Dr. Anirban Bandyopadhyay

Contents

A Space-Time-Topology-Prime, stTS Metric for a Self-operating Mathematical Universe Uses Dodecanion Geometric Algebra of 2-20 D Complex Vectors 1
Pushpendra Singh, Pathik Sahoo, Komal Saxena, Subrata Ghosh, Satyajit Sahu, Kanad Ray, Daisuke Fujita, and Anirban Bandyopadhyay

An Effective Filtering Process for the Noise Suppression in Eye Movement Signals 33
Sergio Mejia-Romero, J. Eduardo Lugo, Delphine Bernardin, and Jocelyn Faubert

Trust IoHT: A Trust Management Model for Internet of Healthcare Things 47
Naznin Hossain Esha, Mst. Rubayat Tasmim, Silvia Huq, Mufti Mahmud, and M. Shamim Kaiser

ACO-Based Control Strategy in Interconnected Thermal Power System for Regulation of Frequency with HAE and UPFC Unit 59
K. Jagatheesan, B. Anand, Nilanjan Dey, Amira S. Ashour, Mahdi Khosravy, and Rajesh Kumar

Cryptosystem Based on Triple Random Phase Encoding with Chaotic Henon Map 73
Archana, Sachin, and Phool Singh

Different Loading of Distributed Generation on IEEE 14-Bus Test System to Find Out the Optimum Size of DG to Allocation in Transmission Network 85
Rakesh Bhadani and K. C. Roy

Kidney Care: Artificial Intelligence-Based Mobile Application for Diagnosing Kidney Disease 99
Zarin Subah Shamma, Israt Jahan Rumman, Ali Mual Raji Saikot, S. M. Salim Reza, Md. Maynul Islam, Mufti Mahmud, and M. Shamim Kaiser

A Framework to Evaluate and Classify the Clinical-Level EEG Signals with Epilepsy 111
Linkon Chowdhury, Bristy Roy Chowdhury, V. Rajinikanth, and Nilanjan Dey

Designing of UWB Monopole Antenna with Triple Band Notch Characteristics at WiMAX/C-Band/WLAN 123
S. K. Vijay, M. R. Ahmad, B. H. Ahmad, S. Rawat, P. Singh, K. Ray, and A. Bandyopadhyay

The Dynamic Performance of Gaze Movement, Using Spectral Decomposition and Phasor Representation 133
Sergio Mejia-Romero, J. Eduardo Lugo, Delphine Bernardin, and Jocelyn Faubert

Novel Hairpin Band-Pass Filter Using Tuning Stub 145
Sonu Jain, Taniya Singh, Ajay Yadav, and M. D. Sharma

Recognition of Faults in Wind-Park-Included Series-Compensated Three-Phase Transmission Line Using Hilbert–Huang Transform 153
Gaurav Kapoor

Simulation of Five-Channel De-multiplexer Using Double-Ring Resonator Photonic Crystal-Based ADF 165
Neha Singh and Krishna Chandra Roy

Optimization of Surface Roughness and Material Removal Rate in Turning of AISI D2 Steel with Coated Carbide Inserts 177
Anil Kumar Yadav and Bhasker Shrivastava

An Approach to Improved MapReduce and Aggregation Pipeline Utilizing NoSQL Technologies 187
Monika and Vishal Shrivastava

Evaluation of Bio-movements Using Nonlinear Dynamics 197
Sergio Mejia-Romero, J. Eduardo Lugo, Delphine Bernardin, and Jocelyn Faubert

An Examination System to Classify the Breast Thermal Images into Early/Acute DCIS Class 209
Nilanjan Dey, V. Rajinikanth, and Aboul Ella Hassanien

Implementation of Hybrid Wind–Solar Energy Conversion Systems 221
Pooja Joshi and K. C. Roy

Accident Prediction Modeling for Yamuna Expressway 241
Parveen Kumar and Jinendra Kumar Jain

Optical Image Encryption Algorithm Based on Chaotic Tinker Bell Map with Random Phase Masks in Fourier Domain 249
Sachin, Archana, and Phool Singh

Fiber Optics Near-Infrared Wavelengths Analysis to Detect the Presence of Liquefied Petroleum Gas 263
H. H. Cerecedo-Núñez, Rosa Ma Rodríguez-Méndez, P. Padilla-Sosa, and J. E. Lugo-Arce

A Novel Approach to Optimize SLAM Using GP-GPU 273
Rohit Mittal, Vibhakar Pathak, and Amit Mithal

Making of Streptavidin Conjugated Crypto-Nanobot: An Advanced Resonance Drug for Cancer Cell Membrane Specificity 281
Anup Singhania, Pathik Sahoo, Kanad Ray, Anirban Bandyopadhyay, and Subrata Ghosh

Performance Evaluation of Fuzzy-Based Hybrid MIMO Architecture for 5G-IoT Communications 289
Fariha Tabassum, A. K. M. Nazrul Islam, and M. Shamim Kaiser

Reducing Frequency Deviation of Two Area System Using Full State Feedback Controller Design 299
Shubham, Sourabh Prakash Roy, and R. K. Mehta

Author Index 311

About the Editors

Dr. Kanad Ray is a Professor and Head of the department of Physics at the Amity School of Applied Sciences Physics Amity University Rajasthan (AUR), Jaipur, India. He has obtained MSc & PhD degrees in Physics from Calcutta University & Jadavpur University, West Bengal, India. In an academic career spanning over 25 years, he has published and presented research papers in several national and international journals and conferences in India and abroad. He has authored a book on the Electromagnetic Field Theory. Prof. Ray's current research areas of interest include cognition, communication, electromagnetic field theory, antenna & wave propagation, microwave, computational biology, and applied physics. He has served as Editor of Springer Book Series such as AISC, LNEE etc. and an Associated Editor of Journal of Integrative Neuroscience published by IOS Press, Netherlands. He has established an MOU between his University and University of Montreal, Canada for various joint research activities. He has also established MOU with National Institute for Materials Science (NIMS), Japan for joint research activities and visits NIMS as a visiting Scientist. He had been visiting Professor to Universiti Teknologi Malaysia (UTM) and Universiti Teknikal Malaysia Melaka (UTeM), Malaysia. He had organized international conference series such as SoCTA, ICOEVCI as General Chair. He is a Senior Member, IEEE and an Executive Committee member of IEEE Rajasthan. He has visited Netherlands, Turkey, China, Czechoslovakia, Russia, Portugal, Finland, Belgium, South Africa, Japan, Malaysia, Thailand, Singapore etc. for various academic missions.

Dr. Krishna Chandra Roy is working as Principal in Kautilya Institute of Technology & Engineering, Jaipur with 22.04 years of academic experience. He received M.Tech in the Electrical Engineering Department from NIT Patna and Ph.D in Electrical Engineering from North Eastern Regional Institute of Science and Technology (Under the MHRD, Govt. of India) Itanagar. He has worked as Principal, Dean (Engg.), PG Coordinator and HOD of various Engineering Colleges and University. He supervised many Ph.D Thesis and M.Tech. Dissertation. He has organized Workshop, FDP, Seminar, Conference and Research program as Coordinator who has fully funded by MHRD and AICTE New Delhi. He has published and presented more than 110 research papers in renowned National and International Journals and conferences. He has published White research papers and patents and also published two books. He is a member of various National and International technical board & committees like International Association of Computer Science & Information Technology, Computer Society of India, Indian Society of Technical Education, Institute of Electronic & Telecommunication Engineering, and Indian Science Congress Association.

Er. Sandeep Kumar Toshniwal working as Registrar & HOD, ECE in Kautilya Institute of Technology & Engineering, Jaipur with 16 years of academic experience. He received M.Tech in Digital Communication from MNIT, Jaipur and pursuing Ph.D in Electronics & Communication from JK Laxmipat University, Jaipur. He supervised many M.Tech. Dissertation. He has organized Workshop, FDP, Seminar, Conference and Research program as Coordinator. He has published and presented about 25 research papers in renowned National and International Journals and conferences.

Dr. Harish Sharma is an Associate professor at Rajasthan Technical University, Kota in Department of Computer Science & Engineering. He has worked at Vardhaman Mahaveer Open University Kota, and Government Engineering College Jhalawar. He received his B.Tech and M.Tech degree in Computer Engg. from Govt. Engineering College, Kota and Rajasthan Technical University, Kota in 2003 and 2009 respectively. He obtained his Ph.D. from ABV - Indian Institute of Information Technology and Management, Gwalior, India.

He is secretary and one of the founder members of Soft Computing Research Society of India. He is a life time member of Cryptology Research Society of India, ISI, Kolkata. He is an Associate Editor of "International Journal of Swarm Intelligence (IJSI)" published by Inderscience. He has also edited special issues of the journals "Memetic Computing" and "Journal of Experimental and Theoretical Artificial Intelligence". His primary area of interest is nature inspired optimization techniques. He has contributed in more than 45 papers published in various international journals and conferences.

Dr. Anirban Bandyopadhyay is a Senior Scientist at the National Institute for Materials Science (NIMS), Tsukuba, Japan. He completed his PhD in Supramolecular Electronics at the Indian Association for the Cultivation of Science (IACS), Kolkata, 2005. From 2005 to 2008, he was an independent researcher, as an ICYS research fellow at the International Center for Young Scientists (ICYS), NIMS, Japan, where he worked on the brain-like bioprocessor building. In 2008, he joined as a permanent scientist at NIMS, working on the cavity resonator model of human brain and design synthesis of brain-like organic jelly. From 2013 to 2014 he was a visiting scientist at the Massachusetts Institute of Technology (MIT), USA. He has received several honors, such as the Hitachi Science and Technology award 2010, Inamori Foundation award 2011–2012, Kurata Foundation Award, Inamori Foundation Fellow (2011–), and Sewa Society international member, Japan. He has patented ten inventions (i) a time crystal model for building an artificial human brain, (ii) geometric-musical language to operate a fractal tape to replace the Turing tape, (iii) fourth circuit element that is not memristor, (iii) cancer & alzheimers drug, (iv) nano-submarine as a working factory & nano-surgeon, (vi) fractal condensation based synthesis, (vii) a thermal noise harvesting chip, (viii) a new generation of molecular rotor, (ix) spontaneous self-programmable synthesis (programmable matter) (x) Fractal grid scanner for dielectric imaging. He has also designed and built multiple machines and technologies, (i) THz-magnetic nano-sensor (ii) a new class of fusion resonator antenna etc. Currently, he is building time crystal based artificial brain using three ways, (i) knots of darkness made of fourth circuit element, (ii) integrated circuit design, (iii) organic supramolecular structure.

A Space-Time-Topology-Prime, stTS Metric for a Self-operating Mathematical Universe Uses Dodecanion Geometric Algebra of 2-20 D Complex Vectors

Pushpendra Singh, Pathik Sahoo, Komal Saxena, Subrata Ghosh, Satyajit Sahu, Kanad Ray, Daisuke Fujita, and Anirban Bandyopadhyay

Abstract Advancing from eight imaginary worlds of octonion algebra, for the first time we introduce dodecanion algebra, a mathematical universe made of twelve imaginary worlds one inside another. The difference between eight and twelve imaginary worlds is that the Fano plane that sets the products of imaginary vectors is

---

P. Singh · K. Ray
Amity School of Applied Science, Amity University, Kant Kalwar, NH-11C, Delhi Highway, Jaipur, Rajasthan 303007, India
e-mail: singhpushpendra548@gmail.com

K. Ray
e-mail: kanadray00@gmail.com

P. Singh · P. Sahoo · K. Saxena · D. Fujita · A. Bandyopadhyay (✉)
International Center for Materials and Nanoarchitectronics (MANA), Research Center for Advanced Measurement and Characterization (RCAMC), National Institute for Materials Science, 1-2-1 Sengen, Tsukuba, Ibaraki 3050047, Japan
e-mail: anirban.bandyo@gmail.com

P. Sahoo
e-mail: 2c.pathik@gmail.com

K. Saxena
e-mail: komal03oct@gmail.com

D. Fujita
e-mail: FUJITA.Daisuke@nims.go.jp

P. Sahoo · S. Ghosh
Chemical Science & Technology Division, CSIR-North East Institute of Science & Technology, Jorhat, Assam 785006, India
e-mail: ocsgin@gmail.com

K. Saxena
Microwave Physics Laboratory, Department of Physics and Computer Science, Dayalbag Educational Institute, Agra, Uttar Pradesh 282005, India

S. Sahu
Department of Physics, Indian Institute of Technology, Jodhpur, Rajasthan 303007, India
e-mail: satyajit@iitj.ac.in

K. Ray et al. (eds.), *Proceedings of International Conference on Data Science and Applications*, Lecture Notes in Networks and Systems 148,
https://doi.org/10.1007/978-981-15-7561-7_1

replaced by a triplet of manifolds that could coexist in three forms. In the proposed algebra product tensors-like quaternion, octonion, dodecanion, and icosanion are deconstructed as a composition of prime dimensional tensors. We propose a generic conformal cylinder of imaginary worlds, similar to modulo or clock arithmetic, using that one could build the group multiplication tables of multinions, which would enable developing the associated algebra. Space-time (st) metric is known, we added two concepts, 15 geometric shapes as topology (T) and 15 primes as symmetry (S) to build a new metric, space-time-topology-prime(stTS) for a self-operating mathematical universe with n nested imaginary worlds. The stTS metric delivers a decision as shape-changing geometry with time, following fractal information theory (FIT) proposed earlier for hypercomputing in the brain. FIT includes two key aspects, the geometric musical language (GML) and the phase prime metric (PPM) that operates using clock architectures spread over 12 dimensions.

Keywords Octonion algebra · Dodecanion algebra · 11 dimension · Imaginary number · Fano plane · Division algebra · Manifold · Conformal · Space-time metric

1 Introduction

Geometric computing where information is converted into geometric shape is advancing rapidly [1]. Geometric shapes are connected to numbers used to present its symmetries. Similar to real numbers (1, 2, 3, 4, 5… n) there are imaginary numbers like quaternions [2] and octonions [3], it's surprising why other imaginary numbers are missing, never explored in the last 200 years history of imaginary numbers. Quaternion curves are widely used in graphics [4], analyzing the higher level codes in DNA [5], quaternions look like electron spins, polarization of photons, hence used is quantum mechanics or any two-state systems. Historically, octonions are the largest imaginary number classes reported to date, they are widely used in astrophysics in modeling information in the black hole [6]. There exists no example of higher dimension numbers, even imaginary numbers with 3, 5, 6, and 7 imaginary axes are missing in the works of literature below the octonions, while 9, 10, 11 dimension complex numbers or vectors we would have to create for reaching the dodecanion which is our aim in this particular work. There are two problems in creating a new algebra, first, generating a multiplication table, and second, determining the sign of the products (+ive or −ive). For inventing a new algebra, constructing a multiplication table is enough, the basic properties of the new algebra could be derived from the table.

Thus far, majority of invented algebras looked into changing the product tensors in the desired way, for a particular purpose, arranged the symmetry of elements in the matrix or tensor. Instead of the arithmetic of complex numbers where we add, subtract, divide or multiply the tensors, when geometric musical language GML [7] binds the corresponding geometric shapes [8], it demands much more than the existing culture of inventing a new algebra. GML requires inserting a geometric shape in the corners of another geometric shape, since it considers, that the corners should be the singularity points if the geometric shape is not the smallest layer in the architecture or innermost system. Therefore, one cannot rewrite tensors by putting the desired terms to zero, in order to get interesting product and symmetry, the elements of the tensors include the details of the singularity points. The tensors are the imaginary numbers of various dimensions and for implementing GML, one has to write the tensor of a given dimension in terms of lower dimensional tensors. Thenceforth, the combination of quaternion algebra, octonion algebra, dodecanion algebra, icosanion algebra and all possible higher order algebras would simply be like providing a platform to superpose various different kinds of dynamics exclusive to a particular algebra to dominate one or another depending on the specific composition. Therefore, the idea to build a new kind of algebra suitable for operating GML is not limited to a particular complex number of a given dimension, preferably, a collection of a large number of complex numbers operating together similar to the natural numbers. Though the development of a complex number series is a generic protocol, the finite number of platonic solids (five) of GML suggests that more than the icosanion, the 20-dimension (20D) complex number is not essential for developing a universal geometric language. What has been tagged as the mathematical recreations thus far might deliver a universal language [9].

Figure 1a shows 15 geometric shapes used to define the geometric musical language, GML. All forms of information are converted in terms of these 15 geometric shapes. One of the examples of using GML is shown in Fig. 1b. A pair of DNA is converted into a quaternion.

2 Introducing the New Concept of Dimension to Fit the Necessity of Fractal Information Theory, FIT

2.1 Fundamental Differences Between the Existing Concept of Dimension and Our Concept of Dimension

In the existing concept of dimension, a distinct dynamic is assigned to a new axis, along that axis only that particular dynamics operate, all other dynamics are silent. A distinct existence of a particular axis or dynamics is the hallmark of this particular concept of dimension. In principle one could add an unlimited number of axis, within the domain of this definition, an upper limit of dimension could not be fixed. It is the combination of independent systems, where the participating systems do not lose their distinct identity. When the systems assemble one inside another, i.e., they assemble within and above, we may assign each imaginary world one unique dynamics, however, they are not independent. The rotation of one axis by 90° does not transfer to another axis, as it happens in the conventional wisdom of dimension. In the nested universe of imaginary worlds one inside another, if two imaginary worlds interact a third imaginary world would get affected (there is a century-old fun,

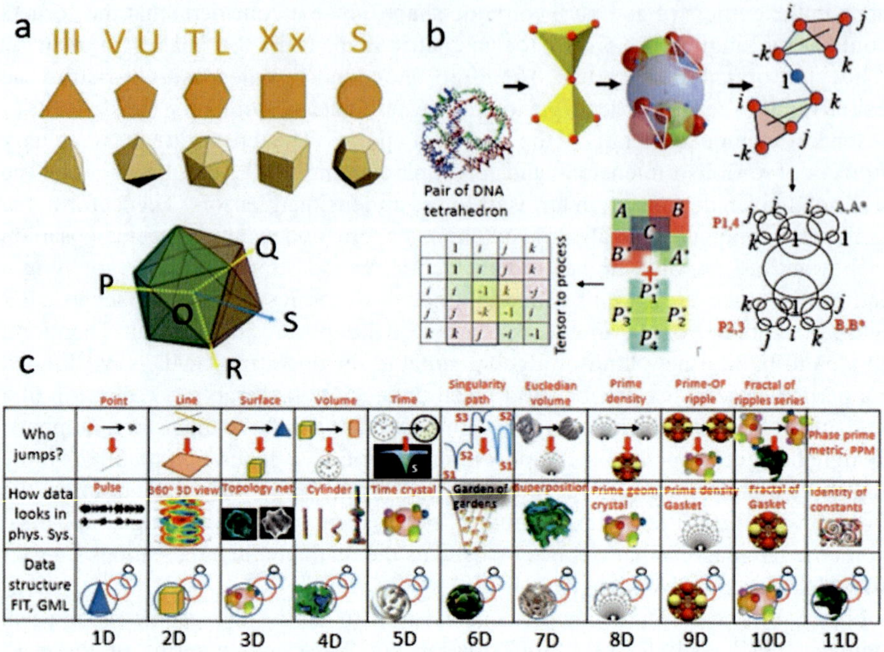

Fig. 1 **a** 15 geometric shapes used in a typical case of GML, 1D (straight I/II, corner V/U, angle T/L, cross X/x, spiral/vortex S), 2D (triangle, square, pentagon, hexagon, circle), and 3D (tetrahedron, cube squares, octahedron, dodecahedron, icosahedron). The icosahedron plot at the bottom of 15 geometric shapes show how three axes from the triangular surface, P, Q and R are three corners, each corner is connected to 5 triangular planes. In an icosahedron there are 12 such axes. The corner is the singularity point where new structures are embedded in FIT-GML (Fractal information theory and Geometric musical language) protocol. S axis comes out of the triangular plane, 20 triangular planes of icosahedron have similar 20 similar axes. **b** We demonstrate how DNA dynamics could be written as tetrahedron, converted to a time crystal and eventually sets of clock arithmetic systems which could be converted into a tensor is shown. **c** A table shows how to understand the concept of dimension. Three rows are there. The first row shows what question to ask, picturising 12 different dimensions. Second row shows how the data might look like in a real physical scenario. The third row shows that how in FIT-GML the information structure looks like (adopted from Ref 17, for details)

parody, jokes with the messy nature of multiplication, e.g., Mad Hatter scene in Alice in wonderland novel was inspired by quaternion mess). This is a new kind of non-rotational, rather clocking relationship between the hypothetical axis of dynamics we observe in the nested imaginary worlds.

Since a pure dynamics-axis relationship does not arise when imaginary worlds nest within and above, the implications are many folds. There is an upper limit of dimension. Figure 1c describes the concept of dimension, where, the addition of a dimension changes the perception of space, time, topology and prime. 1D to 3D describes spatial dimension, 4D to 6D is about time, 7D to 9D it is about topology where the space-time dimensions reside and 9D to 11D is about prime where the topology is regulated. Since prime is not related to any physical significance, a higher to this dimension is not achievable.

However, first, we need to investigate the concept of the dimension of dimensions. Mirror symmetry across the diagonal elements of the representative tensor defines the unit of tensor or local group of imaginary worlds. It means, we consider a 20×20 tensor, we may rewrite the same tensor as 16 number of 5×5 tensors, where 5×5 tensors form local group or distinct clocked relationships between the participating imaginary worlds. If we consider each of the 5×5 tensor is a unit cell then we are primarily dealing with 4×4 tensor while actually, we have a 20-dimensional tensor, icosanion. Furthermore, if we consider 20×20 tensor as 2×10 or 100 number of 2×2 tensor, then, since 2×2 tensors would be the unit cells, we would have actually $10 \times 10 = 100$ tensors at max. Therefore, each of the 2×2 tensor could represent a higher dimension than what we normally consider, each element of 20×20 tensor could hold the information for that dimension (dynamics). Furthermore, 10×10 tensor has four 5×5 tensors, therefore, we have 4×4 tensor representing the dimension of dimensions. Such a hierarchy of dimensions enables encoding higher level dynamics of a system in the same tensor. Since the nested imaginary worlds are linked with each other such that functions cannot represent them, the dimension of dimensions is an approach analogous to a function, mapped at the multiplication table, not by the definition of mathematical functions. At the same time, it allows to apply clock arithmetic [10], which introduces a loop to represent the rank of the constituent tensors, which are primarily the tensors with prime dimensions like 2×2, 3×3 or even 5×5.

Therefore, when we group systems one inside another, a singular host function A represents an imaginary world, however, each element in the domain of that function is not just defined by other functions of a different imaginary world B. This is not one-step process, it continues, say up to Z, i.e., 26 worlds from A to Z assembled one inside the other, forming a mathematical universe. Moreover, these elements are affected by the arrangement and grouping of other imaginary worlds, which is never defined within the definition of the function A. That is not complete. Each element of the host function A, accepts the topological projection and symmetry breaking effects of those topologies from all other imaginary worlds say, A to Z, governed by Phase Prime metric PPM [11]. How other worlds arrange and constitute, affects the product the defines a particular world in this mathematical universe. A simple tutorial presentation is made in Fig. 2a. The number of all possible choices to form a group from a given number of entities is the ordered factor as shown in Fig. 1a. When one increases the given number of entities one by one, and plots the choices, it is the 2D plot. Now, to convert a 2D plot into a 3D architecture that adds another dimension of the contribution of individual primes we simply rotate the 2D plot of Fig. 1a. Prime 2 occurs 50% of all integers, prime 3 occurs purely for 16% cases of all integers. Two primes, 2 and 3 alone covers 66% integers or symmetries possible in a mathematical universe. Now, if we continue our calculation and go on adding the contributions, after reaching 15 primes, we would find that we have reached a contribution of 100% as shown in Fig. 2b. So, if one takes the 2D plot and rotates it by 360°, considering that 15 primes deliver 15 different angles on a polar plot as

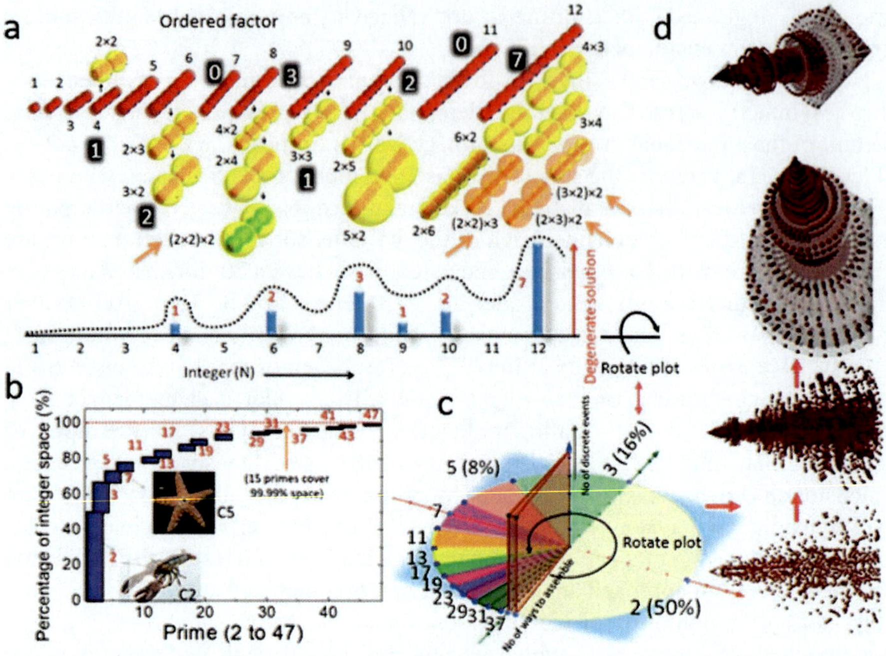

Fig. 2 **a** The development of a phase prime metric or PPM. There are two sub-panels. In the top, we show an array of balls in a single line representing the integers. For each linear arrangement of balls, groups of balls are tagged which could vibrate together as a single phase space. All possible compositions for a single linear array are shown below the line. By changing the order, we get different combinations here, e.g., 2×3 is not equal to 3×2. In the bottom panel we plot the count of group compositions we can make from a single number. This number is also the number of degenerate solutions for the generic oscillations of a string. **b** The contribution for a particular prime in the integer space is counted. For example, prime 2 contribute to 50% of all possible integers in the number system. For each prime while calculating the contribution, only its contribution alone is calculated, for example, 6 could be counted for 3 and 2, we have counted 6 only once for 2, not 3. Similarly, we have counted for 15 primes and reached total contribution of 15 primes to 99.99%. **c** The degeneracy plot of panel **a** is rotated along the integer axis, the total number of rotational angles is 15 and their contributions are plotted in the *XY* plane while the degeneracy is plotted along the *Z*-axis. **d** In the panel (**c**) and the panel (**a**), the continuously decreasing contributions of primes are ignored and all 15 primes are given equal contributions 24 degrees. Then the bottom plot in panel (**a**) is rotated 360 degree to get the plots of panel (**d**) bottom to top (adopted from reference 17, for details)

shown in Fig. 2c, a 3D architecture is found as shown in Fig. 2d. This particular architecture has equal contributions from all primes.

Therefore, "A" described above cannot be a proper function, a defined entity, but the dimension of dimensions enables creating an integrated virtual imaginary world composed of all the imaginary worlds A to Z. The most interesting aspect of this mathematical universe ($\geq 12D$) is that no constituent imaginary world is defined, in other words, no imaginary world exists as a defined mathematical expression, a set of multiplication tables representing the same tensor holds the virtual map that is closest in defining the universe. When we integrate the product tensors in the dimension of dimensions, the linking pathway of dimensions is defined, the architecture of that pathway could be represented by a fractal function, but it leads to the formation of an imaginary hyperspace that is not differentiable. Therefore, in this particular universe of 26 nested worlds A to Z, even the inclusion of all participating imaginary worlds does not make it complete or defined. *The origin of undefined nature is not out of the box contribution, rather, evolution of symmetries* via *PPM, which does not have any physical link to any system anywhere.* Out of the box imaginary worlds (if we study function confined in P, then the rest of the imaginary world in A to Z are out of the box) project superposition of topologies to infinity following an infinite series of connected symmetries in a PPM. Moreover, due to the superposition of symmetries, the PPM start extrapolating the input symmetries, the resultant composition of output symmetries has no link with the projected topologies or even their symmetries. Since PPM is a pattern of ordered factor of integers, when we split tensors as a composition of primes, they would be found infinite times in the PPM, every time in a new combination of primes.

In the literature, when the authors describe dimension, they suggest additional dynamics that needs a separate axis, where its values vary. All the dimensions could be in the real world itself. We do not discard that. However, we have introduced another concept of dimension above where if an element A is made of element B and A does not have any other identity than a composition of B, then, A is an imaginary world for B and B is an imaginary world for A (A + *i*B or B + *i*A). The interaction with the observer decides which one is real and which one is imaginary. Therefore, one could build a network where *Z* is inside *Y*, *Y* is inside *X*, *X* is inside W….C is inside B, B is inside A, to explain the 26 dimensions made of 26 elements in a 26D vector. Therefore, when we build a vector product, the tensor representing the multiplication table (say, 26 × 26 elements), of two systems each made of 26D elements, then, the product affects all the worlds independently and dependently, i.e., modulation is both ways.

When we build a nested architecture A–Z one inside another, tag all layers as imaginary because each layer contributes to others by changing the phase value of the periodic function (imaginary part) and assign a dynamics related to a particular dimension of a system in each imaginary world, we do not change the idea of dimension significantly. What is actually done mathematically is making the subsets of a function undefined, it is not function in a function in a function …, rather, it is making a function a subset of its own map. Research on such undefined functions is limited. What is advantageous here is that the dimension, which earlier demanded a new axis, now acquires a physical substrate or the definition of a virtual system where one could map the whole universe, but not its worlds as distinctly. Therefore, the idea of dimension does not remain an abstract concept, the map of the universe holds its topological representation. However, one has to compensate for that hyperspace projection. When two systems interact or we take a product of the higher dimension vectors, different imaginary worlds interact and affect at different but strictly defined world level. It means imaginary world C could interact with the imaginary world Q and affect the imaginary world T. From the product tensor by following the horizontal and vertical axes of two imaginary worlds one could figure out who would interact to affect whom.

2.2 When the Worlds Are Nested to Form a Mathematical Universe Why There Is an Upper Limit on the Dimension?

PPM has only 15 prime related symmetries governing 99.999999999% of all possible symmetries of the universe. When the elements of higher dimension tensors group, one could notice that unit tensors are all made of prime number of elements. First 12 primes that contribute to 99.99999% (2, 3, 5, 7, 11, 13, 17 contribute 99%) of all possible tensors, Fig. 2b. In reality, using four primes all other prime-related tensors could be rebuilt ($17 = 3 \times 5+2$). Just like for platonic solids, there is no point going beyond icosahedron with 20 planes (20D), because all other solids are made of triangles only, similarly, beyond first eight primes, (19D) splitting tensors with prime tensors does not yield new dynamics. Since icosahedron has 12 corners and our journey through different worlds for our mathematical universe is via corners, 12D is the maximum dimension for the systems assembled "within and above".

If one asks for conventional dynamics (adding a new axis for a distinct dynamics) the upper limit of our universe is 20, it means in any one of the imaginary worlds a system could exhibit 20 distinct dynamics, 20D system. However, when we represent the integrated dynamics of nested worlds, i.e., the dynamics that is not confined within one but multiple worlds, emerge as one unit, then the dimension of that kind of dynamics is limited to 12, i.e., 12D.

3 Construction of a Dodecanion Algebra

3.1 The Concept of Deconstruction of Higher Dimension Complex Numbers with Lower Dimension Complex Numbers

Here we would explore what happens if we try to build higher dimensional complex numbers by adding new imaginary axes one by one and thus building a new kind of number system of multinions. The key point here is that we do not want to discover a new algebra which uses only one kind of complex number, say quaternion algebra, Q; octonion algebra, O or create a new algebra say, dodecanion algebra, D or icosanion algebra, I. Here, we suggest an algebra that uses multiple complex numbers. One could easily build new composite numbers as Q + iO + jD + kI; composition of elementary complex numbers to construct the larger complex numbers would have plethora of applications in geometric algebra. We explore the possibilities here.

The rise of the multiplication table for two complex numbers: Real numbers are on a line, 1D space. While taking the product of two complex numbers, and normalizing the value to 1, one could start from 1 and reach-1 crossing origin at 0, when two perpendicular rotations are carried out across i twice, it means $i^2 = j^2 = k^2 = ijk = -1$. One additional event is the three rotations along three imaginary dimensions ijk. When one tries to multiply two quaternions then the rules are simple. The multiplication is unique in the sense that for cartesian coordinate systems with three orthogonal axes, such results never arise. It is the unique rotation across the real world that enables a situation that two rotations across two imaginary axes deliver a change in another imaginary axis, which in itself tells a unique physical situation. Quaternion products summed up in a tensor (not matrix because three imaginary parts form a vector, which makes it a tensor) is the complete picture of quaternion algebra. To make a new algebra, one has to build such a tensor, since the foundation of algebra is in multiplication, tensor form reveals fundamental criteria.

How to write the multiplication table to construct the algebra of a particular complex number?: Writing the multiplication table is utmost important for inventing a particular kind of algebra. Figure 3 explains how to write a 20×20 tensor, i.e., a pair of 20D complex vectors multiplication table. One complex vector is written along the column and another complex vector is written along the horizontal row. To create an element of this tensor, one has to add the coefficients of column and row and shown in Fig. 3 right. Normally, the three integers are put in a circle and a clocking direction is set. Say the product of A and B is C. Now, often the pair and the row to be multiplied are common but we need to put an alternative to retain symmetry of the arrangement of elements along the diagonals. A + B = C, but normally we make B + C = A, C + A = B. However, if one makes true addition of B + C would be a different integer than A, similarly, C + A would be a different integer than B. If those different additive results are necessary, then, we put two clocks connected to each other with a pair of the common point in between. Now, one such example is shown in Fig. 3. $3 + 4 = 7$, and $4 + 7 = 11$. Now, when we bond the two clocks, we make sure both rotate in the same direction in the overlapping region. Thus, outside, both rotate one opposite to another. In order to write a tensor, it is better to fill the four corner values and then fill the diagonals. Two ways the diagonals could be filled, left to right and right to left as shown in Fig. 3 bottom. Then, the values of the diagonals are identical or increases/decreases by a difference of two. Therefore, it is very easy to right the values, except for one major problem we encounter, we describe it below.

The necessity of modulo arithmetic to fill up the tensor multiplication table: While filling up the right lower half of the tensor along the diagonal, one might encounter a situation where the value of addition is much more than the maximum coefficient value. To resolve the issue, we have introduced a bilayer wheel, once the maximum coefficient is reached, the counting begins from the second layer along the perimeter of the wheel. Moreover, the values are connected using a line inside

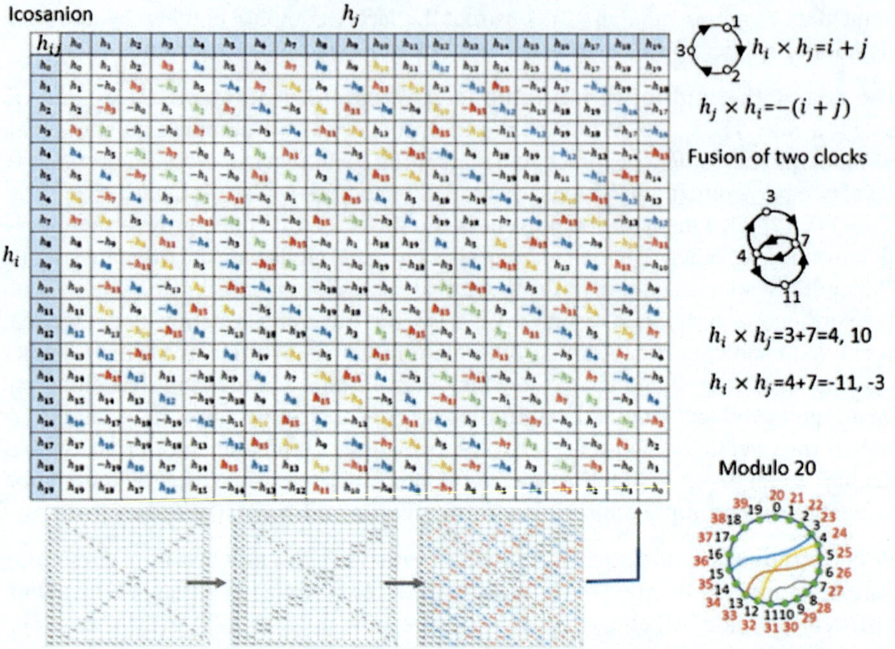

Fig. 3 How to draw an icosanion or 20-dimensional tensor. This is a 20 × 20 tensor where the border values, which are identical are filled up first. Then the diagonals are filled up one by one. The left top to the right bottom diagonals are filled up, near to the diagonal values are identical. The right top to the left bottom diagonals is filled up. Then at a certain gap one could find identical diagonal values. The process is repeated for both the cross-directional diagonals. In the second step, all clocks are written. The upper left triangular region of the tensor has values less than 20, however, the bottom right triangular part has values more than 20. Therefore, clock arithmetic or modulo 20 is used to find the clocks for the bottom right triangular region. Three values make a clock, lower indices to higher are kept as clockwise rotation and an arrow is put to depict the direction

the wheel, since the line carries solution of symmetry, we call it braiding, as shown in Fig. 3, right. The braids have a typical bonding, to generate symmetry along a diagonal, either end of the braid contains integer, one could take either one, as required. The pattern of braids could change for a tensor, it would generate different kinds of tensors.

How to determine the clocking direction for all the elements?: Fig. 4 outlines a table that one may create to set the clocking direction. To build this table one has to start from the lowest coefficient 1 and draw all possible clocks from bottom to top, vice versa. One should increase the coefficient value one by one and create columns. Since coefficients are added to build the multiplication product of the complex vectors, the number of elements in a particular column decreases as we increase the coefficient gaps. Figure 4 bottom right also shows how to build the clocks for a modulo arithmetic wheel.

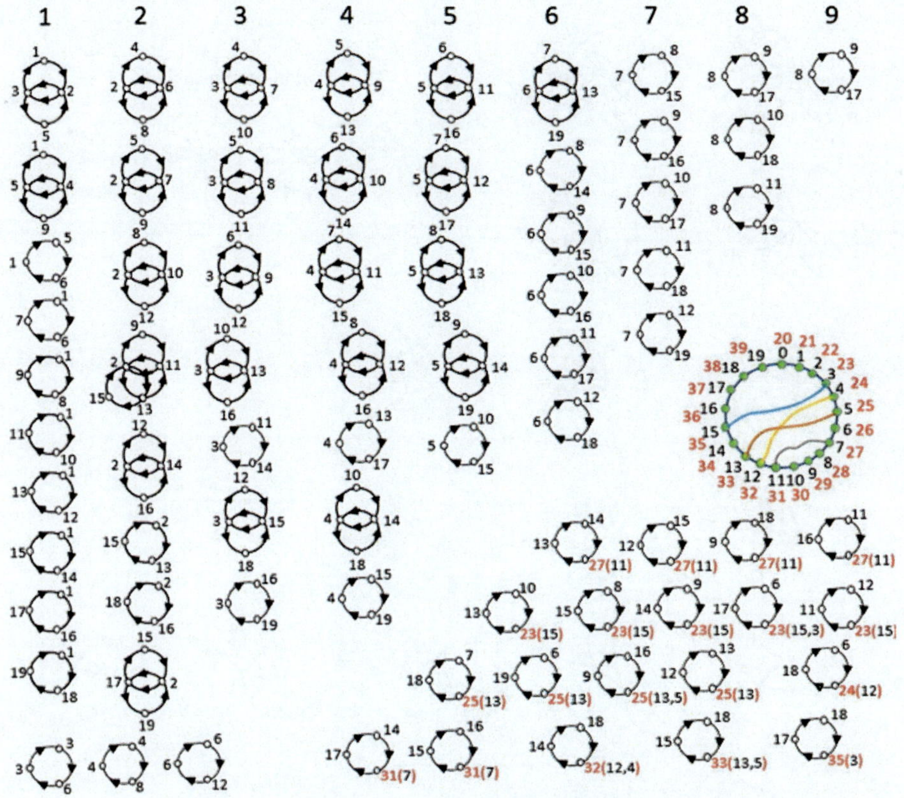

Fig. 4 All possible clocks for icosanion or a 20-dimensional tensor is written here. There are nine columns. Each column has a common point or integer written at the top. An integer represents a coefficient, we write 7 to represent h_7. To the right there is a wheel, numbering 0 to 19 and then 20 to 39. The wheel is modulo arithmetic, there are braiding connecting the integers. These connections represent equivalent values. Below the modulo arithmetic wheel, we find the corresponding clocks

How to build the manifold?: Fig. 4 does not allow us to build manifold, circular clock-like presentation of Fig. 4 certainly delivers an idea of clock-like architecture of the entire multiplication table. Now, to integrate all the clocks we adopt a bilayer clock-like representation of Fig. 4. Here instead of a circular clock, triangular clock is used and instead of one layer of all connected clocks, a bilayer is used. The clock directions never contradict.

One of the interesting aspects of a 20 $\times$ 20 icosanion manifold of Fig. 5 is that the protocol could be followed for any tensor of any dimension. One could cut the paper and glue the common triangles, matching the three numbers or coefficients. The resultant structure would be the manifold that represents the information about the complex vector multiplication. All the circles could be arranged along a cylinder to easily find whom to glue. This particular cylinder is the manifold operator.

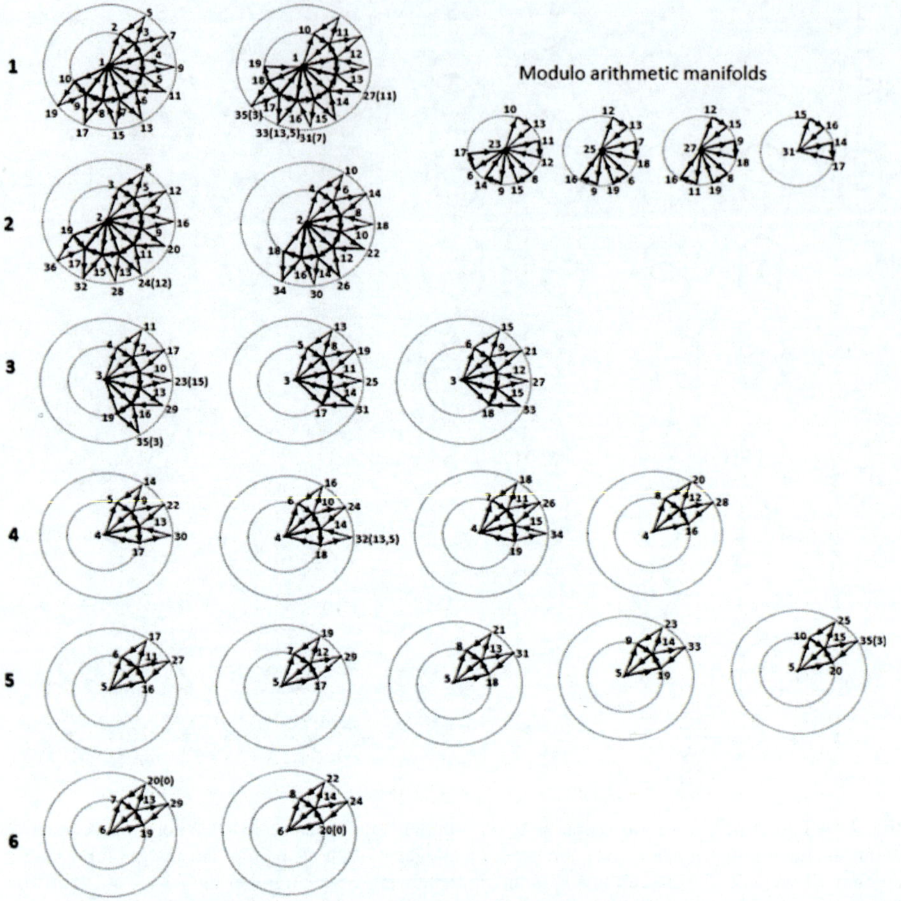

Fig. 5 A table has been created in Fig. 5 analogous to the same plot in Fig. 4. Figure 4 is vertical, a particular common coefficient is arranged vertically but Fig. 5 is horizontal, common coefficients are arranged horizontally. Nine columns of Fig. 4 is shrinked to 6 horizontal rows in Fig. 5. Modulo arithmetic manifolds are shown top right corner. Each plot has two circles. One starts counting from center, and reaches the modulo value. The second layer is started from the first layer, clock directions never contradict. Sum of multiplicative coefficients are shown, the modulo values are noted in parenthesis

The coexistence of three-manifolds: In Fig. 6, we have summarized the journey from quaternion to octonion to dodecanion. This is utmost surprising that the Fano plane that tells us the product and the rotational direction of the clocks was discovered 100 years after the conception of octonion. Similar types of sub-tensors are highlighted with the shades of similar intensity. Quaternion, octonion, and dodecanion have three distinct features of internal spontaneous symmetry breaking. If that mathematical operation is allowed, dodecanion could simultaneously hold two distinct compositions of symmetric arrangements of sub-tensors as outlined in the

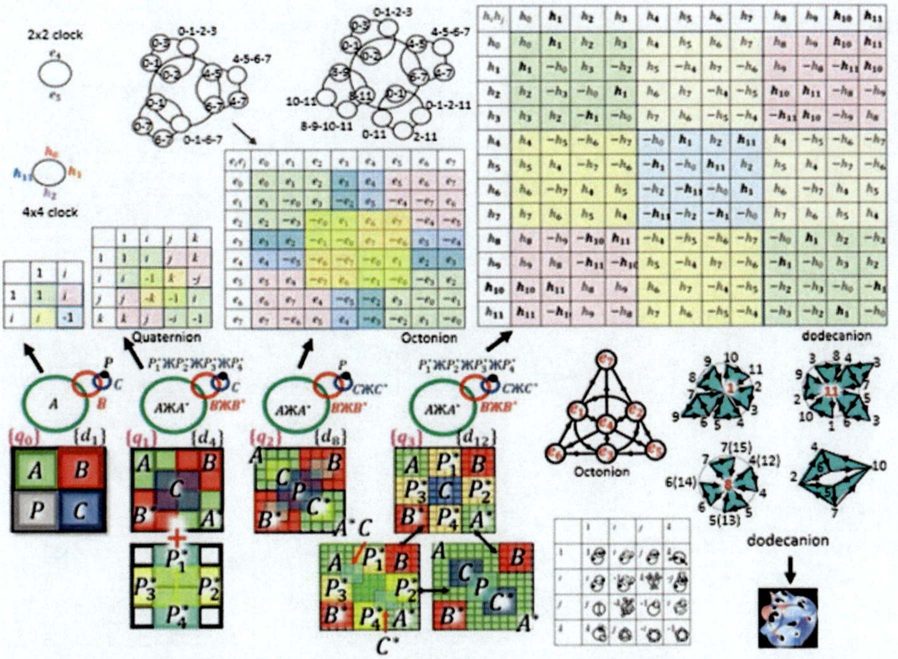

Fig. 6 Four types of tensors are presented here dinion, quaternion, octonion, and dodecanion algebra for the time crystal representation. In four different ways, the multinions are explained. First, time crystal presentation of the multinion, second, matrix representation which are colored as square matrices; third, linguistic presentation, here, we present the tensors as a subset of four clocks, each clock represents a sub-matrix of the entire tensor. Sub-matrices A, B, C, and P. Quaternions show duality and dodecanions show simultaneous coexistence of three tensors, as precursor to self-operating universe. To the bottom right corner Fano plane is shown for quaternions and manifolds are shown for dodecanions. There is a pictorial clock-like presentation of a quaternion at the bottom right, it suggests how a single element in a tensor looks like topologically

bottom row of Fig. 6. There could be simultaneously three distinct compositions of sub-tensors for the dodecanion. It means the values of the multiplication tables may group three different ways, and obviously, three distinct manifolds would arise under certain values of the multiplication table.

Braiding map is the signature of groups: The most fundamental criterion to create a self-operating mathematical operation is that the values of the multiplication table are such that three different group formation is not inhibited. It means when different sets of elements are taken into account, they build C2 symmetry of the diagonals. We have listed braiding for all the complex vectors, 12D, 14D, 15D, 16D, 18D and 20D in Fig. 7a, we have found that the braiding if self-similar, difficult to rearrange the groups. However, if the braids are less in number and overlap while connecting the two integers, i.e., the braids cross each other, then, at those conditions, regrouping of elements of the multiplication table following the three options may become feasible.

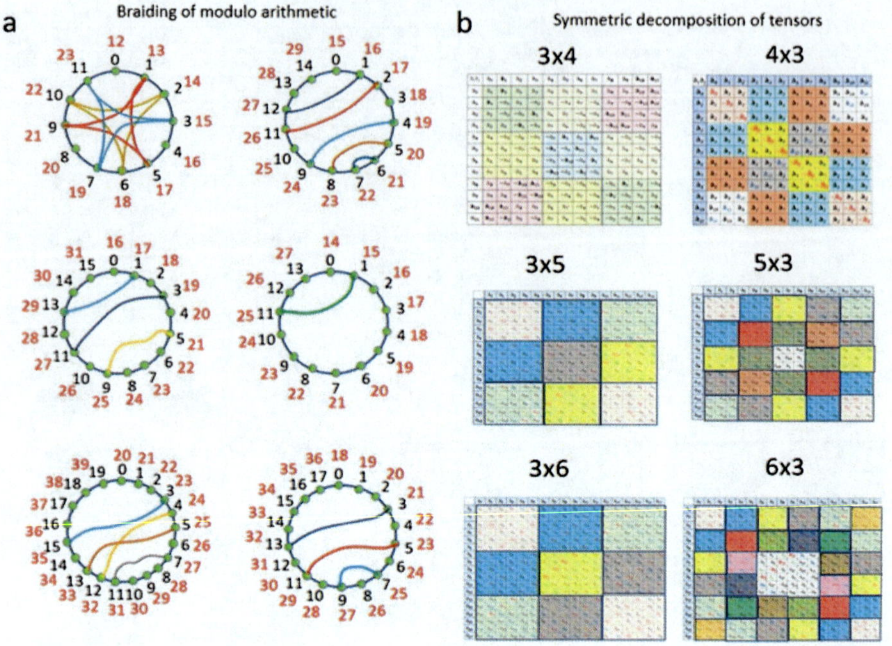

Fig. 7 **a** Braiding of modulo arithmetic used to easily find the coefficients for the lower value tensors are shown. The links depict equivalent values, braiding word is used to represent the connecting line. **b** Two possible symmetric decomposition of tensors is shown for dodecanion (top), pentadecanion (middle) and octodecanion (bottom)

Regrouping of sub-tensors: Fig. 7b shows regrouping of sub-tensors or decomposition of three higher dimensional complex vectors multiplication tables. We show three examples. First, a dodecanion is split into 3×4 and 4×3, it means first the tensor could be divided into sub-tensors of nine 4×4 tensors. Then the same dodecanion tensor is divided into sixteen 3×3 sub-tensors. Second, a pentakaidecanion or pentadecanion, in short, is divided into 3×5 and 5×3 format. Third, octokaidecanion is divided into 3×6 and 6×3 tensors. One may use the term decomposition of tensors. One important criterion to decompose a tensor would be that for dodecanion, 3×3 tensor should be nearly symmetric as well as the 4×4. For that purpose, the value of some elements if made zero, we would find that it's easier to decompose a tensor into various sub-tensors. The nullification of values is similar to the various attempts made by mathematicians while inventing new algebra. Different mathematicians, at different times of the history, used to make most elements of the tensor or multiplication table zero, suitably so that a particularly invented algebra would serve modeling a certain class of physical phenomena. Here our objective is to encode singularity points in the tensor suitably for implementing GML, we discuss it below, in details.

3.2 The Necessity of Polytopes

When we increase the dimension of a complex number from quaternion to octonion to dodecanion, we do not change the form, an octonion is eventually presented like a composition of quaternion and then, a dodecanion is also presented like a composition of quaternion (Fig. 6).It means, when we write the product of two octonions, we could divide the 8×8 tensor elements into five smaller sized quaternions (O = 5Q). While inventing a new algebra, for the last two hundred years, the practice has been deleting the majority of the tensor elements making their value as zero and then, find the multiplication table delivering new structures. Here, when we rewrite the tensor as a composition of other tensors as shown in Fig. 7, then, the number of elementary tensors required to fill the higher dimensional tensors provide us with the corners of geometric shape hidden in it, that is key to operate GML. For example, 8×8 tensor elements when divided into five elements, we get a pyramid or tetragonal 3D structure. These sub-tensors could hold information for the geometric shape that could be embedded in the corner of that pyramid.

It's not that a tensor could be written in terms of elementary tensors following only one way. Here in this report, we confine ourselves in only one possible way. Conventional mathematical literature also follow the same protocol. Therefore, the 12D manifold created in the dodecanion tensor turns quadrilaterally conformal in a cyclic cylinder when geometric information transfers between different architectures of information, be it quaternion 4D, octonion 8D, dodecanion (12D) and hexakaidecanion (16D) or icosahedron (20D). Similarly, hexanion (6D), onunion(9D), dodecanion (12D) and pentakaidecanion (15D) and octakaidecanion (18D) are conformal in triangular topology. The pentanion (5D), decanion (10D), pentakaidecanion (15D), and icosanion (20D) are conformal in pentagonal topology, and hexanion (6D), dodecanion (12D) and octadaidecanion (18D) are conformal in hexagonal topology. We do not go beyond icosanion because icosahedron is the largest platonic solid. All polytopes above 20D are triangular. For a particular dimension we assign a plane or the corner. When dimension means the addition of a new axis, we place the dynamics along the plane, but if it means composite coupling across the imaginary worlds, we get a corner, as a gate to enter the imaginary world. Icosanion holds the largest corner, 12. One could take complex 3D geometric shapes to build all these dimensions, there are plenty of research works done on this aspect [12–14]. The rings of this generic cylinder that helps in writing product tensors of complex numbers represent clock arithmetic or modulo algebra, but links between the rings on this cylinder ensure conformal links between two clocks [15].

3.3 The Correlation Between the Complex Number and the Polytopes

One should note that we assign conformal feature based on the points on the perimeter of the circular sides of the generic cylinder we have built, not the geometric shape of the polytopes. Triangular planes are often used to create polytopes, they are called deltahedrons [13, 16], the number of sides of a deltahedron could vary (4, 6, 8, 10, 12, 14, 16 and 20 faces) thus topological conformity suggests that all are delta or triangle, however, we look at the multiples of 4 (4, 8, 12, 16 and 20) as quadrilateral because, in the multiplication cylinder, the periodicity is regulated by quadrilateral (4) not delta (3).

3.4 Polytopes of Higher Dimensions

If we increase the dimensions, 3, 4, 5, 6, 7, 8, 9, 10, 11, 12 … 18, 19, 20 then the equivalent polytopes would be made of sides 3, 4, 5, 6, 3, 3, 3, 3… The saturation of polytope series in a triangle (3) is interesting, it leads to the platonic solids. Topologically there is no advantage going further up in dimensions. So, we get an upper limit of dimension at 20, icosahedron with 20 faces and 12 corners. We need maximum 12 corners to build a dodecanion algebra. Here are the 15 geometric shapes that are enough to transform information into the universe in terms of geometric shapes. 15 basic geometric shapes are matched (straight I/II, corner V/U, angle T/L, cross X/x, spiral/vortex S), (triangle, square, pentagon, hexagon, circle), (tetrahedron, cube squares, octahedron, dodecahedron, icosahedron), see Fig. 1a. A tetrahedron is formed by placing three equilateral triangles at a vertex (sum of angles at vertex is 180°). It has 4 vertices, 6 edges, and 4 faces. An octahedron is formed by placing four equilateral triangles at each vertex (sum of angles at vertex is 240°). It has 6 vertices, 12 edges, and 8 faces. An icosahedron is formed by placing five equilateral triangles at each vertex (sum of angles at vertex is 300°). It has 12 vertices, 30 edges, and 20 faces. A hexahedron, or cube, is formed by placing three squares at each corner (sum of angles at vertex is 270°). It has 8 vertices, 12 edges, and 6 faces. A dodecahedron is formed by placing three regular pentagons at each vertex (sum of angles at vertex is 324°). It has 20 vertices, 30 edges, and 12 faces.

Quaternions with tetrahedron, octonions with cubes and octahedron, dodecanions with dodecahedron and icosanions with icosahedron play topological mathematics. It means five imaginary complex vectors representing the five platonic solids where the dynamical elements corresponding to each dimension acquires a distinct plane. We discuss below the use of polytope planes and corners of the 3D shapes as singularity points in the Geometric musical language (GML).

Vertices of a polytope are singularity, faces or planes hold information for a typical dimension: A quaternion has four imaginary worlds, one of them, that is being observed or probed is considered real. Each of the four imaginary worlds of a quaternion could expand along a well-defined triangular plane in the tetrahedron. Tetrahedron has four faces and four corners, Each corner is a singularity point by GML and holds a For octonion, there could be two distinct topologies, either each of the 8 planes could be represented each by four sides of a parallelogram/square in a cube, or it could be an octahedron where each of the 8 planes is a triangle. The 12 planes of a dodecahedron could represent 12 distinct dynamics of systems assembled within and above.

3.5 Some Basic Concept of Dodecanion Algebra

The basic algebraic properties of the complex numbers: Moreover, quaternions are non-commutative; it means the order of dimensions is applied effectively to transitions. The order of rotations along *I* axis or *j* axis or *k* axis for two cubes does matter, they would look different. Quaternions are associative with each other but not the dodecanion and higher order terms. If x, y and z denote arbitrary elements of algebra. Associative: $(xy)z = x(yz)$. Dodecanion algebra studies holomorphic or angle- and orientation preserving maps between manifolds.

Writing manifolds of geometric musical language, GML in the phase prime metric, PPM: When dimensions of the multinions are the products of 4, then, one could build a triplet of clocks where each clock is represented by a 2×2 matrix, or a tensor. GML is made of 15 elementary geometric shapes listed in Fig. 1a (one could choose their own geometries to build another form of GML) and ten classes of PPM operate together using manifolds as shown in Fig. 8a. While constructing the multiplication table of imaginary numbers, the manifolds are naturally created as shown in Fig. 5. These manifolds may represent the symmetry of geometric shapes if one writes the elements of tensors using the singularity features (see clock section below for details). Though we add coefficients while constructing a multiplication table, in reality, we take products and the manifolds could be considered as a list of linked products of integers or symmetries. For constructing the PPM, similar product of primes are used as shown in Fig. 2a. Therefore, it is possible to match the products and find the statistically dominating loops of PPM as shown in Fig. 8a. Once the manifold is bridged with the PPM, other linked loops of PPM becomes a set of the manifolds, or the manifolds bind together.

Clocks in the GML and clocks in the PPM: To represent the complete rotation both sides of the diagonal of a tensor should be similar, like a mirror reflection across the diagonal, irrespective of the tensor dimension. One we derive the integers represented by the symmetry of the geometric shape embedded in the local 2×2 matrices; it is trivial to calculate the rotation direction. In Fig. 8b we plot PPM in a polar plot format. In Fig. 2a, ordered factor of an integer is plotted linearly and then we rotate the horizontal axis by 360° to get the PPM. If we bend the entire 2D plot and convert entire horizontal axis into a single point, then, we get a polar plot of the same PPM as shown

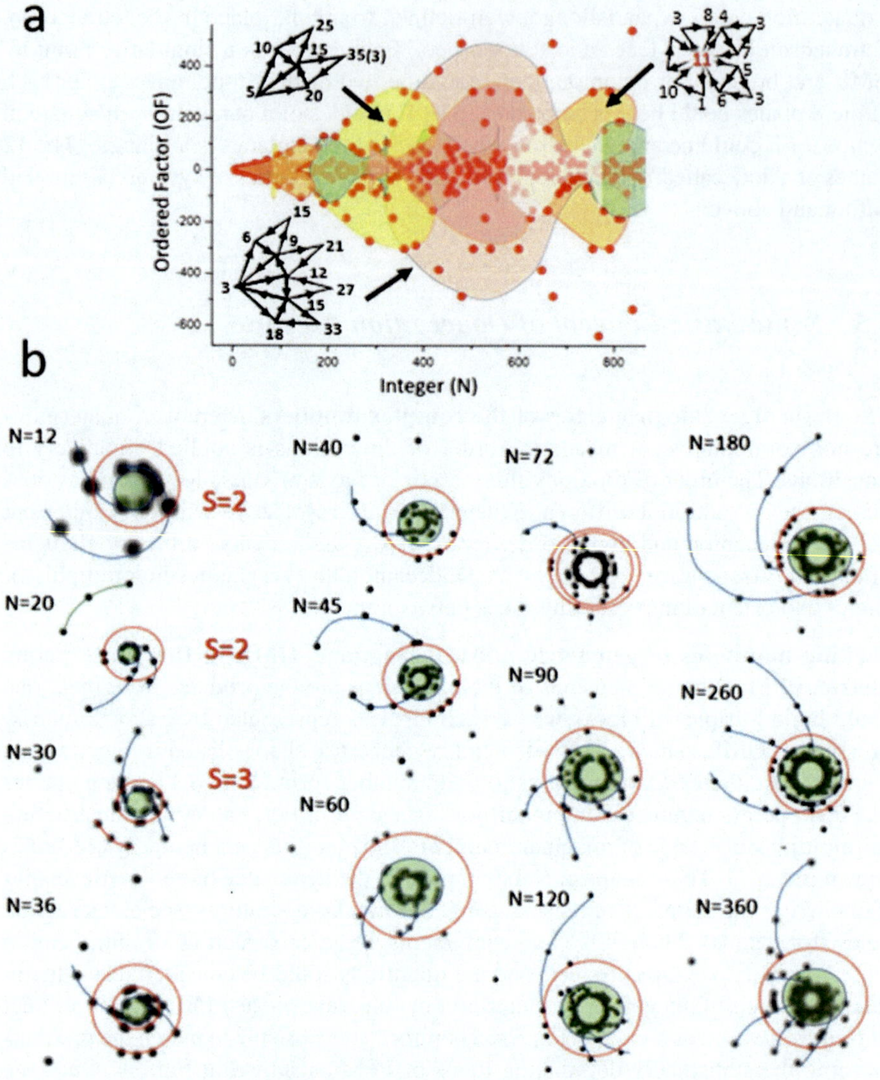

Fig. 8 **a** Ordered factor vs integer N plot, known as a typical phase prime metric. Lines connect the values build distinct areas. Each area represented by a typical manifold is shown using an arrow. **b** Ordered factor is plotted against maximum integer in a polar plot. Say we want to plot $N = 30$, then we create an imaginary circle where each integer is separated by 12° so that total phase is 360° and for each integer we move outward radially, since the distance from center would measure the ordered factor. Connected line shows clockwise and anticlockwise rotation of the minimum distance lines created by connecting the ordered factor of a natural number N

in Fig. 8b. In the example, 13 PPMs are plotted with the increasing number of integers. The plots reveal one of the finest observations in the integer system, the natural classification of integers and their ordered factors in terms of clocking directions. If the order factors of a group of integers would increase with the increasing integer in a particular domain, it would be a clockwise rotation. If the ordered factors decrease with the increasing integers, it would be anticlockwise rotation. The observation of an additional quantization of clocks is evident in the phase prime metric, PPM plot of Fig. 8b, where we have put a circular boundary to isolate a lower level clocking event [17]. The finding would have enormous applications. First, it means, the loops of Fig. 8a, representing the classification of parts of PPM would have favored clock direction. This constraint enables bonding of local clocks of manifolds with the global clocks of PPM. Second, hierarchy of clock integration becomes a fundamental property generated by the integer systems. PPM naturally quantizes a certain set of choices or order factors and integrates the whole system as one unit. For geometric musical language, GML, the elementary 2×2 tensor represents cycloid like systems, i.e., a pair of circles connected and having a relative rotation, means, their symmetries change differently. The quaternion, octonion, dodecanion, hexadecanions (multiple of 4) are most suited to explaining rotation in a 3D system, i.e., effective tools to present a clocking architecture operating on a 3D Euclidean system. However, we cannot rule out the other 2D to 20D tensors listed in the Figs. 9, 10 and 11.

Hierarchical network of clocks set by PPM and the formation of 3D clock architecture: Fig. 8b outlines how if one fixes the maximum integer for a PPM would find a clocking direction, the next step is to really build an architecture of clocks using PPM as shown in Fig. 12. In Fig. 12a, we have shown coding protocol in a system, as one example, where, instead of ordered factor one writes the corresponding integer but keeps the areas covered by the ordered factor intact. Then builds a counter. For example, in Fig. 12b, 3-4-4-3 clock is created, one could build a real physical element with the vibrations similar to the ratios of 3:4:4:3 and that system could represent this particular code. Since in a PPM, there are guest areas built by small ordered factor values in the host areas built by large ordered factor values as shown in Fig. 12b and c, one could rewrite the codes as circles or clocks as shown in Fig. 12b and d. Interconnected clock representation was also described in Fig. 4, however, for dodecanion, $2 \times 2 \times 3$, $2 \times 3 \times 2$ and $3 \times 2 \times 2$ sets deliver, three simultaneously possible clock architectures as shown in Fig. 12d bottom. This is the same criterion of a self-operating universe as described in the Fig. 6 bottom panel where we suggested how one could deconstruct the dodecanion tensor and create three simultaneous tensors.

Octo-decanion h_j

h_{ij}	h_0	h_1	h_2	h_3	h_4	h_5	h_6	h_7	h_8	h_9	h_{10}	h_{11}	h_{12}	h_{13}	h_{14}	h_{15}	h_{16}	h_{17}
h_0	h_0	h_1	h_2	h_3	h_4	h_5	h_6	h_7	h_8	h_9	h_{10}	h_{11}	h_{12}	h_{13}	h_{14}	h_{15}	h_{16}	h_{17}
h_1	h_1	$-h_0$	h_3	$-h_2$	h_5	$-h_4$	h_7	$-h_6$	h_9	$-h_8$	h_{11}	$-h_{10}$	h_{13}	$-h_{12}$	h_{15}	$-h_{14}$	h_{17}	$-h_{16}$
h_2	h_2	$-h_3$	$-h_0$	h_1	h_2	h_7	$-h_4$	$-h_5$	h_6	h_{11}	$-h_8$	$-h_9$	h_{10}	$-h_{15}$	$-h_{12}$	h_{17}	h_{14}	h_{15}
h_3	h_3	h_2	$-h_1$	$-h_0$	h_7	h_2	$-h_3$	$-h_4$	$-h_{11}$	$-h_6$	h_{13}	h_8	h_{15}	$-h_{10}$	h_{17}	$-h_{12}$	h_{15}	h_{14}
h_4	h_4	$-h_5$	$-h_2$	$-h_7$	$-h_0$	h_1	h_2	h_{11}	h_4	$-h_5$	$-h_6$	$-h_{15}$	$-h_8$	h_{17}	h_{14}	h_{15}	$-h_{12}$	$-h_{13}$
h_5	h_5	h_4	$-h_7$	$-h_2$	$-h_1$	$-h_0$	h_{11}	h_2	h_3	h_4	h_{15}	$-h_6$	h_{17}	$-h_8$	$-h_{15}$	h_{14}	h_{11}	$-h_{12}$
h_6	h_6	$-h_7$	h_4	h_3	$-h_2$	$-h_{11}$	$-h_0$	h_1	$-h_2$	h_{15}	h_4	h_{17}	h_{14}	$-h_{15}$	$-h_8$	$-h_9$	$-h_{10}$	$-h_{11}$
h_7	h_7	h_6	h_5	h_4	$-h_{11}$	$-h_2$	$-h_1$	$-h_0$	h_{15}	$-h_2$	h_{17}	h_4	h_{15}	h_{14}	$-h_7$	$-h_8$	$-h_{11}$	$-h_{10}$
h_8	h_8	$-h_9$	$-h_6$	h_{11}	$-h_4$	$-h_3$	h_2	$-h_{15}$	$-h_0$	h_{17}	h_{14}	h_{15}	h_4	h_5	h_6	$-h_{11}$	$-h_8$	$-h_9$
h_9	h_9	h_8	$-h_{11}$	h_6	h_5	$-h_4$	$-h_{15}$	h_2	h_{17}	$-h_0$	h_{15}	$-h_{14}$	$-h_3$	$-h_4$	$-h_{11}$	h_6	h_{13}	h_8
h_{10}	h_{10}	$-h_{11}$	h_8	$-h_{13}$	h_6	$-h_{15}$	$-h_4$	h_{17}	$-h_{14}$	$-h_{15}$	$-h_0$	h_1	$-h_2$	$-h_{11}$	$-h_4$	h_5	h_6	h_7
h_{11}	h_{11}	h_{10}	h_9	$-h_8$	h_{15}	h_6	h_{17}	$-h_4$	$-h_{15}$	h_{14}	$-h_1$	$-h_0$	h_{11}	$-h_2$	h_3	$-h_4$	$-h_7$	h_6
h_{12}	h_{12}	$-h_{13}$	$-h_{10}$	$-h_{15}$	h_8	h_{17}	$-h_{14}$	$-h_{15}$	$-h_4$	h_3	h_2	$-h_{11}$	$-h_0$	h_1	h_2	h_7	$-h_4$	h_5
h_{13}	h_{13}	h_{12}	$-h_{15}$	h_{10}	h_{17}	h_8	h_{15}	$-h_6$	$-h_5$	h_4	h_{11}	h_2	$-h_1$	$-h_0$	h_7	h_2	$-h_3$	h_4
h_{14}	h_{14}	$-h_{15}$	h_{12}	h_{17}	$-h_{14}$	h_{15}	h_8	h_7	$-h_6$	h_{11}	h_4	$-h_3$	$-h_2$	$-h_7$	$-h_0$	h_1	$-h_2$	h_3
h_{15}	h_{15}	h_{14}	h_{17}	h_{12}	$-h_{15}$	$-h_{14}$	h_9	h_8	h_{11}	$-h_6$	$-h_5$	h_4	$-h_7$	$-h_2$	$-h_1$	$-h_0$	h_3	h_2
h_{16}	h_{16}	h_{17}	$-h_{14}$	$-h_{15}$	$-h_{12}$	$-h_{11}$	h_{10}	h_{11}	h_8	$-h_{13}$	$-h_6$	h_7	h_4	h_3	h_2	$-h_3$	$-h_0$	h_1
h_{17}	h_{17}	h_{16}	$-h_{15}$	$-h_{14}$	h_{13}	$-h_{12}$	h_{11}	h_{10}	h_9	$-h_8$	$-h_7$	$-h_6$	h_5	$-h_4$	$-h_3$	h_2	$-h_1$	$-h_0$

h_i

tetra-decanion h_j

h_{ij}	h_0	h_1	h_2	h_3	h_4	h_5	h_6	h_7	h_8	h_9	h_{10}	h_{11}	h_{12}	h_{13}
h_0	h_0	h_1	h_2	h_3	h_4	h_5	h_6	h_7	h_8	h_9	h_{10}	h_{11}	h_{12}	h_{13}
h_1	h_1	$-h_0$	h_3	$-h_2$	h_5	$-h_4$	h_7	$-h_6$	h_9	$-h_8$	h_{11}	$-h_{10}$	h_{13}	$-h_{12}$
h_2	h_2	$-h_3$	$-h_0$	h_1	h_2	h_7	$-h_4$	$-h_5$	h_6	h_{11}	$-h_8$	h_{13}	h_{10}	$-h_{11}$
h_3	h_3	h_2	$-h_1$	$-h_0$	h_7	h_2	$-h_3$	$-h_4$	$-h_{11}$	$-h_6$	h_{13}	h_8	h_{11}	$-h_{10}$
h_4	h_4	$-h_5$	$-h_2$	$-h_7$	$-h_0$	h_1	h_2	h_{11}	h_4	h_{13}	$-h_6$	$-h_{11}$	$-h_8$	h_9
h_5	h_5	h_4	$-h_7$	$-h_2$	$-h_1$	$-h_0$	h_{11}	h_2	h_{13}	h_4	h_{11}	$-h_6$	h_9	$-h_8$
h_6	h_6	$-h_7$	h_4	h_3	$-h_2$	$-h_{11}$	$-h_0$	h_{13}	$-h_2$	h_{11}	h_4	h_5	h_6	$-h_7$
h_7	h_7	h_6	h_5	h_4	$-h_{11}$	$-h_2$	h_{13}	$-h_0$	h_{11}	$-h_2$	h_3	h_4	$-h_7$	h_6
h_8	h_8	$-h_9$	$-h_6$	h_{11}	$-h_4$	h_{13}	h_2	$-h_{11}$	$-h_0$	h_1	h_2	$-h_7$	h_4	h_5
h_9	h_9	h_8	$-h_{11}$	h_6	h_{13}	$-h_4$	$-h_{11}$	h_2	$-h_1$	$-h_0$	$-h_7$	$-h_2$	$-h_5$	$-h_4$
h_{10}	h_{10}	$-h_{11}$	h_8	h_{13}	h_6	$-h_{11}$	$-h_4$	$-h_3$	$-h_2$	h_7	$-h_0$	h_1	$-h_2$	$-h_3$
h_{11}	h_{11}	h_{10}	h_{13}	$-h_8$	h_{11}	h_6	$-h_5$	$-h_4$	h_7	h_2	$-h_1$	$-h_0$	$-h_3$	$-h_2$
h_{12}	h_{12}	h_{13}	$-h_{10}$	$-h_{11}$	h_8	h_9	h_6	h_7	$-h_4$	h_5	h_2	h_3	$-h_0$	h_1
h_{13}	h_{13}	h_{12}	$-h_{11}$	h_{10}	$-h_9$	h_8	h_7	$-h_6$	$-h_5$	h_4	$-h_3$	h_2	$-h_1$	$-h_0$

h_i

Fig. 9 An example of octodecanion (top) and an example of tetradecanion (bottom) hexa-decanion h_j (columns); h_i (rows)

h_{ij}	h_0	h_1	h_2	h_3	h_4	h_5	h_6	h_7	h_8	h_9	h_{10}	h_{11}	h_{12}	h_{13}	h_{14}	h_{15}
h_0	h_0	h_1	h_2	h_3	h_4	h_5	h_6	h_7	h_8	h_9	h_{10}	h_{11}	h_{12}	h_{13}	h_{14}	h_{15}
h_1	h_1	$-h_0$	h_3	$-h_2$	h_5	$-h_4$	h_7	$-h_6$	h_9	$-h_8$	h_{11}	$-h_{10}$	h_{13}	$-h_{12}$	h_{15}	$-h_{14}$
h_2	h_2	$-h_3$	$-h_0$	h_1	h_2	h_7	$-h_4$	$-h_5$	h_6	h_{11}	$-h_8$	$-h_9$	h_{10}	$-h_{15}$	$-h_{12}$	h_{13}
h_3	h_3	h_2	$-h_1$	$-h_0$	h_7	h_2	$-h_3$	$-h_4$	$-h_{11}$	$-h_6$	h_{13}	h_8	h_{15}	$-h_{10}$	h_{13}	$-h_{12}$
h_4	h_4	$-h_5$	$-h_2$	$-h_7$	$-h_0$	h_1	h_2	h_{11}	h_4	$-h_5$	$-h_6$	$-h_{15}$	$-h_8$	h_9	h_{10}	h_{11}
h_5	h_5	h_4	$-h_7$	$-h_2$	$-h_1$	$-h_0$	h_{11}	h_2	h_3	h_4	h_{15}	$-h_6$	h_{13}	$-h_8$	$-h_{11}$	h_{10}
h_6	h_6	$-h_7$	h_4	h_3	$-h_2$	$-h_{11}$	$-h_0$	h_1	$-h_2$	h_{15}	h_4	h_5	h_6	$-h_{11}$	$-h_8$	$-h_9$
h_7	h_7	h_6	h_5	h_4	$-h_{11}$	$-h_2$	$-h_1$	$-h_0$	h_{15}	$-h_2$	h_3	h_4	h_{11}	h_6	$-h_9$	$-h_8$
h_8	h_8	$-h_9$	$-h_6$	h_{11}	$-h_4$	$-h_3$	h_2	$-h_{15}$	$-h_0$	h_1	h_2	h_{11}	h_4	h_5	h_6	$-h_7$
h_9	h_9	h_8	$-h_{11}$	h_6	h_5	$-h_4$	$-h_{15}$	h_2	$-h_1$	$-h_0$	h_{11}	$-h_2$	$-h_3$	$-h_4$	$-h_7$	h_6
h_{10}	h_{10}	$-h_{11}$	h_8	$-h_{13}$	h_6	$-h_{15}$	$-h_4$	h_3	$-h_2$	$-h_{11}$	$-h_0$	h_1	$-h_2$	$-h_7$	$-h_4$	h_5
h_{11}	h_{11}	h_{10}	h_9	$-h_8$	h_{15}	h_6	$-h_5$	$-h_4$	$-h_{11}$	h_2	$-h_1$	$-h_0$	h_7	$-h_2$	h_5	$-h_4$
h_{12}	h_{12}	$-h_{13}$	$-h_{10}$	$-h_{15}$	h_8	h_{13}	$-h_6$	$-h_{11}$	$-h_4$	h_3	h_2	$-h_7$	$-h_0$	h_1	h_2	h_3
h_{13}	h_{13}	h_{12}	$-h_{15}$	h_{10}	$-h_9$	h_8	h_{11}	$-h_6$	$-h_5$	h_4	h_7	h_2	$-h_1$	$-h_0$	h_3	h_2
h_{14}	h_{14}	$-h_{15}$	h_{12}	h_{13}	$-h_{10}$	h_{11}	h_8	h_9	$-h_6$	h_7	h_4	$-h_5$	$-h_2$	$-h_3$	$-h_0$	h_1
h_{15}	h_{15}	h_{14}	h_{13}	h_{12}	$-h_{11}$	$-h_{10}$	h_9	h_8	h_7	$-h_6$	$-h_5$	h_4	$-h_3$	$-h_2$	$-h_1$	$-h_0$

Penta-decanion h_j (columns); h_i (rows)

h_{ij}	h_0	h_1	h_2	h_3	h_4	h_5	h_6	h_7	h_8	h_9	h_{10}	h_{11}	h_{12}	h_{13}	h_{14}
h_0	h_0	h_1	h_2	h_3	h_4	h_5	h_6	h_7	h_8	h_9	h_{10}	h_{11}	h_{12}	h_{13}	h_{14}
h_1	h_1	$-h_0$	h_3	$-h_2$	h_5	$-h_4$	h_7	$-h_6$	h_9	$-h_8$	h_{11}	$-h_{10}$	h_{11}	h_{14}	h_{13}
h_2	h_2	$-h_3$	$-h_0$	h_1	h_2	h_7	$-h_4$	$-h_5$	h_6	h_{11}	$-h_8$	$-h_9$	h_{14}	$-h_{11}$	$-h_{12}$
h_3	h_3	h_2	$-h_1$	$-h_0$	h_7	h_2	$-h_3$	$-h_4$	$-h_{11}$	$-h_6$	h_7	h_{14}	h_9	$-h_{10}$	h_{11}
h_4	h_4	$-h_5$	$-h_2$	$-h_7$	$-h_0$	h_1	h_2	h_{11}	h_4	$-h_5$	h_{14}	$-h_7$	$-h_8$	h_{11}	h_{10}
h_5	h_5	h_4	$-h_7$	$-h_2$	$-h_1$	$-h_0$	h_{11}	h_2	h_3	h_{14}	h_5	$-h_6$	h_{11}	$-h_8$	$-h_9$
h_6	h_6	$-h_7$	h_4	h_3	$-h_2$	$-h_{11}$	$-h_0$	h_1	h_{14}	$-h_3$	h_4	h_{11}	h_6	$-h_9$	$-h_8$
h_7	h_7	h_6	h_5	h_4	$-h_{11}$	$-h_2$	$-h_1$	h_{14}	h_1	$-h_2$	h_{11}	h_4	h_5	h_6	$-h_7$
h_8	h_8	$-h_9$	$-h_6$	h_{11}	$-h_4$	$-h_3$	h_{14}	h_1	$-h_0$	h_{11}	$-h_2$	h_3	h_4	h_7	h_6
h_9	h_9	h_8	$-h_{11}$	h_6	h_5	h_{14}	h_3	h_2	h_{11}	$-h_0$	h_1	$-h_2$	$-h_7$	$-h_4$	h_5
h_{10}	h_{10}	$-h_{11}$	h_8	$-h_7$	h_{14}	$-h_5$	$-h_4$	h_{11}	h_2	$-h_1$	$-h_0$	$-h_7$	$-h_2$	h_5	$-h_4$
h_{11}	h_{11}	h_{10}	h_9	h_{14}	h_7	h_6	h_{11}	$-h_4$	$-h_3$	h_2	h_7	$-h_0$	$-h_1$	$-h_2$	h_3
h_{12}	h_{12}	$-h_{11}$	h_{14}	h_9	h_8	h_{11}	$-h_6$	h_5	$-h_4$	h_7	h_2	h_1	$-h_0$	h_3	h_2
h_{13}	h_{13}	h_{14}	$-h_{11}$	h_{10}	h_{11}	h_8	$-h_9$	$-h_6$	$-h_7$	h_4	h_5	h_2	$-h_3$	$-h_0$	h_1
h_{14}	h_{14}	$-h_{13}$	h_{12}	h_{11}	h_{10}	h_9	h_8	h_7	$-h_6$	h_5	h_4	$-h_3$	$-h_2$	$-h_1$	$-h_0$

Fig. 10 An example of hexadecanion (top) and an example of pentadecanion (bottom)

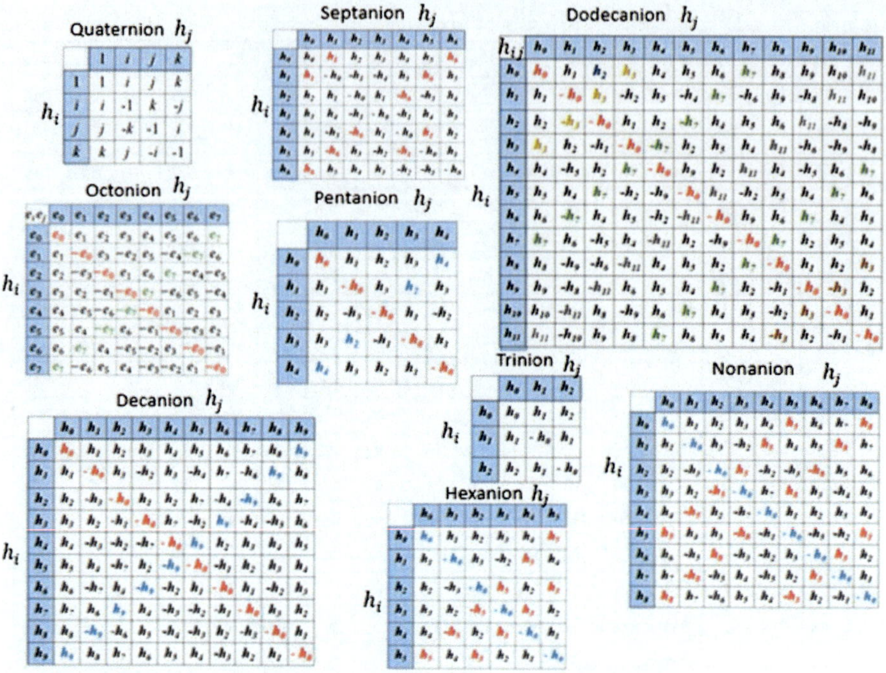

Fig. 11 Different multinions are shown

4 Applications of Dodecanion Geometric Algebra: Construction of Space-Time-Topology-Prime (stTS) Metric

4.1 *Four Fundamental Tensors with a Physical Significance*

Using different compositions of multinions as listed in the Figs. 9, 10 and 11, we could build space-time-topology-prime metric to explain the 11D dynamics of a multi-dimensional system [18, 19] because they are unit tensors which could not be used to define the other multi-dimensional tensors, note that we produce tensors here below 20 ($2 \times 11 > 20$). In GML, we use all spheres or clocks as unitary tensors, even the whole clock architecture or tensor representing entire system is unitary tensor, thus, entire locally and globally, any elementary tensor is unitary.

To build the metric for the self-operating universe, we take four tensors 2, 3, 5 and 7 as fundamental, a 2×2 tensor which we assign as time, 3×3 tensor we assign as space, 5×5 tensor we assign as topology, 7×7 tensor we assign as primes. If an event happens following this universe one has to assign time, space, topology and primes. It means by using 2×2 tensor or two basic values, one real and the other imaginary, we define time, 3×3 tensor we define space, basically three spatial

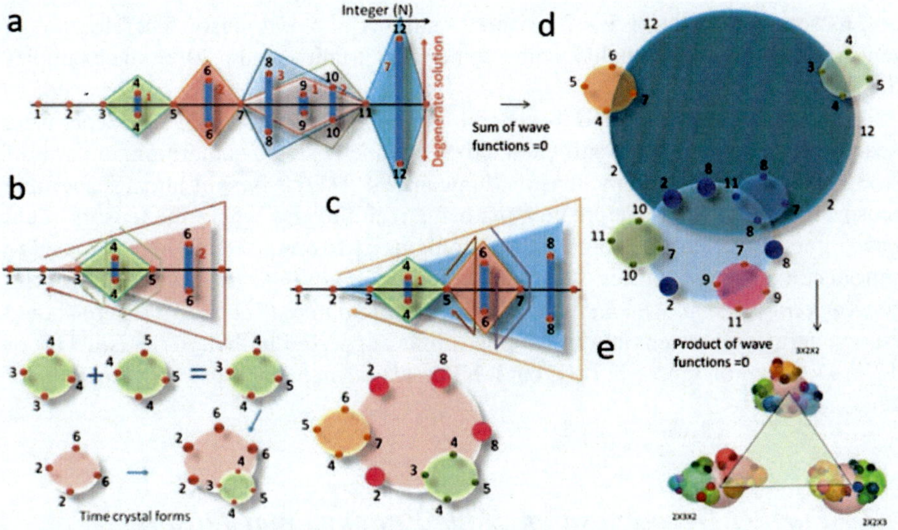

Fig. 12 **a** Integers of ordered factor metric is shown with its corresponding topology in the metric. Integers are plotted in ordered factor metric along the horizontal axis, we expand the integer, instead of putting the value of ordered factor we put the integer itself. **b** We put an arrow around a topology to depict its equivalent clock. **c** Topology inside a topology suggests connected clocks. **d** An example of a clock architecture built from a typical PPM. **e** The clock architecture for a self-operating universe, i.e., a dodecanion

coordinates, 5 × 5 tensor we choose 5 topologies from 15 and using 7 × 7 we choose 7 of the 15 primes that we have selected.

4.2 *Multinion Tensors Could Be Deconstructed in Terms of Prime Tensors*

Now, we could have 2 × 2, 3 × 3, 5 × 5 and 7 × 7 tensors embedded inside other tensors. For example, tensor 4 × 4 could be written as four 2 × 2 tensors. Then 6 × 6 tensor could be written as four 3 × 3 tensors. Similarly, 8 × 8 tensor could be written as 16 number of 2 × 2 tensors or 4 number of 4 × 4 tensors. 9 × 9 tensor could be written as 9 number of 3 × 3 tensors. 10 could be written as four 5 × 5 tensors or 25 number of 2 × 2 tensors. 12 × 12 tensor could be written as 16 number of 3 × 3 tensor, or 9 number of 4 × 4 tensor. 14 × 14 tensor could be written as 4 number of 7 × 7 tensor or 49 number of 2 × 2 tensors. 15 × 15 could be written as 9 number of 5 × 5 tensor or 25 number of 3 × 3 tensor. 16 could be rewritten as 64 number of 2 × 2 tensor, or 16 number of 4 × 4 tensors or 4 number of 8 × 8 tensor. 18 × 18 tensor could be rewritten as 36 number of 3 × 3 tensor, 4 number of 9 × 9 tensor, 81 number of 2 × 2 tensor. 20 × 20 tensor could be rewritten as 100 number of 2

× 2 tensor, 16 number of 5 × 5 tensor, 25 number of 4 × 4 tensor. Therefore, most tensors could show the duality, some could show triality below 20. Some examples are there in Fig. 7b.

Earlier in astrophysics and in general relativity, the fundamental concepts were space and time which were sufficient to be explained using a quaternion, in terms of 3 × 3 space and a 1 × 3 time tensors. It means a 4 × 4 tensor split into a space-time tensor. However, when we create a mathematical universe where the tensors could have dimension 2–20 for holding the topologies up to anicosanion, here we need to introduce two more entities. That's why now *space-time to be replaced by space-time-topology-prime, which could be used to build a new 11D metric.* There should be no confusion between multiple upper limits suggested in the article, is it 11D or 12D? Dodecanion tensor is 12D, but it holds 11D dynamics as listed in the Table of Fig. 1c.

4.3 Further Decomposition of the Fundamental Tensors

String theory that promised to explain what is there in the vacuum, suggested that using warp factor one could combine 4D tensor with an 8D tensor and build an 11D universe [19]. This is a clever idea when we have only two types of tensors. Here we change the protocol a bit differently, we create a catalog of tensors of all dimensions from 2D to 20D and use the composition of those tensors to represent the mathematical universe of PPM. For 9D worlds, we need only one kind of tensor, spatial 3 × 3, a triplet of spatial hyperplanes could warp fractally and couple to build 9D. Nested hyperplane feature requires 9D. For the 6D world, it is either 2 × 3, means a pair of spatial hyperplanes warp, or 3 × 2, means three imaginary worlds one inside another build a time crystal architecture. 5D tensor one could write in two ways, as a fundamental tensor, or sum of two types of tensors 3 × 3 and 3 × 2 or 5 × 2. When it is a fundamental tensor then it represents a combination of 5 topologies. However, when 5D tensor represents a combination of multiple values, similar to 4 × 4 tensors or quaternions we get a space-time metric, where 3 or 5 imaginary worlds are nested. For the 7D tensor, when its fundamental, PPM dynamics related to five different topologies would activate. However, if it could be 5 + 2, 6 + 1, 3 + 4, if 7D tensor is a composite tensor of topology + time (when a geometric shape changes with time, 5 + 2); warped pair of hyperspaces + time (6 + 1), Space + space-time. Thus, even the fundamental tensors 5 and 7 could be divided into space and time parts. The decomposition of higher dimensional tensors in terms of smaller dimensional tensors is listed in Table 1.

The inclusion of topology and prime in the dual concept of space and time enables geometric language features GML and the self-operational features of PPM to be included as a fundamental feature in the constituents that build the universe. It would have severe ripple effects in the formulation of associated theories.

Table 1 A summary of multinion algebra de-construction. From 2D to 20D we can split large dimensional tensors using smaller dimension tensors

Dimen sion	Tensor	Split protocol	Code of tensor
2D	2x2	Prime. (time, t)	h_1, h_2
3D	3x3	Prime. (space, s)	h_3
4D	4x4	2x2, 3+1	$h_2^2, h_{3,1}$
5D	5x5	Prime, 3+2. (Topology, T)	$h_5, h_{3,2}$
6D	6x6	2x3, 3x2, 5+1	$h_3^2, h_2^3, h_{5,1}$
7D	7x7	Prime, 5+2, 2x3+1, (Symmetry, S)	$h_7, h_{5,2}, h_{3,1}^2$
8D	8x8	2x4, 4x2, 5+3, 2x3+2	$h_2^4, h_4^2, h_{5,3}, h_{3,2}^2$
9D	9x9	3x3, 5+4, 7+2	$h_3^3, h_{5,3,1}, h_{7,2}$
10D	10x10	2x5,5x2, 7+3, 3x3+1	$h_2^5, h_5^2, h_{7,3}, h_{3,1}^3$
11D	11x11	Prime, 2x5+1, 3x3+2	$h_{11}, h_{5,1}^2, h_{3,2}^3, h_{7,3,1}$
12D	12x12	3x4,4x3,5x2+2, 7+5,	$h_3^4, h_4^3, h_{5,2}^2, h_{7,5}$
13D	13x13	Prime, 3x4+1, 5x2+3, 2x3+7, 3x3+4	$h_{13}, h_{3,1}^4, h_{5,3}^2, h_{2,7}^3$, $h_{3,4}^3$…4 means $h_{3,1}$..
14D	14x14	2x7,7x2,3x3+5,5x2+4, 3x4+2, 2x4+2x3	$h_2^7, h_7^2, h_{3,5}^3, h_{5,4}^2, h_{3,2}^4$ $h_{2,2}^{4,3}$
15D	15x15	3x5,5x3,7x2+1,6x2+3	$h_3^5, h_5^3, h_{7,1}^2, h_{6,1}^2$
16D	16x16	3x5+1,2x7+2,11+5,13+3	$h_{3,1}^5, h_{7,2}^2, h_{11,5}, h_{13,3}$
17D	17x17	Prime, 3x5+2,2x7+3, 2x6+5,	$h_{17}, h_{3,2}^5, h_{7,2}^2, h_{2,5}^6$
18D	18x18	3x6,6x3,2x9,9x2,4x4+2, 2x7+4, 13+5, 11+7	$h_3^6, h_6^3, h_9^2, h_2^9, h_{4,2}^4$ $h_{2,4}^7, h_{11,7}, h_{13,5}$
19D	19x19	Prime, 3x6+1, 17+2, 13+6, 11+8,12+7,2x9+1	$h_{19}, h_{3,1}^6, h_{17,2}, h_{13,6}$ $h_{11,8}, h_{12,7}, h_{2,1}^9$
20D	20x20	2x10, 4x5, 3x6+2,2x7+6, 11+9, 13+7, 1+19,3+17	$h_2^{10}, h_4^5, h_{3,2}^6, h_{7,6}^2$, $h_{11,9}, h_{13,7}, h_{1,19}, h_{3,17}$

4.4 Multi-dimension Could Be Achieved in Various Ways: Our "Within and Above" Assembly Could Process Quantum Effect and Minkowski Space-Time in Two Nested Worlds

Often in the literature, we find that octonions are used in quantum mechanics, to explain eight distinct dynamics in 8 different dimensions. Minkowski space-time is realized using 6×6 and 10×10 tensors, in all these applications we should note that there is only one imaginary time dimension, different dimensions are used to explore different dynamics. When we describe dodecanion, with 11 imaginary worlds nested one inside another we explore a kind of dynamics that communicates across different imaginary worlds. We repeatedly note that in our formulation we need only 2×2 tensor, one real and one imaginary world to explain quantum mechanics and Minkowski space-time. For Minkowski space-time we need only 2×2 tensor, one real and one imaginary world, just like quantum mechanics, a pair of the real or imaginary world could share space and time parts of the tensor, together and separately simultaneously.

4.5 Supersymmetry (SUSY), Partner Hamiltonian and Warped 11D Space-Time Metric to 11D Space-Time-Topology-Prime Metric or 2-3-5-7 Metric

The supersymmetry concept was introduced to bring together Bosons and Fermions, time crystal in GML does the same, including the rotational angle. Two kinds of spins in SUSY was applied to branes and strings to constitute the vacuum of the universe in String theory. The string is a mapping from 2d space-time to 1D-dimension space time or 1D extendable wire. Here, due to GML, the 11D transition from 1D, is very well-defined and there is no short-cut like string to connect higher dimensions. In GML, 1D is point, 2D is line, 3D is an area, 4D is time, 5D is singularity, 6D is singularity pathway, 7D is time crystal, 8D is 15 topological hyperspaces holding time crystals, 9D is statistical dominance of 15 primes, 10D is 10 classes of phase prime metrics, 11D is the metric of metrics that selects the metrics itself. Therefore, higher dimensions in the GML-PPM protocol does not require a typical structure to connect, instead of remaining within a system inside one could build SUSY was thus far limited to space-time concept only. Therein 10D heterotic string on the space-time background is

$$\mathrm{d}s^2 = -\mathrm{d}x_0^2 + \mathrm{d}x_1^2 + \mathrm{d}x_2^2 + \mathrm{d}s_X^2 \tag{1}$$

$$ds^2 = \Delta^{-1}\left(-\mathrm{d}x_0^2 + \mathrm{d}x_1^2 + \mathrm{d}x_2^2\right) + \Delta^{\frac{1}{2}}\mathrm{d}s_X^2 \tag{2}$$

The second equation is the 11D space-time metric, wherein,s_X is the Calabi-Yau 4-fold metric. Here we get $\Delta(y)$ as wrap factor, which integrates the 4D Minkowski space with the 3D or 8D (Eq. 2) space, to build an 11D metric. Here the metric measures spatial distance, for self-assembly of systems within and above we could measure the space $\mathrm{d}s^2(h_3)$, time $\mathrm{d}t^2(h_{1,2})$, topology $\mathrm{d}T^2(h_5)$ and symmetry $\mathrm{d}S^2(h_7)$. Here the combinations of dual, triple and quad features are:

Space-time $st(h_3, h_{1,2})$, with 3 + 1/2 = 5/6 dimensions; Space-topology $sT(h_3, h_5)$, with 3 + 5=8 dimensions; time-symmetry $tS(h_{1,2}, h_7)$, with ½ + 7=8/9 dimensions; topology-symmetry $TS(h_5, h_7)$, with 5 + 7=12 dimensions; $stST(h_3, h_{1,2}, h_5, h_7)$, with 3 + 1/2 + 5+7 = 16 or 17 dimensions. Space-symmetry $sS(h_3, h_7)$, with 3 + 7=10 dimensions; Space-topology $tT(h_{1,2}, h_5)$, with ½ + 5=6/7 dimensions; space-symmetry-time $sSt(h_3, h_7, h_{1,2})$ with 3 + 7+1/2 = 11/12 dimensions; space-time-topology $stT(h_3, h_{1,2}, h_5)$ with 2 + 1/2 + 5=8/9 dimensions; space-symmetry-topology $sST(h_3, h_7, h_5)$ with 3 + 7+5 = 15 dimensions; symmetry-time-topology $tST(h_{1,2}, h_7, h_5)$ with ½ + 7+5 = 13/14 dimensions; symmetry-space-time $sSt(h_3, h_7, h_{1,2})$ 3 + 7+1/2 = 11/12 dimensions. Please find the list of decompositions of 2D to 12D tensors in the Table 1.

Since icosahedron has 12 corners, for the "within-and-above" universe we would have a maximum of 12 dimensions and since icosahedron has 20 planes we would have 20 dimensions when dimension means adding a new dynamics. Hence, $stST$ would be confined in one imaginary world, i.e., it would represent the projected and the feedback time crystal "to and from" the PPM.

The metric representing the time crystal in $stST = S2T2$ universe of 12 nested worlds is given by

$$\begin{aligned}\mathcal{H} = {} & \Delta_{PPM1} st(h_3, h_{1,2}) + \Delta_{PPM2} sT(h_3, h_5) + \Delta_{PPM3} tS(h_{1,2}, h_7) \\ & + \Delta_{PPM4} TS(h_5, h_7) + \Delta_{PPM5} sS(h_3, h_7) + \Delta_{PPM6} tT(h_{1,2}, h_5) \\ & + \Delta_{PPM7} sST(h_3, h_7, h_5) + \Delta_{PPM8} stT(h_3, h_{1,2}, h_5) \\ & + \Delta_{PPM9} tST(h_{1,2}, h_7, h_5) + \Delta_{PPM10} sSt(h_3, h_7, h_{1,2}) \\ & + \Delta_{Project-Feedback} stST(h_3, h_{1,2}, h_5, h_7). \end{aligned} \tag{3}$$

Possible applications of space-time-topology-prime metric in elementary physics: The hierarchy problem in physics is that the weak force is 10^{24} times stronger than gravity. Quantum field theory (QFT) is a theoretical framework that combines classical field theory, special relativity, and quantum mechanics. QFT treats particles as excited states (also called quanta) of their underlying fields, whereas in GML we consider a generic composition of the time crystals represents the interacting systems, while the interaction is governed by PPM. Space-time-topology-Symmetry (S2T2) metric implements PPM, no need for Lagrangian (dynamic state of a system) and Feynman diagram, time crystal structures take care of both. In GML, one has to match topologies and bind the time crystals, i.e., apply ℋ (Appendix I for details), one could find the future dynamic states of the particles. In the perturbation theory of quantum mechanics, one starts with the known solution and perturbs it to reach to the solution of a complex quantum scenario, which is not required here for time crystals, because, we start from the simplest clock and build an architecture of clocks. PPM guides to fuse the time crystals and expand, thus, no need to implement trial and error to fit observation. Quantum electrodynamics (QED), a theory that predicts light-matter interaction (1927, Dirac coined QED), adds additional terms with the free electromagnetic field first terms is an additional interaction expression between electric current density and the second term is electromagnetic vector potential. Since weak interaction is 10^{24} times smaller than gravity, it is difficult to add additional terms related to gravity, herein, however, that problem does not arise if we put the forces in different imaginary worlds of the universe. How do the clocks are assembled, based on that the power of tens is achieved in a universe of nested worlds by GML? For example, to understand the planks constant GML takes two imaginary worlds, in one it explains power 10 (10^{10}) and in another the power -35 (anti-clock rotation is negative, 10^{-35}). In GML the fundamental constant becomes a topological constraint that is embedded in the time crystal of an atom or nanoscale system. Geometric constraints are fundamental to the 15 geometries selected in Fig. 1 and the 10 different classes of PPM [17]. Time cycles make quantum and classical computing equivalent [20].

Quantization of space, time, topology and prime in Gravity: Now, one operation of these tensors is geometry, could try solving the gravitational quantum problems, renormalization problems [21], which were historical hallmarks for the development of new protocols in physics that could solve long-unsolved mysteries. Time crystal with clockwise and anticlockwise rotating singularity spheres [22, 23] could represent spin network [24] or spin foam, to provide space-time architecture required in quantum. Loop quantum gravity quantizes area, so GML which quantizes space, topology and quantized symmetry quantize time, thus, Space-time-topology-Symmetry (S2T2) is quantized.

4.6 Higher Dimension in the Brain

Recently 11D structures are found in the brain, and they were linked to human cognition and consciousness [25]. Eight imaginary worlds are not sufficient to model information processing in the human brain because at least 12 brain components are there with 12 nested layered hardware assembled within and above. If the first 15 primes used in building the sides of 15 geometric shapes that combine to build every possible geometric structure of the universe then, using the protocol as a language one could construct 11D tensors. The operation of these tensors is geometric algebra. One of the critical features of dodecanion tensor is three compositions of the 12 universes as a group of $2 \times 2 \times 3$, $3 \times 2 \times 2$, and $2 \times 3 \times 2$. Here $2 \times 2 \times 3$ means take two universes as the supreme host, put 2 universes as guests inside the supreme host and in each pair of guest universes put three more universes. Then use dodecanion tensors to build how the interaction of different imaginary worlds affects new universes. $2 \times$

2×3, $3 \times 2 \times 2$, and $2 \times 3 \times 2$ grouping is a trigonal, i.e., topological clocking or coexistence of three choices, which is not part of the dodecanion tensor. Thus, for the first time, when we reach 12 universes in a single tensor a hierarchical topological operator is born, any hardware nested with 12 imaginary worlds within and above would show such behavior.

A complete model of the human brain using time crystal architecture: While modeling the human brain and building a time crystal architecture [26], Singh et al. [26] and Ghosh et al. [27] found 12 layers one inside another explain life cycles with 10^9 s (100 years) to 10^{-16} s faster oscillation of atomic bonds in amines and that 10^{25} order changes in time scale could be analyzed using GML. The hierarchy problem for the universe is not much different.

5 Conclusion and Future

There is one clear message from this work, quaternion and octonions are not the lone pairs in the series of the imaginary world of complex tensors, one could create dimensions of any kind and that is easy to build. There is no need to confine into the domain of quaternion and octonion alone. Many different kinds of algebra have been formulated within the domain of quaternion and octonion, we propose a new kind of algebra which is not limited to dodecanion or icosanion, instead it's a combination of tensors from dinion (2×2), trinion (3×3), quaternion (4×4), pentanion (5×5), hexanion (6×6), heptanion, octonion, nanonion, decanion, mono-decanion (11D), dodecanion (12D), tridecanion (13D), tetradecanion (14D), pentadecanion (15D), sexadecanion (16D), septadecanion (17D), octodecanion (18D), nanodecanion (19D) and icosanion (20). These are 19 different tensors and their internal compositions as listed in Table 1, which means inside an icosanion one may put pentanion and other tensors, could eventually generate a new algebra. Inventing a new algebra means defining the product of tensors, because addition and subtraction are straightforward. If one wants, of course following our product writing trick described here, could build higher level tensors and associated algebra, since our objective is to build an algebra for applying GML's 15 topologies, we restrict ourselves to dimension 20.

In summary, a set of 19 classes of tensors powered by space-time-topology-prime concept could build new metrics for astrophysics, partner Hamiltonians for elementary particle physics, quantum gravity and solid-state physics, where SUSY and quantum field theory finds difficulties in holding renormalizability intact. Gauge invariance or dynamic state invariance across the 11D universe (one inside another) is ensured by conformal transfer in GML-PPM. If we want to transmit a triangle across the imaginary worlds then, the three angles of a triangle would remain constant we squeeze or expand the triangle to fit it in different imaginary worlds. Thus, the difference between different imaginary worlds is the diameter of the maximum and minimum operating circle or sphere where we incorporate the 15 geometric shapes of GML. Therefore, the conformal feature is always ensured in FIT-GML-PPM formulations described here.

Acknowledgements Authors acknowledge the Asian office of Aerospace R&D (AOARD) a part of United States Air Force (USAF) for the Grant no. FA2386-16-1-0003 (2016–2019) on the electromagnetic resonance based communication and intelligence of biomaterials.

Appendix

While others would continue to build new algebras using addition, subtraction, multiplication and division of these complex numbers, we advocate operating a new function ※ that is used frequently in the fractal information theory (FIT), which is a combination of geometric musical language (GML) and phase prime metric, (PPM). The new function ※ looks into the topological symmetry of the participating elements of tensors, wherein all elements are geometric shapes. How interacting geometric shapes would bond together building a new geometric shape, FIT is a systematic study for that purpose. Therefore, when we write Q ※ iO ※ jD ※ kI, it consolidates that the self-similar geometries initiate bonding of wide ranges of geometric shapes in a complex 3D architecture. If all geometric shapes of the architecture are connected to clocks or modulo (modulo = number of corners of a geometric shape), the tensor gets an application in physics. The architecture explores all possible dynamics among the participant complex numbers.

References

1. Corrochano B (2010) Geometric computing for wavelet transforms, robot vision, learning, control and action. Springer, Heidelberg, Chapter 6, pp 149–183
2. Rozenfel′d AB (1988) A history of non-euclidean geometry: evolution of the concept of a geometric space. Springer, Heidelberg, p 373 (1988)
3. Conway JH, Smith DA (2003) On quaternions and octonions: their geometry, arithmetic, and symmetry, p 9. ISBN 1-56881-134-9
4. Shoemake K (1985) Animating rotation with quaternion curves. Comput Graphics 19(3):245–254
5. Shu JJ, Ouw LS (2004) Pairwise alignment of the DNA sequence using hypercomplex number representation. Bull Math Biol 66(5):1423–1438
6. Leron B, Duminda D, Duff MJ, Hajar E, Williams R (2009) Black holes, qubits and octonions. Phys Rep 471(3–4):113–219
7. Ghosh et al (2019) JP-2017-150171, World patent WO 2019/026983
8. Preparata F, Hong SJ. Convex hulls of finite sets of points in two and three dimensions. In: Manacher G, Graham SL (eds) CACM, vol 20, issue 2, p 88
9. Gardner M (1992) Fractal music, hypercards, and more: mathematical recreations from scientific american. W. H. Freeman, New York, pp 40, 53, and 58–60
10. Cormen TH, Charles EL, Ronald LR, Stein C (2001) Introduction to algorithms, 2nd edn. MIT Press and McGraw-Hill, pp. 862–868. ISBN 0-262-03293-7. Section 31.3: Modular arithmetic

11. Reddy S et al (2018) A brain-like computer made of time crystal: could a metric of prime alone replace a user and alleviate programming forever? Stud Comput Intell 761:1–44
12. Bewersdorff J (2005) Asymmetric dice: are they worth anything? In: Luck, logic, white lies: the mathematics of games. A K Peters, Wellesley, MA, pp 33–36
13. Cundy HM, Rollett A (1989) 3.11. Deltahedra. Mathematical models, 3rd edn. Tarquin Pub., Stradbroke, England, pp 142–144
14. Trigg CW (1978) An infinite class of deltahedra. Math Mag 51(1):55–57 (JSTOR 2689647)
15. Lasenby A (2005) Recent applications of conformal geometric algebra. Computer algebra and geometric algebra with applications. In: Li H, Olver PJ, Sommer G (eds). IWMM 2004, GIAE 2004. Lecture Notes in Computer Science, vol 3519. Springer, Heidelberg
16. Pugh A (1976) Polyhedra: a visual approach. University of California Press, Berkeley, California. ISBN 0-520-03056-7, pp 35–36
17. Bandyopadhyay A (2020) Nanobrain. The making of an artificial brain from a time crystal, 1st edn. CRC Press, March 16, 2020 (Forthcoming), ISBN 9781439875490 - CAT# K13502
18. Kac VG, Moody RV, Wakimoto M (1988) On E10. Differential geometrical methods in theoretical physics (Como, 1987). NATO Adv Sci Inst Ser C Math Phys Sci 250. Kluwer Acad Publ, Dordrecht, pp 109–128. MR 0981374
19. West P (2001) E11 and M theory. Classical Quant Gravity 18(21):4443–4460
20. Watrous J, Aaronson S (2009) Closed time like curves make quantum and classical computing equivalent. Proc R Soc A Math Phys Eng Sci 465(2102):631
21. Hamber HW (2009) Quantum gravitation—the Feynman path integral approach. Springer Nature. ISBN 978-3-540-85292-6
22. Penrose R (1971) Angular momentum: an approach to combinatorial spacetime. In Bastin T (ed) Quantum theory and beyond. Cambridge University Press, Cambridge
23. Penrose R (1969) Applications of negative dimensional tensors: combinatorial mathematics and its applications. In: Welsh DJA (ed) (Proc. Conf., Oxford, 1969), Academic Press, pp. 221–244, esp. p. 241. On the origins of twistor theory in: gravitation and geometry, a Volume in Honour of I. Robinson, Biblipolis, Naples 1987
24. Oeckl R (2003) Generalized lattice gauge theory, spin foams and state sum invariants. J Geometry Phys 46(3–4):308–354
25. Reimann MW et al (2017) Cliques of neurons bound into cavities provide a missing link between structure and function. Front Comput Neurosci, 12 June 2017. https://doi.org/10.3389/fncom.2017.00048
26. Singh P, Ray K, Fujita D, Bandyopadhyay A (2019) Complete dielectric resonator model of human brain from mri data: a journey from connectome neural branching to single protein. In: Ray K, Sharan S, Rawat S, Jain S, Srivastava S, Bandyopadhyay A (eds) Engineering vibration, communication and information processing. Lecture Notes in Electrical Engineering, vol 478. Springer, Singapore
27. Ghosh S, Sahu S, Fujita D, Bandyopadhyay A (2014) Design and operation of a brain like computer: a new class of frequency-fractal computing using wireless communication in a supramolecular organic, inorganic systems. Information 5:28–99

An Effective Filtering Process for the Noise Suppression in Eye Movement Signals

Sergio Mejia-Romero, J. Eduardo Lugo, Delphine Bernardin, and Jocelyn Faubert

Abstract All eye movement measurement contains some level of noise; denoising requires the use of digital filters and subsequent measurement processing. The elimination of noise in the eye movement signals commonly uses some linear time-invariant filter, such as the Butterworth, Gauss, or Savitzky-Golay filters. However, in high noise signals, the low-pass part of the filters can significantly distort and alter speeds saccades. This can be remedied by adding a high pass component, but this introduces its artifacts, especially false oscillations. Some of the problems can be improved using the mean filter, the Kalman filter, or the Bilateral filter. Although they are efficient, they still share the main feature: the resulting signal contains the remaining noise. Our proposed method is using the Wiener filter to denoise signal, and we will consider the estimation of the noise power spectrum calculated using an artificial eye. The results indicate that the method eliminates noise efficiently.

Keywords Eye tracker · Denoising · Reconstruction signal · AR model · Wiener filter

1 Introduction

Eye-tracking technique is widely used in neurosciences to know how a person performs the visual exploration of their environment during daily life or in a specific task, this technique allows you to create a record of the sequence of eye movement when you pass from one point of interest to another. Raw data always containing noise, derivate from saccadic oscillations, and smooth pursuit [1]. The

S. Mejia-Romero (✉) · J. Eduardo Lugo · D. Bernardin · J. Faubert
FaubertLab, School of Optometry, Université de Montréal, Montréal, QC, Canadá
e-mail: sergio.mejia.romero@umontreal.ca

D. Bernardin
Essilor Canada Ltd., Montréal, QC, Canadá

K. Ray et al. (eds.), *Proceedings of International Conference on Data Science and Applications*, Lecture Notes in Networks and Systems 148,
https://doi.org/10.1007/978-981-15-7561-7_2

correct acquisition of data is a critical step in fields such as life science research. The data collected efficiently guarantees reliability and ensures that the data analysis is prepared correctly [2]. Registered data must be obtained with optimal performance and with minimal error.

The position of the eyes is always changing. The saccades movements are abrupt and rapid changes in the direction of the gaze when the eye changes from one fixation point to another. These movements are rapid and vary between 30 ms and 120 ms [3] and are intended to place new points of interest in the visual scene in the fovea. The period in which the eyes are relatively stationary capturing the information is called fixation. The fixation last on overage typically range between 150 and 600 ms [3].

During fixation, the eye constantly oscillates with low amplitude movements, generally ranging from 2 to 120 arc-minutes [3], and they are completely imperceptible under normal conditions. This type of movement is necessary to maintain the projected image on the photoreceptors of the retina. Without these microsaccades, staring at a point would cause cessation of stimuli sent to the brain, since the rods and cones only respond to changes in luminance.

Total noise is the combination of several noises, but mainly to the vibration and additional noise inherent in the recording method (for example, image sensor noise and electromagnetic noise). The noise level may vary depending on the environment and equipment, from around 0.01° to around 0.3°.

The level of noise and speed variability of eye movements make it hard to determine if the subject is looking at one point or several points around a small area. The goal of signal treatment is to separate the real signal from the signal with noise and determine the least possible number of false fixations estimates as possible.

Here, we present the Wiener filter using the noise power spectrum estimated using an artificial eye. The approach is novel in that it differs in concept from the traditional workflow using a window to estimate the noise power spectrum. Instead, it integrates the noise power spectrum parameters for the Wiener filter.

2 Method

2.1 Eye Signal

Data is collected using SMI commercial eye-tracking, software for gaze analysis [4]. The eye tracker was properly calibrated before use; the user is asked to fixate tree calibration points shown on the screen. The data acquisition is made at 120 Hz. Horizontal and vertical gaze point is acquired using the technique described by SMI [4]. The resolution was 0.5 min of arc. The position of the eyes with respect to the calibration screen is estimated and the raw signal was smoothed with a symmetrical digital low-pass filter (−3 db at 60 Hz).

The estimated coordinates of the pupil position in the frame, are estimated using the coordinates (*xr, yr*) from the eye-tracking system and are transferred to the reference frame using the position of the projection of the subject pupil on the eye tracker camera (*x, y*). Where the coordinates in the image plane are the result of the linear transformation applied by the software according to the first initial calibration of the three points.

The center of the pupil, on the camera, this is known as an arbitrary reference system. If the camera does not move, then the minimum variation of the recorded point represents an eye movement [5], each data point is transformed to the arbitrary frame of reference in order to always be a positive integer ranging from 0 to 1240 pixels units. The linear transformation between the pupil coordinates $p(t)$ on the data provided by the tracker system is expressed by:

$$p(t) \in \{0, 1, 2 \ldots, \text{res}\}, \tag{1}$$

where $p(t)$ is the horizontal and vertical position. "res" are the camera reference, (0,0) it is the position upper left corner and (res, res) represents the lower right corner.

If we consider that the acquisition system is linear and when this system with impulse response $h(t)$ is activated with an arbitrary input position signal $p(t)$, its response, $g(t)$, can be computed straightforwardly:

$$g(t) = \int h(\sigma) * p(t)\, \mathrm{d}\sigma. \tag{2}$$

Alternatively, in the Fourier domain,

$$G(\omega) = H(\omega)P(\omega). \tag{3}$$

where ω is the angular frequency; $G(\omega)$ is the FT of $g(t)$; $H(\omega)$ is the FT representation of the impulse response $h(t)$, and $P(\omega)$ is the input signal in Fourier representation.

If the noise is not ignored in Eq. (2) and it is assumed additive and signal independent, where $s(t)$ signal detected with noise, $g(t)$ is the detected signal, $n(t)$ is the unwanted additive noise. Since the signal detected is obtained using a linear system in amplitude, the model can be expressed as:

$$s(t, k) = g(t, k) + n(t, k), \tag{4}$$

where $\{t = 0, 1, 2\ldots, N-1\}$ is the discrete-time index, $\{k = 1, 2\ldots N\}$ is the frame number and N is the length of the frame.

Using the Fourier transform, Eq. (4) can be rewritten as:

$$S(\omega, k) = G(\omega, k) + N(\omega, k), \tag{5}$$

where $S(\omega, k)$, $G(\omega, k)$ and $N(\omega, k)$ are the Fourier transforms of $s(t, k)$, $g(t, k)$ and $n(t, k)$, respectively, in here ω is the angular frequency index of sampling.

$$S(\omega, k) = \sum_{n-\infty}^{\infty} y(n)\omega(k-n)\mathrm{e}^{-j\omega n}, \tag{6}$$

where $\omega(n)$ is the work window by k samples. Multiplying both sides of Eq. (5) by their complex conjugates, we get:

$$S(\omega, k)|^2 = |G(\omega_2 k)|^2 + |N(\omega, k)|^2 + 2|G(\omega, k)N(\omega, k)|, \tag{7}$$

where $|G(\omega, k)|^2$ is the power spectral of G. The $|N(\omega, k)|^2$ and $|\ G(\omega, k)\ N(\omega, k)|$ can't be obtained directly so they need to be approximated as:

$$\begin{gathered} E\{\ S(\omega, k)|^2\} = E\{|G(\omega, k)|^2\} + E\{|N(\omega, k)|^2\}\ldots \\ +2E\{|G(\omega, k)N(\omega, k)|\}, \end{gathered} \tag{8}$$

where $E\{.\}$ represents the expectation operator. As the additive noise is assumed to be zero-mean and uncorrelated, the $|G(\omega, k)\ N(\omega, k)|$ reduces to zero.

$$E\{\ S(\omega, k)|^2\} = E\{|G(\omega, k)|^2\} + E\{|N(\omega, k)|^2\} \tag{9}$$

Normally, $E\{N(\omega, k)|^2\}$ the estimate is made during most stable periods and is denoted by $\acute{P}n(\omega, k)$. Therefore, the estimated of the clean signal power spectrum can be expressed as:

$$P_g(\omega, k) = P_s(\omega, k) - P_n(\omega, k), \tag{10}$$

here $Pg(\omega, k)$ is referred to as the estimated signal power spectrum, $Ps(w, k)$ corresponds to the noise signal power spectrum, and $Pn(\omega, k)$ is the estimated noise power spectrum from stable signal periods [6].

The objective of denoising in the frequency domain is then to find an optimal gain $H(\omega)$ at each frequency (ω) that would attenuate the noise as much as possible with little distortion of the desired signal.

The spectral subtraction Eq. (10) can be written as:

$$P_g(\omega, k) = P_s(\omega, k)1 - \left\lceil \frac{P_n(\omega, k)}{P_s(\omega, k)} \right\rceil, \tag{11}$$

where $1 - \left\lceil \frac{P_n(\omega,k)}{P_s(\omega,k)} \right\rceil$ is the filter gain, which is real.

2.2 Wiener Filter

Wiener filter is used over noised signals. The response of this filter is linear, stable and efficient when the noise assumption is approximately equal to that of the system. This filter is based on minimizing the differences between the output signals and the actual signals [7–10]. The filter assumes that both noise and signal are second-order functions so it cannot rebuild the frequency components that are suppressed.

The goal of the Wiener filter is to estimate the signal without noise by filtering the signal using a sample of the signal with noise. Below, we show the model that describes the Wiener filter:

$$g(t,k) \to G\left(e^{j\sigma}\right) \to +n(t,k) \to s(t,k) \to H\left(e^{j\sigma}\right) \to y(t,k). \quad (12)$$

The signal $g(t, k)$ is detected by the linear time-invariant system (LTI) $G(e^{j\sigma})$ and contains additive noise $n(t, k)$, resulting in the signal detected as $s(t, k)= h(k)* g(t, k) + n(t, k)$, and as a reconstructed signal $s(t, k)$ after using the filter $H(e^{j\sigma})$ to make an estimate $y(t, k)$ from $g(t, k)$.

The goal of the Wiener filter is to estimate the LTI system $H(e^{j\sigma})$ such that the resulting signal $y(t, k)$ matches $g(t, k)$ as best as possible using the error signal.

We define the error signal (ε) to the result of the estimated signal and desired signal at a frequency (ω) as:

$$\varepsilon(\omega,k) = Y(\omega,k) - G(\omega,k). \quad (13)$$

This error can also be put into the form the quadratic mean error $\varepsilon(k)$, as:

$$E\{|e[k]|^2\} = 1/2\pi \int_{-\pi}^{\pi} \varphi_{eef}\left(e^{j\Omega}\right)d\Omega = \varphi_{ee[K=0]}, \quad (14)$$

where we get the minimization of the MSE derived from the raw signal $g(t, k)$ and the estimate signal $y(t, k)$.

The Wiener filter method assumes that the cross-power spectral density $\Phi sg(e^{j\Omega})$ from the observed signal $s(t, k)$ and the original signal $g(t, k)$, and the observed signal $\Phi ss(e^{j\Omega})$ are known.

The filter works by looking for the minimized MSE $E\{|e[k]|^2\}$ using as a reference to the transfer function (TF) $H(e^{j\sigma})$.

$$H\left(e^{j\sigma}\right) = \left[\Phi gs\left(e^{j\Omega}\right)/\Phi_{SS}\left(e^{j\Omega}\right)\right]. \quad (15)$$

Using the Wiener filter is not necessary to know the actual distortion process. We need to know the power spectral density $\Phi gs(e^{j\Omega})$ and $\Phi ss(e^{j\Omega})$ to make an approximate of $g(t, k)$ from $s(t, k)$ with a minimum *MSE*.

These PSDs can be estimated from the PSDs from the raw signal $\Phi gs(e^{j\Omega})$ as well as the transfer function $G(e^{j\Omega})$ of the distorting system.

Assuming that the noise function $n(k)$ is not directly correlated with $g(t, k)$ the PSD $\Phi gs(e^{j\Omega})$ can be derived as:

$$\Phi gs\left(e^{j\Omega}\right) = \Phi gg\left(e^{j\Omega}\right) \cdot G\left(e^{-j\Omega}\right), \tag{16}$$

and

$$\Phi ss\left(e^{j\Omega}\right) = \Phi gg\left(e^{j\Omega}\right) \cdot \left|G\left(e^{-j\Omega}\right)\right|^2 + \Phi nn\left(e^{j\Omega}\right), \tag{17}$$

Using the Eqs. 16 and 17 in the general function Wiener filter, the filter has the form:

$$\hat{Y}\left(e^{j\Omega}\right) = \frac{\Phi gg\left(e^{j\Omega}\right) \cdot G\left(e^{-j\Omega}\right)}{\Phi gg\left(e^{j\Omega}\right) \cdot \left|G\left(e^{j\Omega}\right)\right|^2 + \Phi nn\left(e^{j\Omega}\right)}, \tag{18}$$

where $\hat{Y}(\omega)$ represent frequency domain form of the reconstructing signal, $G(\omega)$ is the Fourier representation of optimal gain, $\Phi nn(e^{j\Omega})$ noise signal spectral density, $\Phi gg(e^{j\Omega})$ spectral density of signal with noise.

Equation 18 can be rewritten by introducing the signal to noise ratio (SNR) between the original signal and the noise as:

$$H\left(e^{j\Omega}\right) = \frac{1}{G\left(e^{j\Omega}\right)} \frac{\left|G\left(e^{j\Omega}\right)\right|}{\left|G\left(e^{j\Omega}\right)\right|^2 + \frac{1}{\mathrm{SNR}\left(e^{j\Omega}\right)}}. \tag{19}$$

If there is no noise in the signal, i.e., $\Phi nn(e^{j\Omega}) = 0$, the expression between SNR is equal to 1. Therefore, the Wiener filter is the result of the inverse system to the distortion system:

$$H\left(e^{j\Omega}\right) = 1/G\left(e^{j\Omega}\right), \tag{20}$$

Ideally, if the distortion system is a passthrough $G(e^{j\Omega}) = 1$, the filter is:

$$H\left(e^{j\Omega}\right) = \frac{\Phi gg\left(e^{j\Omega}\right)}{\Phi gg\left(e^{j\Omega}\right) \cdot + \Phi nn\left(e^{j\Omega}\right)}. \tag{21}$$

From Eq. 21, we can make two observations, One, if the noise estimate is low, then the noise in the estimated signal will be maintained, and two, if the noise estimate is high, it will possibly cause the loss of the signal edges.

Noise level estimation is the critical part of the improvement method using the Wiener filter because the quality of the restored signal depends on the precise estimate of the noise power spectrum.

To estimate the PSD of the noise signal and propose an optimal Wiener filter, we use a static eye signal.

The only form to obtain the signal of an eye without movements is to use an artificial eye which must produce the reflexes of the cornea producing a signal in the fully stationary eye tracker [11].

2.3 *Material*

The eye signal acquisition was using SMI eye-tracking [12] is done at 120 Hz. The measurement of gaze point in time $g(t)$ is done using normalized Cartesians-coordinates reference, where (0, 0) is the position upper left corner.

In order to carry out the signal analysis test, we have developed using MATLAB, a tool that manages the eye-tracking data, read and export the raw data of the eye tracker and analyze the gaze direction.

According to the requirements of SMI eye-tracking, before running a test, the tracker is calibrated with three points to several locations of the calibration plane, thus obtaining accurate eye gaze data. After this calibration test, the participant follows the eighteen point on stimuli (Fig. 1).

During the experimental test, the participant observes the grid on the screen and made one fixation in each dot, we can show one test in Fig. 2.

For an estimate of the noise power spectrum, an artificial eye, was used, and the experimental procedure was simple:

- We calibrate the eye tracker system with a human subject in the usual way so we can start recording eye movements data.
- Then, we replace the subject with artificial eyes ensuring the position where the human eyes would have been and verify that the eye-tracking detects the reflection of the artificial eyes, make sure that the artificial eyes gaze positions are within the calibration area, and then recording begins.

Fig. 1 Stimuli designed and implemented for recording eye gaze data

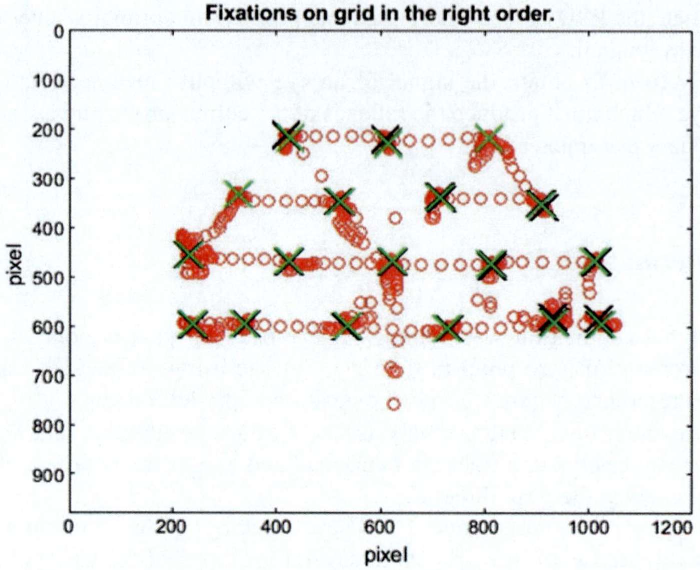

Fig. 2 Denoising gaze pattern on the stimuli tests

- We export the raw data samples, the distance of the eye tracker to the calibration plane is recorded to calculate the eye movement of the samples in visual degrees.

Also, we have made an evaluation of eye movement at known speeds and frequencies Table 1. A total of four different movements: vertical (Fig. 3a) and horizontal (Fig. 3b) movement, as well as rectangular (Fig. 3c) movement and inverted triangle (Fig. 3d) have been used in our propose.

Each movement test has a different speed; for this, present a different frequency spectral distribution.

In order to estimate spatial precision from the Wiener filter, the standard deviation, and the SNR on all filtered series of each sequence were computed and then the average standard deviation was calculated for each task.

Table 1 Speeds and frequencies by movement

Movement	Test		
	Speed movement (m/s)	Frequency of movement (Hz)	Estimate frequency (Hz)
Vertical	0.40	0.40	0.454
Horizontal	0.25	0.70	0.656
Square	0.30	0.40	0.452
Triangle	0.50	0.20	0.207

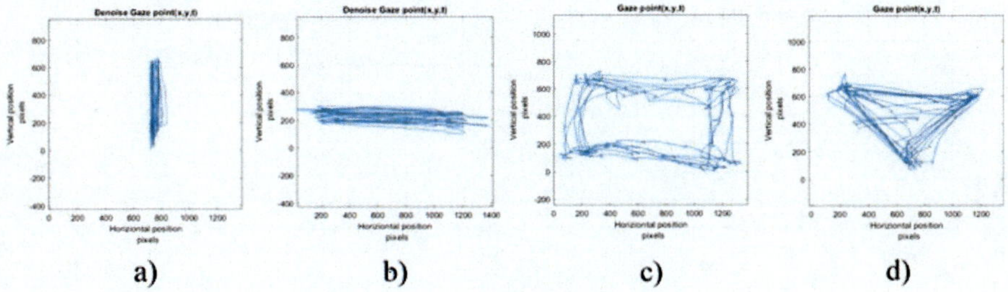

Fig. 3 Gaze pattern, on stimuli tests

With the signal recorded data, precision for the eye signal [1] was calculated based on the function of the angular distances θi between successive data samples (x_i, y_i) to $(x_i + 1, y_i + 1)$, of the form:

$$\theta_{\mathrm{RMS}} = \sqrt{\frac{1}{n}\sum_{i=1}^{n}\theta_i^2}, \tag{22}$$

moreover, the SD was calculated as:

$$\theta_{\mathrm{SD}} = \sqrt{\frac{1}{n}\sum_{i=1}^{n}(x_i - \mu_x)^2 + (y_i - \mu_y)^2}, \tag{23}$$

where μ_x and μ_y are the means of *n* simple locations.

3 Results

This section shows the results of noise elimination performance using the Wiener filter when we use the PSD of the estimated noise from the system using an artificial eye in static motion.

In Fig. 4, we show the time series of the artificial eye on static motion and the corresponding PSD of the noise signal.

Table 2 shows the spatial accuracy of the unfiltered and filtered time series using a mean filter (Fig. 5).

The purpose of this methodology is to evaluate the Wiener filter capacity to filter the noisy data acquired in the experiments.

Table 3 shows the performance of each filter according to its SNR. The spectral subtraction filter achieves a better SNR response than with the median filter for the four-motion series; this performance closely matches the Wiener filter for all series.

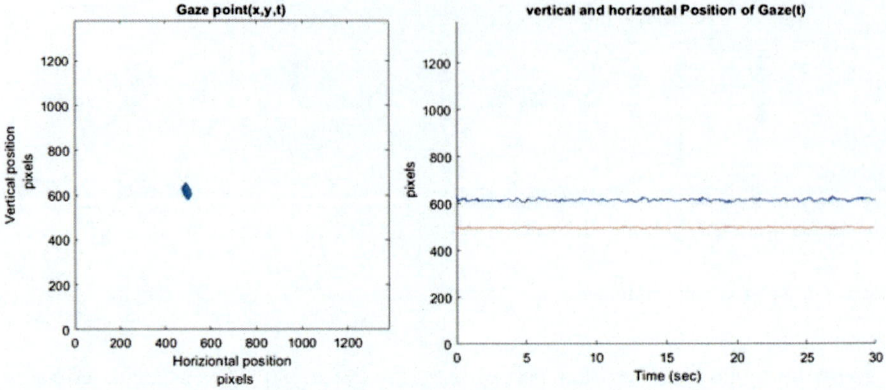

Fig. 4 Artificial eye looking a point on the calibration grid during 10 min, mean position in vertical axis = 613 pixels, and mean position in horizontal axis = 495 pixels

Table 2 Spatial accuracy

Measure	Spatial precision by the artificial eye	
	Unfiltered	Mean filter
Standard deviation of signal	0.08	0.03
Mean horizontal position	495	493
Mean vertical position	613	614
RMS	2.34	0.75

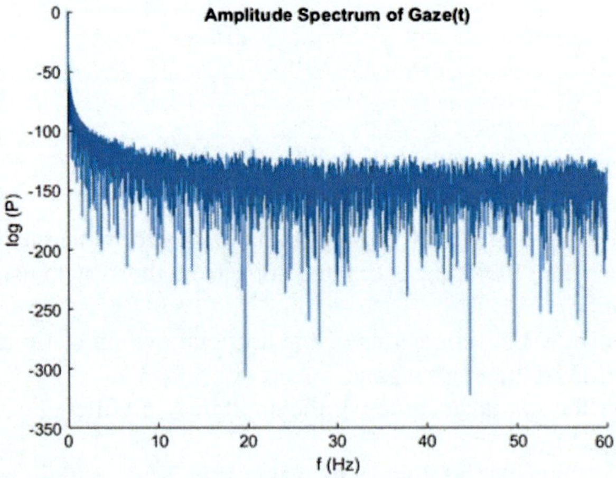

Fig. 5 Frequency responses by the system when an artificial eye is used

Table 3 SNR response

Method	SNR of the time series (dB)				
	Vertical	Horizontal	Rectangle	Triangle	Static
Spectral subtraction	68.18	65.18	66.33	67.29	81.98
Median filter	59.12	49.5	55.76	59.65	83.65
Wiener filter	73.25	75.60	79.69	78.09	84.95
Wiener filter propose	77.12	77.94	82.92	81.30	85.05

The spectral subtraction method and the median filter were also able to eliminate noise properly for the artificial eye. However, the spectral subtraction method and the median of the filter depend on the filter parameters, so all the values of the series change and produce low performance compared to the Wiener filter method.

Nevertheless, the Wiener filter, using the parameters of directly measured errors of the system by the proposed method, shows an average of ~12.13 dB higher for all the series.

In another experiment, the Wiener filter was tested on the experimental time series gotten from the eye tracker (see Fig. 6).

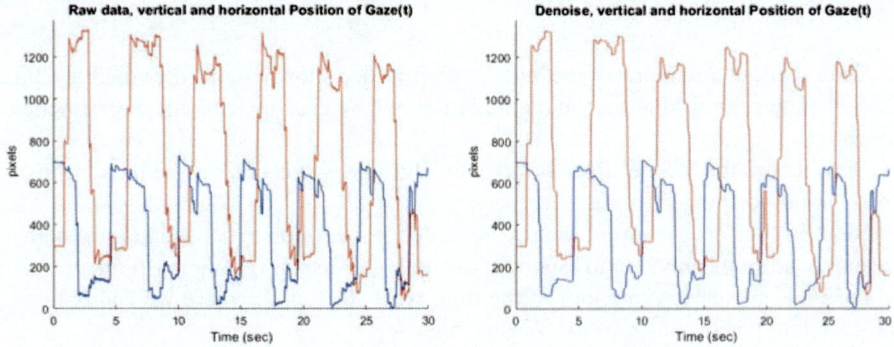

Fig. 6 Gaze pattern, **a** shows the experimental raw data, and on **b** the denoised signal with the Wiener filter is displayed

Table 4 SNR results using different filter

Method	SNR of the time series
	Vertical
Spectral subtraction	63.18
Median filter	49.12
Wiener filter	67.12
Wiener filter	87.12

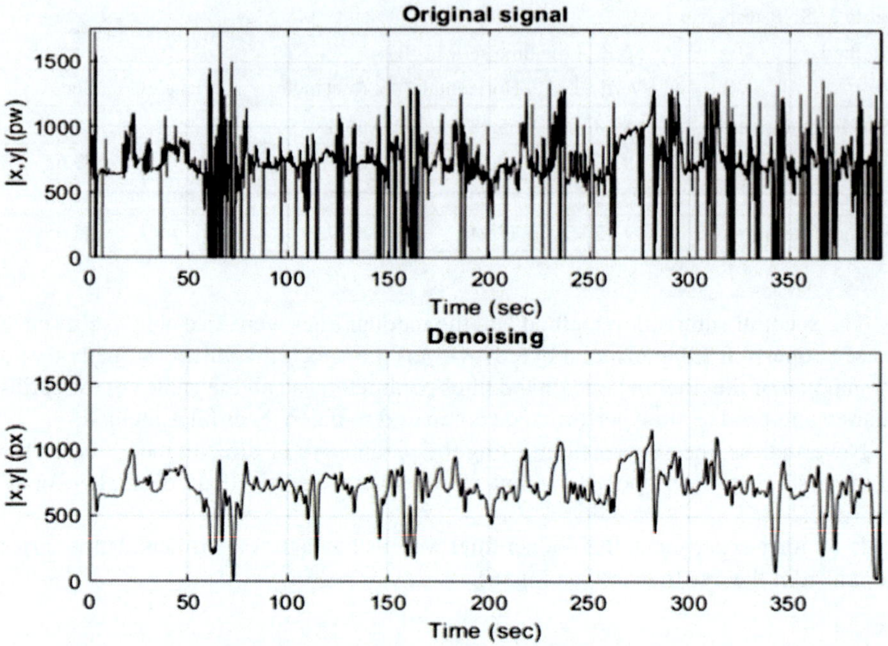

Fig. 7 Eye-tracking activity for a random participant during ~400 s in the test data. (Top) the raw data is displayed, (bottom) the data after using the Wiener filter

The excellent efficiency of the Wiener filter is also shown with experimental data. Table 4 shows the results for the median filter and Wiener filter on the experimental series.

Once again, the Wiener filter effectively eliminates eye-blink as well as intrinsic noise, not modifying the data.

Next, in Fig. 7, we present the raw recorded series of an experimental test under normal conditions and its corresponding signal after using the Wiener filter.

In Fig. 8, we present a zoom of the time series of Fig. 7, where we can observe the excellent performance of the Wiener filter.

4 Conclusions

In this work, the Wiener filtering algorithm was implemented to estimate the noise power spectrum using an artificial eye, and for filtering the eye movement signal degraded by additive noise.

For the evaluation of the Wiener filtering performance, SNR measurements were made in four different noisy signals of eye movement.

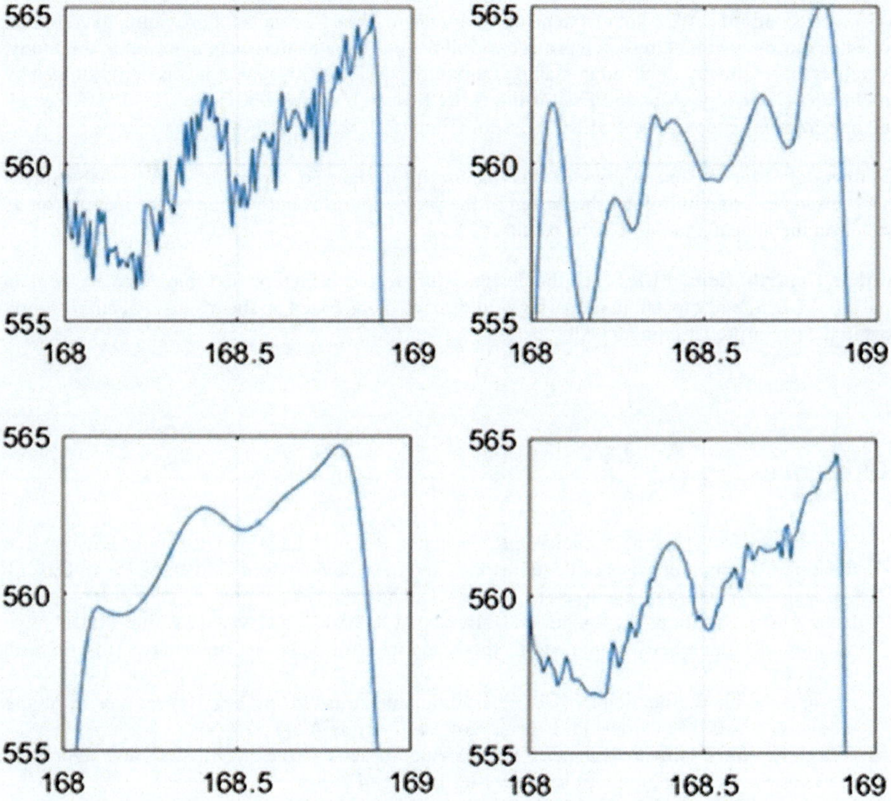

Fig. 8 **a** the raw data, **b** spectral subtraction filter (SNR = 2.77 dB), **c** mean filter (SNR = 0.07 dB) and **d** wiener filter (SNR = 65.50 dB)

The results show that the signal estimated using Wiener filtering has a better SNR in addition to a lower RMS. It has been shown that the implementation of the Wiener filter can be used in the treatment of the eye tracker signal to reduce and replace noisy data.

The results showed that the Wiener filter using this methodology is better compared to a conventional Wiener filtering process and it has better performance than the mean filter or spectral subtraction filter.

Our proposed method considers system noise, from which the filter coefficients are calculated. This property allows the filter to be implemented in real-time using a level of base signal noise, different from other filters. The procedure presented here is not limited to eye tracker, this can be used in similar signals, such as head movement or even pupillary dilation.

Other studies should investigate where the proposed Wiener filtering method could be implemented in real-time with various input devices.

Acknowledgements We want to extend our gratitude to Jesse Michaels for providing the raw data collected during his Ph.D. research project and all the participants that were involved in this study, as well as those directly involved in making this study possible. This research was partly funded by an NSERC Discovery grant and Essilor Industrial Research Chair (IRCPJ 305729-13), Research and development cooperative NSERC - Essilor Grant (CRDPJ 533187-2018), Prompt.

Conflicts of Interest The authors of this manuscript declare no conflict of interest. We express that the sponsors had no role in the design of the study, analysis and writing of the manuscript as well as in the decision to publish the results.

Author Contributions M-R.S. led the design of the research method and implemented the data analysis. M-R.S. participated in preparing of the conclusions based on the results. All authors took part in the paper preparation and edition.

References

1. Holmqvist K, Nystrom M, Andersson R, Dewhurst R, Jarodzka H, van de Weijer J (2011) Eye tracking. A comprehensive guide to methods and measures. Oxford University Press, Oxford
2. Caroenter 1977 libro de eye tracking Andrew Duchowski
3. Holmqvist K, Nystrom M, Andersson R, Dewhurst R, Jarodzka H, van de Weijer J (2011) Eye tracking. A comprehensive guide to methods and measures. Oxford University Press, Oxford, pp 35–40
4. Eason G, Noble B, Sneddon IN (1955) On certain integrals of Lipschitz-Hankel type involving products of bessel functions. Phil. Trans Roy Soc London A247:529–551
5. Hua H, Krishnaswamy P, Rolland JP (2006) Video-based eyetracking methods and algorithms in head-mounted displays. Opt Express 14(10):4328–4350
6. Upadhyay N, Karmakar A (2012) The spectral subtractive-type algorithms for enhancing speech in noisy environments. In: IEEE international conference on recent advances in information technology, ISM Dhanbad, India, pp 841–847
7. Bingyin X, Bao C (2014) Wiener filtering based speech enhancement with weighted denoising auto-encoder and noise classification. Speech Commun 60:13–29
8. MA Abd El-Fattah (2014) Speech enhancement with an adaptive Wiener filter. Int J Speech Technol 17(1):53–64
9. Loizou PC (2013) Speech Enhancement: theory and practice, 2nd edn. CRC Press
10. Haykin S (2003) Adaptive filter theory, 4th edn. Prentice-Hall, Upper Saddle River
11. Abramov I, Harris CM (1984) Artificial eye for assessing corneal reflection eye trackers. Behav Res Methods, Instrum Comput 16:437–438
12. SensoMotoric Instruments and Noldus Information Technology combine eye tracking and video analysis. Noldus. Retrieved April 2, 2014

Trust IoHT: A Trust Management Model for Internet of Healthcare Things

Naznin Hossain Esha, Mst. Rubayat Tasmim, Silvia Huq, Mufti Mahmud, and M. Shamim Kaiser

Abstract Internet of healthcare things (IoHT) creates a network where healthcare sensors, smart healthcare objects and services linked independently. This raises a concern of security and integrity of the interconnected devices and services in IoHT. This paper presents a trust management model for IoHT using fuzzy logic. The fuzzy inference system will consider packet loss rate, aggregated transmission rate, end to end delay, received signal strength indicator, drained battery power and packet integrity of node as input parameters and identify trustworthiness (trusted/maliciousness) of a node. In addition, Chi-squared test is also employed to detect the fault node in the IoHT. The performance evaluation reveals that more than 92% accuracy has been achieved for the proposed model.

Keywords IoHT · Fuzzy logic · FIS · Faulty · Malicious · Trustworthy

1 Introduction

Internet of things (IoT) interconnects a collection of sensors and computing systems to collect data from an environment. The data are then analyzed and converted into actionable insights which can be used for various purposes. While IoT is used to

N. H. Esha · Mst. R. Tasmim · S. Huq · M. S. Kaiser (✉)
Institute of Information Technology, Jahangirnagar University, Dhaka 1342, Bangladesh
e-mail: mskaiser@juniv.edu

N. H. Esha
e-mail: naznin.esha@gmail.com

Mst. R. Tasmim
e-mail: rubayat6442@gmail.com

S. Huq
e-mail: huq.silvia1952@gmail.com

M. Mahmud
School of Science & Technology, Nottingham Trent University, Nottingham, UK
e-mail: mufti.mahmud@ntu.ac.uk

K. Ray et al. (eds.), *Proceedings of International Conference on Data Science and Applications*, Lecture Notes in Networks and Systems 148,
https://doi.org/10.1007/978-981-15-7561-7_3

connect medical devices, patients, doctors, paramedics, nurses, then such system can be termed as Internet of healthcare things (IoHT) [1]. A fully connected IoHT can enable practitioners to provide one by one customized data for treatments. If patient's records are made available, the provider can view a comprehensive medical history and compared the condition with similar patients. The more data and record available regarding current treatments, medications, and patient's history, the higher care the patient will receive. The devices are used in IoHT have the potential to improve patient care, fuel innovative medical research and enhance healthcare system efficiency. At the same time, however, the cybersecurity risks are real and growing. However, as this network has openness character the risk is increasing very quickly. Therefore, it is required to be identify whether the communicating entities (sensors and medical devices) are trustworthy or not [2, 3].

In medical field, the data are sensible and valuable. Nowadays, they are vulnerable to attacks. So security is main concern here. Criminal attacks on healthcare data can be threat to a patient's life. For real-time consultation, data must be secure as well as accurate. That's why authentication, authorization and data integrity and privacy should be ensured in an IoHT system. So an IoHT system must be based on trust, security and privacy of healthcare entities. Reliable and secure trust management model is necessary for a secure and faultless IoHT system.

The contribution in this work is to propose a trust management system to secure IoHT and provide safe communication to patients through different kinds of communicating device. The process of ensuring authenticity of entities in an IoHT system is served by trust management architectures. Before joining the network, the trustworthiness of a new node will be checked by the trust management system. Chi-squared test is employed to detect the fault node in the IoHT. The fuzzy inference system will consider packet loss rate, aggregated transmission rate, end to end delay, received signal strength indicator (RSSI), drained battery power and packet integrity of node as input parameters and identify trustworthiness (trusted/maliciousness) of a node. The trust value of a new node is set to 0.5. For positive interaction, the value will increase and the value will reduce for negative interaction.

The work is organized as follows: Sect. 2 presents related work based on trust-based IoHT system. The proposed system architecture is in Sect. 3. Section 4 presents numerical analysis. Finally, fuzzy-based trust management model is concluded in the last section.

2 Related Work

This section gives an overview of trust-based trust management system proposed by many researchers.

RFID telecare medicine information system has been proposed in [4] which solved the security and privacy problems over insecure healthcare network environments by using many cryptographic techniques. In this technique, the attack has been estimated based on misuse of the timestamp technique which indicated a fault. Here, requestresponse messages have been generated using one way hash function. This method is applied for distant areas and low population density. Besides, RFID authentication procedures are still complex.

Geo-distribution clouds for e-health monitoring system, proposed in [5], it uses an e-health monitoring system which have minimum service delay and privacy preservation by exploiting geo-distributed cloud. In this system, it focuses on the resource allocation program that allows the distributed cloud servers to commonly assign the servers to the requested users under the load balance condition. Besides, the traffic analysis attack is mostly reduced. The main aim of this approach is resource allocation. It can only reduce delay. It cannot handle fault and malicious node.

Middleware solutions for e-healthcare system security, proposed in [6], a middleware compound is used in this approach to secure data/information over network. It focuses security warning on e-healthcare system where an hacker can access both data and network using masquerade attack to protect security and privacy of patients' information. Masquerade attack can be used to the system and patients' data are in danger. Though it uses as middleware to solve problems but middleware can be compromised easily. Middleware itself is vulnerable and easy to hack.

Security and privacy mechanism for IoHT, proposed in [7], it uses a novel security and privacy method to handle insecurity issues. This method is used here to initiate trust in IoT application market. Interactive vector uses here that indicated trustworthiness of application. It becomes vulnerable because of its simplicity. It uses an interactive vector that uses as primary indicator in application marketplaces. This vector does not check nodes' details parameters and their condition. It cannot identify fault and malicious node.

Authentication protocol for information system, proposed in [8], it explores the features of modern information system, problems with traditional authentication protocols and establishes authentication. Authentication suggests checking the introduced users identification and proving its validity. User authentication is must in informational system. Traditional authentication methods have some disadvantage that do not count for typical system structure and thus do not allow to use its components efficiently.

Data privacy protection mechanism, proposed in [9], a shape of the technique analysis of the methodologies and approaches that are being used in present days to cope with the notable issue of privacy. It increases the efficiency, convenience and cost performance for healthcare sector. The security and privacy of the gathered data from devices are in critical state because of transmission to a cloud or while stored in clouds are vital unsolved trouble [10]. Sometimes, it creates problem while handling higher exposer of sensitive data.

Joint authentication as well as and privacy preservation protocol has been proposed for TMIS [11] which is a lightweight authentication protocol for privacy issues. Authors claimed that the proposed protocol can able to protect common security threats. Besides, it introduced an advanced version of its protocol for cloud where the security and privacy have been ensured through cloud. However, the authentication time is high as it is done via cloud.

3 System Model

In an IoHT system, there are several smart devices named heart rate monitor, smart watch, smart bed, blood pressure monitor, etc. These collect data from patient's body sent to appropriate destination via Internet gateway. Patients come to system network via IoHT things(IoHT devices) through gateway. In this gateway, we have developed a trust model to detect fault and malicious node. So that unauthorized and harmful access would not destroy data and system.

3.1 Parameters

These parameters are considered to identify fault node and malicious node so that only trustworthy node can remain in the system.

Aggregate Transmission Rate: The data transfer rate (DTR) is the quantity of digital data which is moved from one place to another in a specified time. It can be defined as the speed of travel of a given amount of data from one place to another.

Packet Loss Rate: Packet loss rate is measured as a percentage of packets lost with respect to packets sent. In general, less than 0.05% packet loss is "good" and 1–1.5% is "acceptable".

End to End Delay (E2E): E2E delay is a time which included processing and propagation delay of a packet to be transmitted from source node to destination node.

Received Signal Strength Indicator—RSSI It is an estimated measurement of how good a node can hear, detect and receive signals from any access point or from a specific router. A signal is indicated through RSSI. It is that helps you determine and know if a signal is sufficient to establish a wireless connection.

Drain Power: A malicious node can waste the battery by performing unnecessarily operations [12]. The amount of energy is stored in the battery known as power capacity which reduces with working process.

Packet Integrity: When user send and receive data over the wireless network, there are possibilities where the data can get altered, corrupted or modified. It can be accidental, purposely done with evil intention. Whatever the case is receiver must verify and figure out if the data is corrupted or altered.

3.2 Fault Node Detection

Node faults mainly occurred due to fabrication, environmental factors, sensor(s)/ communication module(s) failure, adversary attacks and massive drain-out of battery power. In this case, the collected sensor data defers/uncorrelated to the typical characteristics of that sensor node. The faulty nodes required to be identified as these are great threat to IoHT system as it hampered real-data transmit.

So fault node must be detected and rejected from network. Here, Chi-square test method is used for fault node detection.

At first, input of parameters are taken: Packet loss rate, transmission rate, end to end delay as observed value O. We set our expected value E.

From Chi-square test,

$$X^2 = \sum_{i=1}^{n} \frac{(O - E)^2}{E} \tag{1}$$

3.3 *Malicious Node Detection*

A node is said to be malicious when it refuse to provide a desired service to its neighbor node in the network [13]. During the route selection, a malicious node must be excluded from the routing path [14]. As these are threat to an IoHT system.

Here, we have introduced a fuzzy logic-based malicious node detection using FIS. Fuzzy logic is an approach to computing based on multi-valued logic instead of the usual "true or false" (1 or 0) Boolean logic on which the modern computer relies. Fuzzy logic is based on human perceive. It is used in those situations where the available information/data is in form of partial truth that makes decision process very complicated. The degree of truth is represented by membership function in fuzzy logic. FLC controller is used here for decision making for detection malicious node. Here, Takagi-Sugeno FIS is used to identify trustworthy node [15].

Figure 1 views the six input parameters into FIS. Three descriptors for each input. From fuzzy fication, we get three output- fault nodes, malicious nodes and trustworthy nodes.

Table 1 shows some rule bases for our system. Each input parameter has three descriptor—high, medium and low.

Figure 2 shows block diagram of our proposed trust model.

3.4 *Algorithm*

Figure 3 shows the algorithm for this trust model for identifying trustworthy node.

4 Numerical Result

The simulation results of the proposed model is presented in this section.

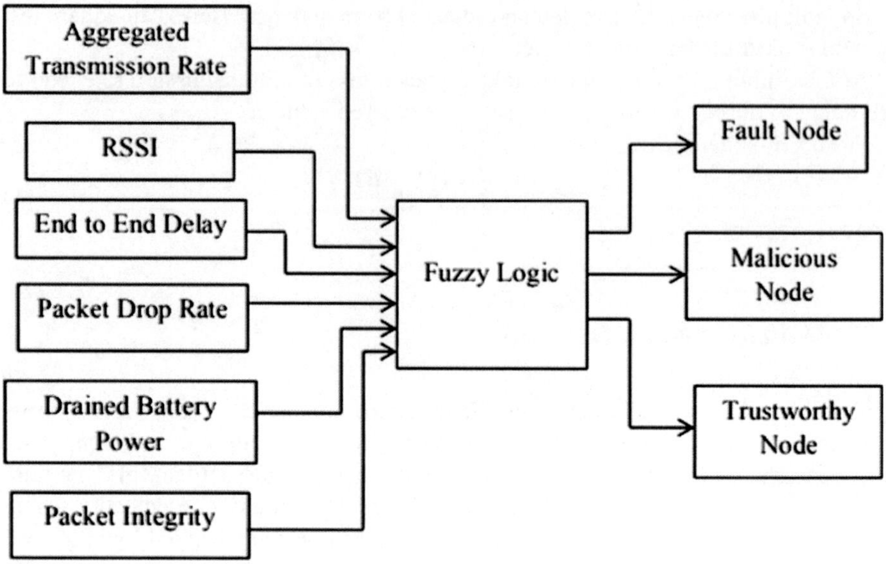

Fig. 1 Block diagram for identifying node

Table 1 Rule base for fuzzy logic

No.	R	$RSSI$	D_{E2E}	PDR	DE	PI	Detection
1	H	H	H	L	M	H	Faulty
2	H	H	M	H	M	L	Malicious
3	H	H	M	M	H	L	Malicious
4	H	H	M	L	H	L	Malicious
5	H	H	M	L	M	H	Trustworthy
6	H	M	L	L	M	H	Trustworthy
7	H	M	L	L	L	L	Malicious
8	H	M	H	L	L	L	Faulty
9	H	L	M	H	M	L	Malicious
10	H	L	M	M	H	L	Malicious
11	H	L	M	L	M	H	Trustworthy
12	H	L	L	L	M	H	Trustworthy
13	H	L	L	L	L	L	Faulty
14	M	H	M	H	M	L	Malicious
15	M	H	M	L	M	H	Trustworthy

Legend R—Aggregate Transmission Rate, $RSSI$—Received signal strength indicator, D_{E2E}—End to end delay, DE—Drained battery, PDR—Packet drop rate, PI—Packet integrity

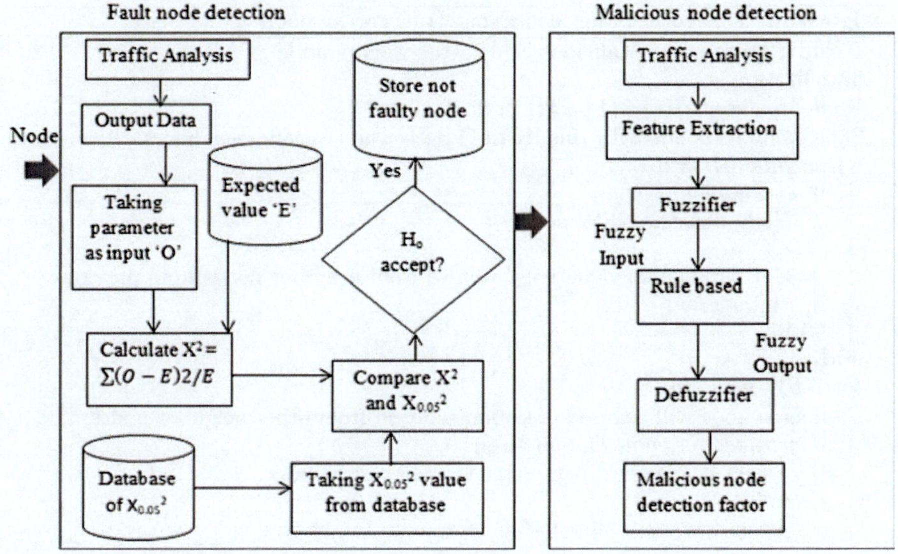

Fig. 2 Block diagram of trust model

4.1 Data Set

KDD means knowledge discovery in databases which is a well-known benchmark in terms of intrusion detection techniques. The NSL-KDD data set with 42 attributes is used in this system.

Table 2 shows the mapping of KDD data set with the FIS input. As we know, in KDD data set, there are 42 features. For our system, these features are mapped with selected dedicated features to detect trustworthy node which has been considered in our model.

4.2 Simulation Results

Figure 4 shows input and output membership functions where triangular membership function was considered for each linguistic term (low, medium and high).

Figure 5 shows the 3D surface view of the rule-based system.

Figure 6a shows the confusion matrix where columns represent true classes while rows represent the classifier's predictions with all correct classifications along the upper-left to lower-right diagonal. Here, 42, 30, 22 are correctly classified as trusted node (level of trustworthy of a node), malicious node and faulty node, respectively. Figure 6b shows the predicted performance in terms of percentage.

Algorithm 1: Algorithm for identifying Trustworthy node

```
Result: Fault node, Malicious Node, Trustworthy node
initialization;
Node broadcasts HELLO packet periodically;
Some node responds with this HELLO packet and become neighbors;
while interaction do
    if no interaction then
        Node turns as NEW node
    else
        Get parameters value(trust values) from neighbor nodes from previous
          interaction
    end
end
for NEW node do
    Source node will ask node recommendation from other neighbor nodes;
    if get node recommendation then
        collect it as nodes' parameter values(trust values)
    else
        set nodes' trust value to 0.5
    end
end
Calculate trustworthiness;
for node 1:n do
    Find k(not faulty node) from Chi square test;
    Repeat fault node
end
for node 1:k do
    Find t(trustworthiness) from FIS;
    Repeat malicious node
end
```

Fig. 3 Algorithm for identifying trustworthy node

Table 2 KDD dataset feature relation with system input

FIS input	KDD data set
R	Service,src_bytes,dst_bytes,protocol type
$RSSI$	Duration,srv_count,count
D_{E2E}	Duration,count,flag,srv_count,diff_srv_rate,num _failed_login
PDR	Serror_rate,rerror_rate,srv_serror_rate,srv_rerror_ rate
DE	Duration
PI	src_bytes,dst_bytes

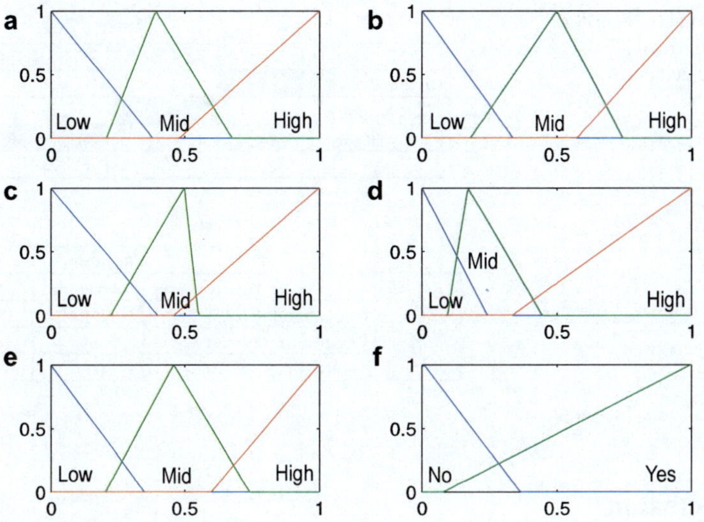

Fig. 4 Fuzzy logic membership functions **a** transmission rate, **b** RSSI, **c** delay, **d** packet drop rate, **e** drained battery and **f** packet integrity

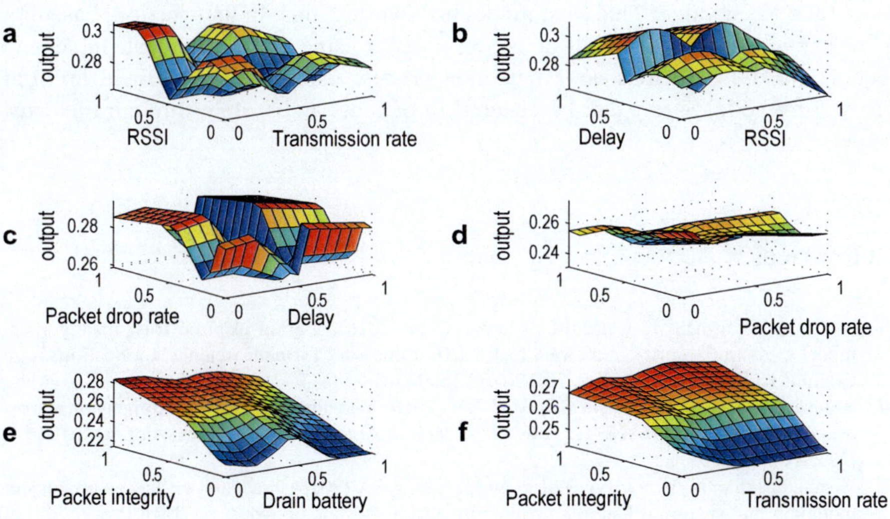

Fig. 5 Surface view of rule-based system

Fig. 6 Confusion matrix **a** number of correctly classified, **b** accuracy

Classified Data	Reference Data			
	Trusted node	Mal. Node	Faulty Node	
Trusted node	42	1	2	45
Mal. Node	1	30	0	31
Faulty Node	2	0	22	24
	45	31	24	100

a

Classified	Reference		
	Trusted node	Mal Node	Faulty Node
Trusted node	0.93	0.02	0.04
Mal Node	0.03	0.97	0.00
Faulty Node	0.08	0.00	0.92

b

5 Conclusion

This paper has introduced a new trust model for IoHT. It selects safe node for IoHT based on packet loss rate, transmission rate, end to end delay, packet integrity, drain power and RSSI value. This trust model has two sections. In first section, based on these parameters, we did fault detection process using Chi-square test. In second section, we did malicious node detection process using fuzzy logic and find out trustworthy node. Safety can be ensured to patients and system through this trust model.

References

1. Mufti Mahmud MS, Kaiser MM, Rahman et al (2018) A brain-inspired trust management model to assure security in a cloud based IoT framework for neuroscience applications. ogn Comput 10:864. https://doi.org/10.1007/s12559-018-9543-3
2. Antesar M, Shabut M, Kaiser S, Dahal KP (2018) Wenbing chen, a multidimensional trust evaluation model for MANETs. J Network Comput Appl 123:32–41. https://doi.org/10.1016/j.jnca.2018.07.008
3. Afsana F, Jahan N, Sunny FA, Kaiser MS, Mamun SA (2015) Trust and energy aware cluster modeling and spectrum handoff for cognitive radio ad-hoc network. In: 2015 Proceeding of ICEEICT, Dhaka, pp 1–6. https://doi.org/10.1109/ICEEICT.2015.7307489
4. Benssalah M, Djeddou M, Drouiche K (2016) Dual cooperative RFID-telecare medicine information system authentication protocol for healthcare environments. Secur Commun Networks 9(18):4924–4948
5. Shen Q, Liang X, Shen X, Lin X, Luo HY (2014) Exploiting geo-distributed clouds for a E-health monitoring system with minimum service delay and privacy preservation. IEEE J Biomed Health Inform 18(2):430–439
6. Bruce N, Sain M, Lee HJ (2014) A support middleware solution for e-healthcare system security. In: Proceeding of 16th international conference on advanced communication technology, pp 44–47

7. Kang K, Pang Z, Wang C (2013) Security and privacy mechanism for health internet of things. J China Univ Posts and Telecommun 20:64–68
8. Saito T, Wen W, Mizoguchi F (1999) Analysis of authentication protocol by parameterized ban logic. Technical report, ISEC
9. Singh N, Singh AK (2018) Data privacy protection mechanisms in cloud. Data Sci Eng 3(1):24–39
10. Sun W, Cai Z, Li Y, Liu F, Fang S, Wang G (2018) Security and privacy in the medical internet of things: a review. Secur Commun Networks. https://doi.org/10.1155/2018/5978636
11. Li C-T, Shih D-H, Wang C-C (2018) Cloud-assisted mutual authentication and privacy preservation protocol for telecare medical information systems. Comput Methods Programs Biomed 157:191–203
12. Rana KG, Yongquan C, Azeem M, Yu H (2017) Detection of malicious node in wireless Ad Hoc network by using acknowledgement based approach. Proceeding of ICCNS 2017, pp 76–80
13. "What is Malicious Node | IGI Global." [Online]. Available: https://www.igi-global.com/dictionary/a-novel-secure-routing-protocol-in-manet/33926. Accessed: 28 Oct 2019
14. Mishra D, Jain YK, Agrawal S (2009) Behavior analysis of malicious node in the different routing algorithms in mobile Ad Hoc network (MANET). In: ICACCT, pp 621–623
15. Kaiser MS et al (2016) A neuro-fuzzy control system based on feature extraction of surface electromyogram signal for solar-powered wheelchair. Cogn Comput 8:946–954

ACO-Based Control Strategy in Interconnected Thermal Power System for Regulation of Frequency with HAE and UPFC Unit

K. Jagatheesan, B. Anand, Nilanjan Dey, Amira S. Ashour, Mahdi Khosravy, and Rajesh Kumar

Abstract Increased quality is an essential requirement in thermal power systems that can be guaranteed by balancing both the load demand and power generation. Accordingly, in an interconnected thermal system, the controller gain values of Proportional Integral Derivative (PID) in load frequency control (LFC) is tuned by using Ant Colony Optimization (ACO) in the proposed system. Besides, a secondary PID controller is employed to regulate the system's frequency and power flow using a tie-line between the interconnected systems through an abrupt load surplus. One percent step load perturbation in area 1 (1% SLP) is considered while determining the optimal controller's gain values. A comparative study was conducted to evaluate the proposed ACO based tuned controller gains against the article Swarm Optimization (PSO) and Genetic Algorithm (GA) technique tuned controller performance. Additionally, an enhancement in the performance of the system is achieved using a unified power flow controller (UPFC) flexible alternating current transmission system devices unit in series with normal tie-line. Furthermore, a hydrogen generative aqua electrolyzer (HAE) energy storage unit is incorporated into the investigated power

K. Jagatheesan
Department of EEE, Paavai Engineering College, Namakkal, Tamil Nadu, India

B. Anand
Department of EIE, Hindusthan College of Engineering and Technology, Coimbatore, Tamil Nadu, India

N. Dey
Department of CSE, Techno India College of Technology, Kolkatta, West Bengal 740000, India

A. S. Ashour (✉)
Department of EEC, Faculty of Engineering, Tanta University, Tanta, Egypt
e-mail: amirasashour@yahoo.com

M. Khosravy
Media Integrated Communication Lab, Graduate School of Engineering, Osaka University, Osaka, Japan

R. Kumar
Department of EE, Malaviya National Institute of Technology Jaipur, Jaipur, India

K. Ray et al. (eds.), *Proceedings of International Conference on Data Science and Applications*, Lecture Notes in Networks and Systems 148,
https://doi.org/10.1007/978-981-15-7561-7_4

system. Finally, the simulation results obviously evident that the proposed system's response improves the time domain measurement parameter values with minimum oscillations.

Keywords Power management · Power quality · Energy regulation · Power system · Controller · Ant colony optimization

1 Introduction

High power quality is required for consistency in frequency, voltage, and high reliability. This is achieved by matching the load demand with power generation. However, in large industrial plants, the operating point is varying continuously leading to a complex balancing process of the total load demand with power generation. Any unexpected load disturbance produces fluctuations in the tie-line power exchange and system frequency. The control of the system's frequency oscillations in each area and power exchanges across the connected areas within the particular limit identifies LFC [1–4].

In the connected power system, several researchers conducted different control approaches for the LFC to support the system's frequency and flow of tie-line power exchange deviations within the already scheduled value, and under normal loading situations as well as a small load disturbance [1–38].

The LFC crisis in power systems is rectified by implementing proper control techniques in power system. Various controlling methods and optimization procedures can be applied, such as the classical controllers [5, 24], fuzzy logic controller [6], PSO [7–9], bacterial foraging technique [10–12], cuckoo search (CS) technique [13, 14], artificial bee colony technique [15], ant colony optimization technique [16, 17], GA [18, 19], artificial neural network method [20, 21], imperialist competitive algorithm [22], firefly algorithm [23, 24], bat algorithm [25], plant biology algorithm [26, 27], conventional controller [28], variable structure control [29], discrete mode [2–4], optimal control theory [30].

Saikia et al. [11] proposed a bacterial foraging technique for tuning the classical controller's gain values. These controllers are integral (I), proportional-integral (PI), PID controller, integral double derivative (IDD) controller, and fuzzy integral double derivative (FIDD) controllers are introduced. Such controllers can be used in three-area hydrothermal power (thermal—Thermal—Hydro) systems. In [22], an imperialist competitive algorithm was implemented to design the fractional order PID controller (FOPID) in three-area hydrothermal (non-reheat, reheat, hydro) power systems with different loading conditions in each area. In [31], evolutionary computational techniques, namely the BF and PSO were presented to adjust the PID controller's parameter in three equal reheat thermal power systems.

Recently, for controlling the load frequency, several nature-inspired computational techniques were developed and effectively employed in interconnected multi-area power systems for adjusting the gain values of the controller. The proportional

Integral plus controller (PI+) was designed by applying the beta wavelet neural network (BWNN) technique in [30]. The LFC parameters were tuned in organized power systems with Redox Flow Battery (RFB) energy storage unit and Hydrogen Aqua Analyzer (HAE) energy storage unit. Fuzzy-PID controller was designed by implementing teaching learning-based optimization (TLBO) technique in [36, 37], for AGC of unequal interconnected two thermal power systems. A comparative study with simulated annealing, Lozi map based chaotic search (PS optimization algorithm (LCOA), and genetic algorithm (GA) established the superiority of this method. In [38], a minority charge carrier with PI controller was applied in AGC of the hydrothermal power system. A new modified harmony search algorithm tuned controller (PID) was functioned to solve the LFC matter in the hydrothermal power system connected two-area power system. Additionally, the integral time absolute error (ITAE) function was implemented during the optimization of controller gain values [39].

In a thermal power system, a multi-area interconnected unit was analyzed by applying dual-mode gain scheduling based PI controller [40] to solve the LFC issue. The controller gain values in the PI controller was tuned by using the Bat inspired algorithm. The used objective function was the integral square error. In [41], a CS algorithm designed 2-DOF (two degrees of freedom) controllers were considered to overcome the AGC problem in a multi-area power generating system. A hybrid Differential Evolution–Pattern Search (DE-PS) algorithm designed controller is applied for LFC of deregulated power system with Unified Power Flow Control (UPFS) and Redox Flow Battery (RFB) to improve the response of system under crisis situation [42]. Coordinative optimization control is applied into microgrid based predictive control in [43] and Facebook messenger Chatbot enabled home control system is implemented based on architectural fretwork in [44].

Based on the preceding studies, the current study evaluated the performance of the ACO tuned controller in LFC in an interconnected thermal system with a 1% step load disturbance. Furthermore, ACO is applied for optimizing the controller's gain values, namely I, P, and D. Finally, UPFC/HAE energy storage unit is used to enhance the response of the interconnected power system. Thus, the current work presents regulation of system's frequency in the thermal area interconnected two-area power system with ACO technique for tuning of PID controller with non-linearity effect, UPFC, and HAE unit. Also, the performance of the system is examined compared to the GA-, and PSO—optimization techniques tuned PID controller performance. The aim of this proposed work is to improve the performance by adding energy storage units and FACTS devices connected parallel to the tie line of the interconnected power system.

The remaining sections are Sect. 2, which presents the interconnected two-area power system model with thermal system among UPFC and HAE, followed by Sect. 3 wherein the method for designing and tuning the PID controller gain values using GA, PSO and ACO techniques are offered. Then, the performance and effectiveness of GA, PSO, and ACO tuned controller are analyzed with and without UPFC and HAE unit in Sect. 4. Lastly, Sect. 5 presents the conclusions of the proposed work.

2 Investigated Power System

The Simulink design of interconnected two-area thermal power system is exposed in Fig. 1, where the parameters of the connected system have the reported values in [6, 24]. Thermal power incorporates governor, reheater, turbine, speed governor and generator. The two areas are interconnected using tie-line. During nominal loading conditions, the individual load disturbance with the parameters limit of the system is considered. After unexpected load disturbance conditions in any of the interconnected power system, the surplus power is transmitted via tie-line to maintain its stability. In this work, the PID controller is applied to standardize the power system parameters with the help of the secondary loop.

In this proposed work, a flexible alternating current transmission system (FACTS) device, including the UPFC is connected parallel to tie-line. Hydrogen Aqua Analyzer (HAE) energy storage unit considered in area-2 for improving the behavior of the system during emergency load demand conditions. The FACTS device having the capacity is used to control and regulate the power flow within the connected power generating system via tie-line. It improves the dynamic behavior of the power system and system getting support for voltage balance [37]. In this work, UPFC FACTS device is considered and connected shunt to the existing tie-line. The transfer function (TF) of UPFC is given by:

$$G_{\text{UPFC}}(s) = \frac{1}{1 + sT_{\text{UPFC}}} \quad (1)$$

where T_{UPFC} time constant of UPFC.

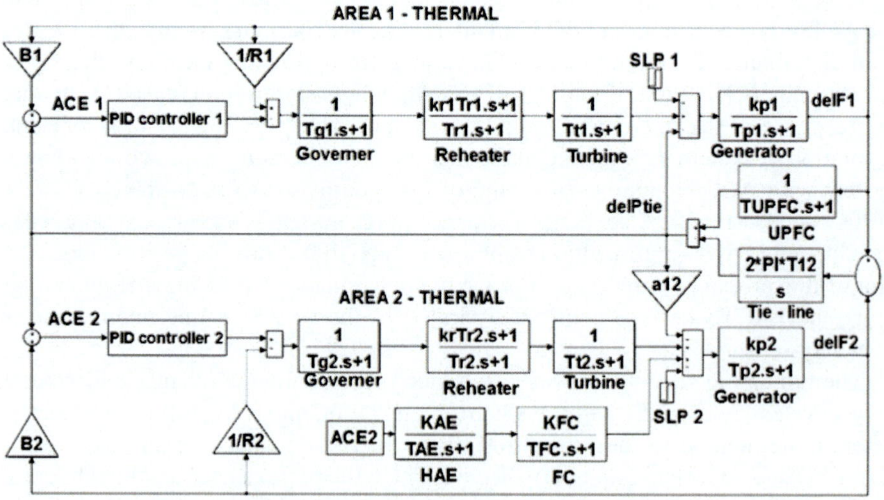

Fig. 1 Two-area connected thermal power system with UPFC and HAE

In recent energy scenarios, the hydrogen is used as a popular alternative energy source. The fuel of stationary generating units is replaced by hydrogen. An electrolyzer unit is used to convert the stored hydrogen from given input electric energy by decomposing the water and hydrogen molecules in the storage system [35]. The energy conversion system converts hydrogen into chemical energy by the support of fuel cells. The aqua electrolyzer transfer function model is given in:

$$G_{\mathrm{AE}}(s) = \frac{K_{\mathrm{AE}}}{1 + sT_{\mathrm{AE}}} \tag{2}$$

The fuel cells convert the chemical energy of fuel. The TF model is given by:

$$G_{\mathrm{FC}}(s) = \frac{K_{\mathrm{FC}}}{1 + sT_{\mathrm{FC}}} \tag{3}$$

The UPFC and HAE are connected into two-area interconnected thermal power plant to get better dynamic behavior of the system during sudden load demand.

3 Design Methodology for Tuning of PID Controller

3.1 PID Controller

A PID controller has a great impact on solving the LFC concern in multi-area connected power system. It is used to realize better response of the system, namely minimum settling time, high over- and under-shoot by eliminating steady-state error [11, 15] at the time of emergency load demand. The transfer function of the controller is given as:

$$G(s) = K_p + \frac{K_i}{s} + K_{d^s} \tag{4}$$

In this, K_p, K_i, and K_d are the *P*, *I* and *D* controller gain values, respectively. The PID controller generated signals are given by:

$$u_1 = K_{p1}\left(\mathrm{ACE}_1 + \frac{1}{sT_{i1}}\mathrm{ACE}_1 + sT_{d1}\mathrm{ACE}_1\right) \tag{5}$$

$$u_2 = K_{p2}\left(\mathrm{ACE}_2 + \frac{1}{sT_{i2}}\mathrm{ACE}_2 + sT_{d2}\mathrm{ACE}_2\right) \tag{6}$$

where u_1 and u_2 are the control input signals of area 1 and 2, respectively; K_{p1} and K_{p2} are proportional gain values of area 1 and 2, respectively; K_{i1} and K_{i2} are the integral gain values of area 1 and 2, respectively; K_{d1} and K_{d2} are the two areas derivative gain values, respectively; ACE_1, ACE_2 are area control error values of area 1 and 2, respectively; T_{i1}, T_{i2} are the integral time constant of area 1 and 2 and T_{d1}, T_{d2} are the derivative time constant of area 1 and area 2. The suitable choice of PID controller gain values is more crucial. Many heuristic algorithms are problem based and definite method. Other side, a metaheuristic algorithm is high level problem-free algorithmic frame-work and it gives a set of strategies to design a heuristic algorithm for optimization issues. Thus, in the current study, the gain values of controllers are optimized by applying ACO optimization algorithm. The limits of controller gain values are $K_{p\min}, K_{i\min}, K_{d\min} \le K_p, K_i, K_d \le K_{p\max}, K_{i\max}, K_{d\max}$, where the minimum limit and maximum limit of gain values of the controller gain values are 0 and 1, respectively. In this proposed work, ITAE cost function is applied [8] during the optimization of gain values of the controller, which is depicted as follows:

$$\mathrm{ITAE} = \int_0^t t.|\,\mathrm{ACE}\,|\mathrm{d}t \tag{7}$$

3.2 *Ant Colony Optimization*

Trial and error tuning methods take a long time to develop and design a proper controller, along with low accuracy. To overcome these drawbacks, several optimization techniques are framed and effectively executed for solving LFC crisis in a connected power generating system. The ACO optimization method is applied in the present work to solve LFC crisis. The major advantage of using ACO is its positive feedback for finding good solutions, avoiding premature convergence by implementing distributed computation and it gives the support for finding an acceptable solution based on its easy levels in the searching process. These are the more crucial reasons for using ACO optimization technique. The foraging behaviour of real ants was the main source for the development of the ACO algorithm, where the solutions of the optimization problem depending on the number of real artificial ants. The ants' information is an exchange based on these solutions through a common scheme, where the reminiscent of an ant adopted by real ants [16, 17, 45]. The natural foraging characters of real ants are inspired by several searchers. It is used to solve several discrete optimization-based problems. During the food penetrating process, ants are searching for their food source. Initially, all the ants are randomly wandered around the surrounding of the nest. After exploring all ants, the food sources are determined as soon as possible. After reaching the food source, the ants evaluate the amount of food source and its quality. At the time of the return trip, the ant carries some food to the nest and deposits few pheromone chemical trials to the ground. Pheromone trial depends on food source quality and quantity, which is used to guide for finding food sources of other ants. The communication is indirect via chemical trials to enable finding the shortest path across nest and food sources. The real ants' behaviour stimulated scientists to find the proper solutions for many optimization tasks [16, 17, 45]. Then, the ants will start their tour and the process of calculating the path length is done. Based on the shortest path, the optimal gain values for PID controllers are tuned using the ACO technique. The ACO algorithm is utilized to solve combinatorial optimization problems. It has desirable characteristics, such as versatility, robustness and population-based approach [11, 15]. Since the proper PID controllers' design for the power system is required to solve the LFC problem by optimizing K_p, K_i and K_d. The proposed work optimized these gain values in the PID controller using ACO technique [16, 17, 45]. During the tour time, the maximum value of the iteration is verified. If it reaches the cost function is evaluated. The performance index value of cost function is reached as per predetermined criterion the controller gain values are displayed else the process again takes place. In another case, the maximum iteration is not reached means a new tour has been taken place and repeats the above process once again. The steps taken by the ACO optimization method applied for designing PID controller gain values are:

Step 1: Start
Step 2: Initialize ACO parameters
Step 3: Run the model and update values of probability and pheromone
Step 4: Evaluate optimal gain values of K_p, K_i, K_d
Step 5: Check maximum iteration reached or not
Step 6: If yes STOP: If no go to step 3.

4 Simulation Result and Discussion

In this work, two areas of thermal system response are examined by considering ACO tuned controller with and without UPFC with HAE unit in the analyzed interconnected power system by pertaining 1% SLP load demand in area 1 of analyzed power system in Fig. 1. The optimal controller gain values are obtained using the GA, PSO and ACO methods [28] for a considered thermal power system with the following two different scenarios of the same analyzed power system as reported in Table 1.

The performance of the interconnected thermal system is analyzed in the following scenarios one the same underwork thermal power system by implementing different technique tuned controllers and energy storage unit:

Scenario 1: Using GA, PSO and ACO PID controller
Scenario 2: Using GA, PSO and ACO PID controller with UPFC and HAE unit.

Scenario 1: Using GA, PSO and ACO PID Controller The response comparisons of PID controller gain values are tuned using ACO, GA, and PSO are clearly illustrated in Fig. 2.

Figure 2 shows the response evaluation of oscillations in area 1 which is clearly evident to the superiority of ACO-PID controller over GA and tuned controller. The

Table 1 Controller gain values for different scenarios

Gain values	Without considering HAE and UPFC unit			With HAE and UPFC unit		
	GA	PSO	ACO	GA	PSO	ACO
Kp1	0.7922	0.9375	0.91	0.7922	0.989	0.91
Ki1	0.9595	0.9988	0.98	0.9595	0.9997	0.99
Kd1	0.3922	0.2967	0.09	0.0357	0.4643	0.02
Kp2	0.7431	0.407	0.9	0.6557	0.8426	0.98
Ki2	0.6555	0.8279	0.94	0.8491	0.947	0.90
Kd2	0.1712	0.4498	0.84	0.1715	0.2968	0.60
Performance indices	0.4258	0.3965	0.3097	0.4456	0.4016	0.3306

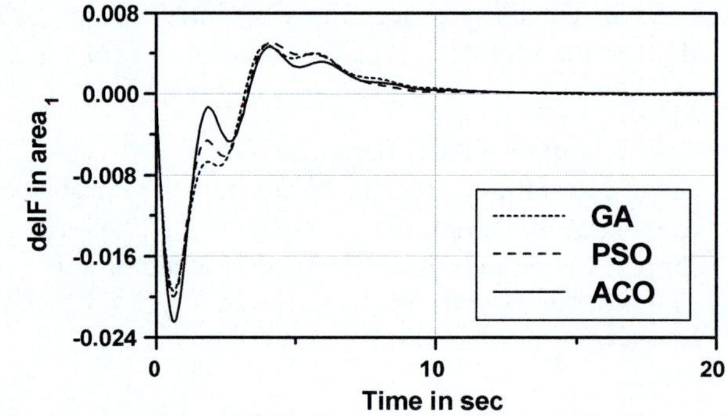

Fig. 2 Response evaluation of frequency oscillations in area 1

specification of time-domain parameters of this corresponding scenario is depicted in Table 2. It is clearly showing the ACO-PID controller yield better-controlled response over GA and PSO based controller with respect to minimum settling time.

Scenario 2: Using GA, PSO and ACO PID controller with UPFC and HAE Unit In this scenario, the proposed system is equipped with the UPFC unit in parallel to the tie-line and HAE energy storage unit in area 2. Figure 3 demonstrates the response

Table 2 Value of Settling Time (ST), Peak Overshoot (PO) and Peak Undershoot (PU) in response for Figs. 2 and 3 with GA, PSO and ACO controller

Time-domain specification	Without considering HAE and UPFC unit			With HAE and UPFC unit		
	GA	PSO	ACO	GA	PSO	ACO
ST (s)	18	17	15.5	15	14	12.5
PO (p.u.)	0.003	0.022	0.026	0.005	0.005	0.004
PU (p.u.)	0.075	0.084	0.073	0.019	0.02	0.022

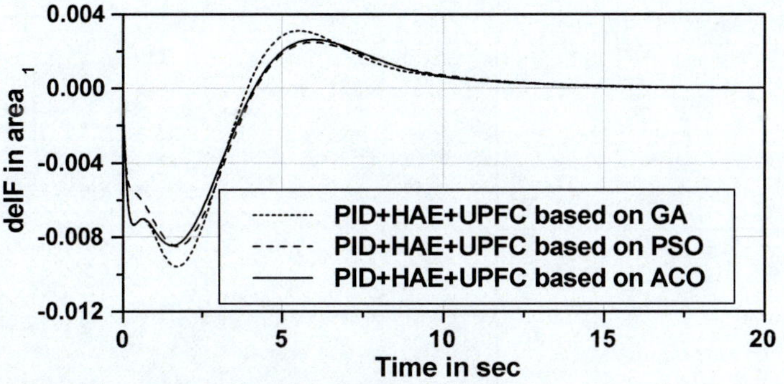

Fig. 3 Response evaluation of oscillations in area 1 frequency

of deviations in the frequency of area 1 in the examined power system using GA, PSO and ACO-PID controllers.

Figure 3 illustrates that the deviations in the power system frequency response with UPFC and HAE units. In addition, the numerical values of time-domain parameters of this scenario are depicted in Table 2 for the responses in Figs. 2 through 3. The peak overshoot (PO), settling time (ST) and peak undershoot (PU) using ACO- and GA-, PSO- tuned PID controller are depicted in Table 2.

Comparison of the bar chart of settling time without considering energy storage unit and UPFC unit analyzed system is depicted in Fig. 4. Similarly, settling comparisons with HAE unit and UPFC unit are given in Fig. 5.

Table 2 and Figs. 4 and 5 proves the superiority of the proposed power system with UPFC and HAE compared to the power system without considering HAE and UPFC unit. By analyzing Figs. 2 through 3 and the settling time, peak over-/under-shoot

Fig. 4 Without considering HAE and UPFC unit

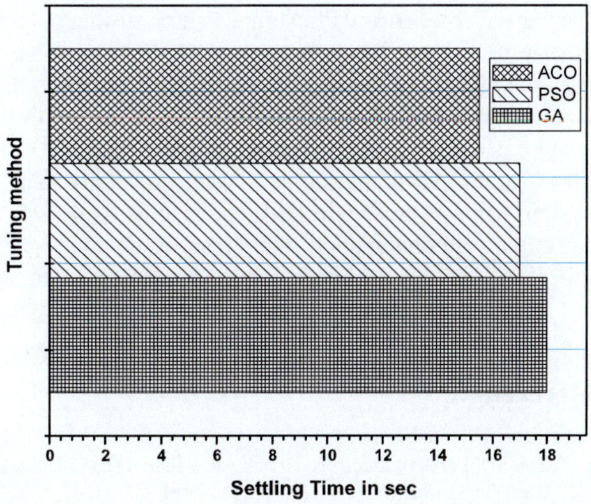

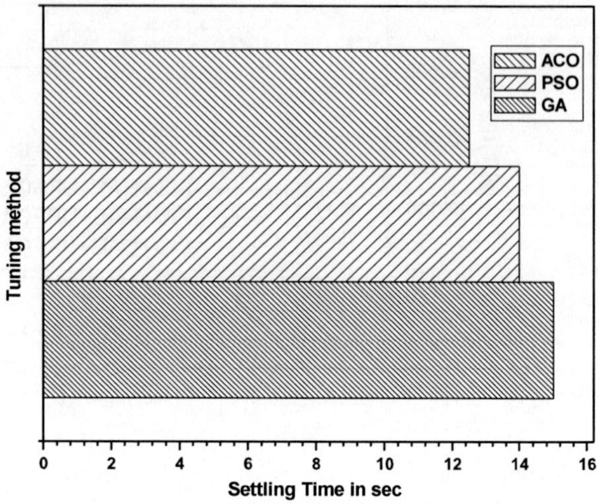

Fig. 5 With HAE and UPFC unit

and performance indices of each technique in Table 2 of scenarios 1 through 2, we establish the result that the proposed ACO- PID controller and ACO-PID with UPFC and HAE unit improve the overall system performance in comparison to using GA and PSO tuned PID controllers for all scenarios of the same power system.

5 Conclusion

In the presented work, the dynamic behavior analysis of two-area thermal connected system is analyzed. The ACO-PID controller response is evaluated and compared to the one with the GA and PSO tuned PID controller response. The simulation outcomes clearly reveal that the proposed ACO-PID controller provides better performance than the PID controller using GA and PSO. Furthermore, the proposed power system is fitted with UPFC and HAE units with the ACO-PID controller. The proposed system performance was compared to GA and PSO PID response with considering UPFC and HAE unit equipped system response. The result clearly shows that the system has a better-controlled response as well as fast settling time, smaller peak overshoot and undershoot (many responses) and higher performance indices during emergency load disturbance condition of the system.

References

1. Elgerd OI (1970) Electric energy system theory: an introduction. Tata Mc-Graw Hill Publishing Company Limited, New York pp 315–389

2. Hari L, Kothari ML, Nandha J (1991) Optimum selection of speed regulation parameters for automatic generation control in discrete mode considering generation rate constraints. IEE Proc-C 13:401–406
3. Nanda J, Kothari ML, Satsangi PS (1983) Automatic generation control of an interconnected hydrothermal system in continuous and discrete modes considering generation rate constraints. IEE Proc 130:17–27
4. Kotari ML, Nanda J, Kothari DP, Das D (1989) Discrete -mode automatic generation control of a two –area reheat thermal system with new area control error. IEEE Trans Power Syst 4:730–738
5. Nandha J, Mishra S (2010) A novel classical controller for Automatic generation control in thermal and hydro thermal systems. PEDES, pp 1–6
6. Anand B, Ebenezer Jeyakumar A (2009) Load Frequency control with fuzzy logic controller considering non-linearities and boiler dynamics. ACSE 8:15–20
7. Ebrahim MA, Mostafa HE, Gawish SA, Bendary FM (2009) Design of decentralized load frequency based-PID controller using stochastic Particle swarm optimization technique. In: International conference on electric power and energy conversion system, pp 1–6
8. Chakraborty S, Samanta S, Mukherjee A, Dey N, Chaudhuri SS (2013) Particle swarm optimization based parameter optimization technique in medical information hiding. In: 2013 IEEE International conference on computational intelligence and computing research (ICCIC), Madurai
9. Khosravy M, Gupta N, Patel N, Senjyu T, Duque CA (2020) Particle swarm optimization of morphological filters for electrocardiogram baseline drift estimation. In: Dey N, Ashour AS, Bhattacharyya S (eds) Applied nature-inspired computing: algorithms and case studies. Springer, Singapore, pp 1–21
10. Ali ES, Abd-Elazim SM (2011) Bacteria foraging optimization algorithm based load frequency controller for interconnected power system. Electric Power Energy Syst 33:633–688
11. Saikia LC, Sinha N, Nanda J (2010) Maiden application of bacterial foraging based fuzzy IDD controller in AGC of a multi-area hydrothermal system. Electric Power Energy Syst 45:98–106
12. Paramasivam B, Chidambaram IA (2010) Bacterial foraging optimization based load frequency control of interconnected power systems with static synchronous series compensator. Int J Latest Trends Comput 1:7–13
13. Dey N, Samanta S, Yang X-Sh, Chaudhri SS, Das A (2013) Optimization of scaling factors in electrocardiogram signal watermarking using cuckoo search. Int J Bio-Inspired Comput (IJBIC) 5:315–326
14. Kumar SR, Ganapathy S (2013) Cuckoo search optimization algorithm based load frequency control of interconnected power systems with GDB nonlinearity and SMES units. Int J Eng Inventions 2:23–28
15. Paramasivam B, Chidambaram IA (2015) ABC algorithm based load-frequency controller for an interconnected power system considering nonlinearities and coordinated with UPFC and RFB. Int J Eng Innovative Technol 1:1–11
16. Omar M, Solimn M, Abdelghany AM, Bendary F (2013) Optimal tuning of PID controllers for hydrothermal load frequency control using ant colony optimization. Int J Electr Eng Inf 5:348–356
17. Hsiao Y-T, Chuang C-L, Chien C-C (2004) Ant colony optimization for designing of PID controllers. In: IEEE International symposium on computer aided control systems design, pp 321–326
18. Chidambaram IA, Paramasivam B (2009) Genetic algorithm based decentralized controller for load-frequency control of interconnected power systems with RFB considering TCPS in the tie-line. Int J Electron Eng Res 1:299–312
19. Gupta N, Khosravy M, Patel N, Senjyu T (2018) A bi-level evolutionary optimization for coordinated transmission expansion planning. IEEE Access 6:48455–48477
20. Shayeghi H, Shayanfor HA (2006) Application of ANN technique based μ–synthesis to load frequency control of interconnected power and energy systems. Int J Electr Power Energy Syst 28:503–511

21. Samanta S, Ahmed SK, Salem M. A, Nath SS, Dey N, Chowdhury SS (2014) Heraldic features based automated glaucoma classification using back propagation neural network. In: The 2014 international conference on frontiers of intelligent computing: theory and applications (FICTA), Special session: advanced research in 'computer vision, image and video processing, pp 14–15
22. Taher SA, Fini MH, Aliabadi SF (2014) Fractional order PID controller design for LFC in electric power systems using imperialist competitive algorithm. Ain Shams Eng J 5:121–135
23. Naidu K, Mokhlis H, Bakar AHA (2013) Application of firefly algorithm (FA) based optimization in load frequency control for interconnected reheat thermal power systems. In: Proceedings of IEEE jordan conference on applied electrical engineering and computing technologies (AEECT)
24. Dey N, Samanta S, Chakraborty S, Das A, Chaudhuri SS, Suri JS (2014) Firefly algorithm for optimization of scaling factors during embedding of manifold medical information: an application in ophthalmology imaging. J Med Imaging Health Inf 4:384–394
25. Moraes CA, De Oliveira EJ, Khosravy M, Oliveira LW, Honório LM, Pinto MF (2020) A Hybrid bat-inspired algorithm for power transmission expansion planning on a practical brazilian network. In: Dey N, Ashour AS, Bhattacharyya S (eds) Applied nature-inspired computing: algorithms and case studies. Springer, Singapore, pp 71–95
26. Gupta N, Khosravy M, Patel N, Sethi IK (2018) Evolutionary optimization based on biological evolution in plants. Procedia Comput Sci 126:146–155 (Elsevier)
27. Gupta N, Khosravy M, Mahela OP, Patel N (2020) Plants biology inspired genetics algorithm: superior efficiency to firefly optimizer. In: Applications of firefly algorithm and its variants, from springer tracts in nature-inspired computing (STNIC), Springer International Publishing, in press
28. Jagatheesan K, Anand B (2012) Dynamic performance of multi-area hydro thermal power systems with integral controller considering various performance indices methods. In: Proceedings of the IEEE international conference of emerging trends in science, engineering and technology (INCOSET), pp 474–478
29. Das S, Kothari ML, Kothari DP, Nanda J (1991) Variable structure control strategy to automatic generation control of interconnected reheat thermal system. IEE proceedings-D 138:579–585
30. Kotari ML, Nanda J (1988) Application of optimal control strategy to automatic generation control of a hydrothermal system. IEE Proc 135:268–274
31. Roy R, Bhatt P, Ghoshal SP (2010) Evolutionary computation based three—area automatic generation control. Expert Syst Appl 37:5913–5924
32. Kundur P (1994) Power system stability and control. Tata Mc-Graw Hill Publishing Company limited, New Delhi, India
33. Gopal M (2003) Digital controls and state variable methods. Tata Mc-Graw Hill Publishing company limited, 2nd edition, New Delhi, India
34. Nagrath J, Kothari DP (1994) Power system engineering. Tata Mc-Graw Hill Publishing Company limited, New Delhi, India
35. Francis R, Chidambaram IA (2015) Optimized PI + load-frequency controller using BWNN approach for an interconnected reheat power system with RFB and hydrogen electrolyser units. Electric power and Energy Syst 67:381–392
36. Bouchekaraa HREH, Abidob MA, Boucherma M (2014) Optimal power flow using teaching-learning-based optimization technique. Electric Power and Energy Syst 114:49–59
37. Sahu BK, Pati S, Mohanty PK, Panda S (2015) Teaching-learning based optimization algorithm based fuzzy-PID controller for automatic generation control of multi-area power system. Appl Soft Comput 27:240–249
38. Nanda J, Sreedhar M, Dasgupta A (2015) A new technique in hydrothermal interconnected automatic generation control system by minority charge carrier inspired algorithm. Electric Power Energy Syst 68:259–268
39. Shivaie M, Kazemi MG, Ameli MT (2015) A modified harmony search algorithm for solving load-frequency control of non-linear interconnected hydrothermal power systems. Sustain Energy Technol Assessments 10:53–62

40. Sathya MR, Ansari MT (2015) load frequency control using bat inspired algorithm based dual mode gain scheduling of PI controller for interconnected power system. Electric Power Energy Syst 64:365–374
41. Dash P, Saikia LC, Sinha N (2015) Comparison of performance of several FACTS devices using Cuckoo search algorithm optimized 2DOF controllers in multi-area AGC. Electric Power Energy Syst 65:316–324
42. Sahu RK, Gorripotu TS, Panda S (2015) A hybrid DE-PS algorithm for load frequency control under deregulated power system with UPFC and RFB. Ain Shams Eng J 6:893–911
43. Hu C, Bi L, Piao Z, Wen C, Hou L (2018) Coordinative optimization control of microgrid based on model predictive control. Int J Ambient Comput Intell (IJACI) 9(3):57–75
44. Aina S, Okegbile SD, Makanju P, Oluwaranti AI (2019) An architectural framework for facebook messenger Chatbot enabled home appliance control system. Int J Ambient Comput Intell (IJACI) 10(2):18–33
45. Jagatheesan K, Anand B, Dey N (2015) Automatic generation control of Thermal-Thermal-Hydro power systems with PID controller using ant colony optimization. Int J Ser Sci Manage Eng Technol (IJSSMET) 6(2):18–34

Cryptosystem Based on Triple Random Phase Encoding with Chaotic Henon Map

Archana, Sachin, and Phool Singh

Abstract Well-known optical image encryption technique double random phase encoding scheme has been shown vulnerable to basic attacks. Thereafter triple random phase encoding scheme is proposed to endure basic attacks. Recently, triple random phase encoding scheme is also shown vulnerable by using deep learning based attack. In this paper, an image encryption scheme based on triple random phase encoding with chaotic Henon map in Fourier domain is proposed. Henon map is used to strengthen the image encryption scheme. Henon map has two parameters and two initial conditions which are highly sensitive to its original value. Experiments were carried out on grayscale images to validate the proposed scheme. Statistical attacks such as information entropy, histogram and 3-D plot analysis are successfully endured by the scheme. Performances against noise and occlusion attacks show the robustness of the scheme. The key sensitivity results indicate that the proposed scheme is highly secure.

Keywords Chaotic Henon map · Image encryption · Double random phase encoding · Triple random phase encoding

1 Introduction

With the population explosion, privacy and security of communication networks have become a big challenge. During the data transmission on social media, security information is an issue. In several applications like an online banking system, biometric signature, one time paid (OTP), identity cards, military applications, confidential video conferences, confidential messages, etc. security is a must. An unauthorized use of data is a serious problem. Throughout transmission, hackers try to hack the

Archana (✉) · Sachin
Department of Mathematics, Central University of Haryana, Mahendergarh 123031, India
e-mail: archana191216@cuh.ac.in

P. Singh
Department of Mathematics, SOET, Central University of Haryana, Mahendergarh 123031, India

K. Ray et al. (eds.), *Proceedings of International Conference on Data Science and Applications*, Lecture Notes in Networks and Systems 148,
https://doi.org/10.1007/978-981-15-7561-7_5

data by use of the internet or other communication media. Data security in term of encryption is well known. With an increase of computational power, digital algorithm for image encryption arises with many limitations. Therefore, optical image encryption or hybrid encryption techniques are more suitable [1].

Double random phase encoding (DRPE) is one of the pioneering and popular optical image encryption technique proposed by Refregier and Javidi in 1995 [2]. This technique encrypts an input image to a stationary white noised encrypted image by use of two random phase masks under 4f optical system. One random phase mask is used in the spatial domain and other one in the Fourier plane. Researchers have proved that DRPE scheme is robust against the brute-force attack, but vulnerable to some basic attacks such as chosen-plaintext attack (CPA) [3], known-plaintext attack (KPA) [4], ciphertext-only attack (COA) [5], chosen-ciphertext attack (CCA) [6]. To resist these basic attacks, researchers have proposed fractional domain in DRPE such as fractional Fourier domain [7], fractional wavelet domain [8], gyrator domain [9], fractional Mellin domain [10], and fractional Hartley domain [11, 12]. Some also used chaotic maps to endure these attacks. Recently, some image encryption algorithm are also proposed based on polarization [13], ptychography [14], virtual optics [15], diffractive imaging [16], photon counting [17], digital holography [18] and ghost imaging [19].

To enhance the security level of DRPE scheme, Ahouzi et al. added a third random phase mask to standard DRPE scheme and termed their scheme as triple random phase encoding (TRPE) scheme [20]. They have shown endurance of TRPE to basic attacks in their work. But recently, cryptanalysis of random phase encoding based optical cryptosystem via deep learning [21] shown that DRPE and TRPE scheme can be cracked by using deep learning (DL) strategy. This motivated us to propose a new secure cryptosystem based on TRPE scheme.

The relationship between cryptosystem and chaotic map is very close. Chaotic map has properties such as control parameters, pseudo-random behavior, sensitivity to the change in the initial conditions. Chaotic system has unpredictable behavior that resembles noise. The strength of chaotic system is based on the effectiveness of produce a pseudo-random sequence. If an initial value is wrong then the chaotic system totally haphazard results. DRPE has been studied with many chaotic systems such as Baker map [22], Affine map [23], Arnold map [24], Logistic map [25], Jigsaw map [26], Lorenz map [27] etc.

In this paper, we have applied Henon map in frequency domain in triple random phase encoding. The rest of the paper is organized as follows: in Sect. 2, TRPE scheme and Henon map are described; in Sect. 3, the proposed scheme is described. Result and discussion are given in Sect. 4. Finally, in the last section, the main conclusions of the work.

1.1 Triple Random Phase Encoding

Triple random phase encoding (TRPE) scheme uses three-phase masks to encrypted an image. Let us consider $f(x, y)$ be the input image. First, an input image $f(x, y)$ is bonded to a random phase mask (RPM1). Thereafter Fourier transform is applied on it as in the case of double random phase encoding. The resulting image bonded with second random phase mask (RPM2). Then Fourier transform (FT) is again applied to make it $4f$ system. Finally, a third random phase mask (RPM3) bonded with a resulting image to get the encrypted image. Mathematically, this process is given by:

$$e_{\mathrm{TRPE}}(x, y) = \mathrm{FT}(\mathrm{FT}(f(x, y) \times \mathrm{RPM1}) \times \mathrm{RPM2}) \times \mathrm{RPM3}$$

where $\mathrm{RPM1} = \exp\{2\pi i\varphi_1(x, y)\}$, $\mathrm{RPM2} = \exp\{2\pi i\varphi_2(u, v)\}$, $\mathrm{RPM3} = \exp\{2\pi i\varphi_3(x, y)\}$, (x, y) and (u, v) stands for the spatial-domain and frequency-domain coordinates, respectively. $\varphi 1$, $\varphi 2$, $\varphi 3$ are randomly generated function having values between [0, 1].

Decryption process for TRPE scheme is the reverse process of the encryption. First, encrypted image $e_{\mathrm{TRPE}}(x, y)$ is bonded with the complex conjugate of a random phase mask third (RPM3*) and then, inverse Fourier transform (IFT) is performed. The obtained result is then bonded with the complex conjugate of second random phase mask (RPM2*) and again IFT is applied to the resulting image, followed by an absolute operation. Mathematically, this process is given by:

$$\mathrm{f}(x, y) = \mathrm{abs}\big(\mathrm{IFT}(\mathrm{IFT}(e_{\mathrm{TRPE}}(x, y)) \times \mathrm{RPM3}^*\big) \times \mathrm{RPM2}^*\big)$$

where "abs" stands for absolute operation of the function. For the sake of clarity, schemes for the encryption and decryption are shown in Fig. 1a, b, respectively.

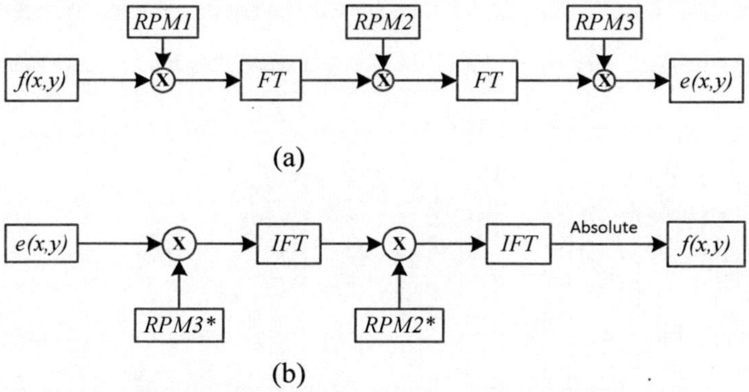

Fig. 1 Flowchart for TRPE scheme **a** encryption process, **b** decryption process

1.2 Henon Map

The Henon map [28] introduced by Michel Henon in 1976 is based on a discrete-time dynamical system. Mathematically, two-dimensional Henon map can be defined as

$$x_{i+1} = 1 - ax_i^2 + y_i$$

$$y_i = bx_i$$

with the initial point (x_1, y_1). The Henon map depends upon two parameters, a and b. Henon map is applied on the grayscale image of Cameraman (Fig. 2a) and the corresponding Henon transformed and retrieved image of Cameraman is given in Fig. 2b, c. Parameters value $a = 1.4$, $b = 0.3$ and initial conditions $x(1) = 0.1354477$, $y(1) = 0.1894063$ for Henon map, are considered in this simulation. This map generates pseudo-random sequence shown in Fig. 3. The time series plot with respect to value x, and the time series plot with respect to the value y shown in Fig. 3a, b respectively. Figure 3c shows the bifurcation diagram of Henon map. It is shown that bifurcation of Henon map is highly periodic. For this bifurcation diagram, parameter a varies from (0, 1.4) and $b = 0.75$.

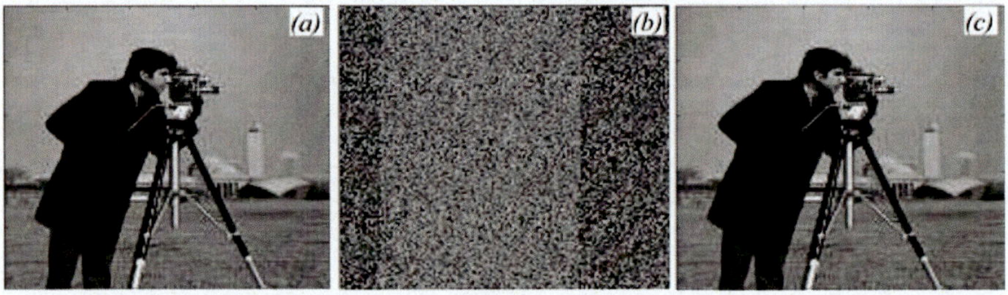

Fig. 2 **a** Input image, **b** encrypted image, **c** decrypted image of Cameraman using Henon map

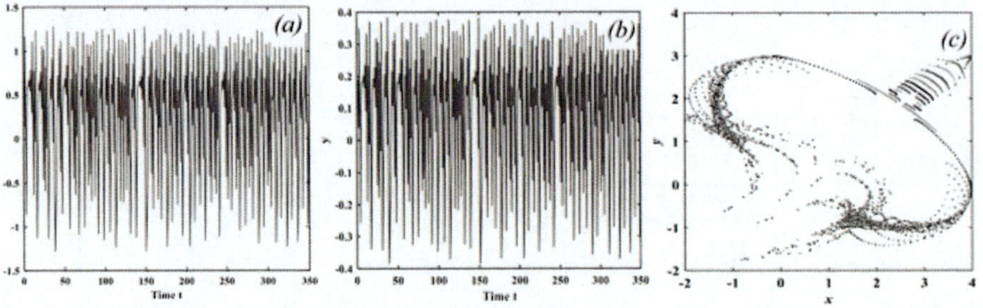

Fig. 3 Time series plot with respect to **a** x, **b** y respectively of Henon map, **c** bifurcation diagram of Henon map

2 Proposed Scheme

We propose an image encryption scheme by using of TRPE scheme with chaotic Henon map in Fourier domain. In the proposed scheme, first, we take the input image of the Cameraman with size $M \times N$, where $M = 256$ and $N = 256$. Input image is bonded with $RPM1$. Fourier transform is applied to the fused image. The resulting image is subjected to Henon map followed by bonding with $RPM2$, and another Fourier transform is applied to it. Finally, RPM3 is bonded with a resulting image to get the encrypted image.

Decryption process for the proposed scheme is the inverse process of encryption process. First, the encrypted image is bonded with RPM3* and then IFT is performed. The resulting image is bonded with RPM2* and followed by the inverse of Henon map and again on the resulting image, IFT is performed. Finally, apply the absolute operation on the resulting image to get the decrypted image. Figure 4 shows the flowchart of the encryption and decryption process for the proposed scheme and Fig. 5 shows the validation of the proposed scheme on the grayscale image of Cameraman.

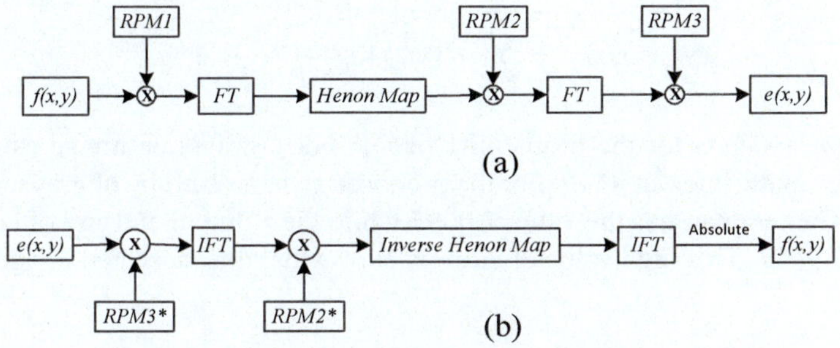

Fig. 4 Flowchart for proposed scheme **a** encryption process, **b** decryption process

Fig. 5 Validation results **a** input image, **b** encrypted image, **c** decrypted image of Cameraman for the proposed scheme

3 Results and Discussion

Many experiments have been conducted on different types of grayscale images to validate the proposed scheme. However, the result section shows the result corresponding to the grayscale image of the Cameraman of size 256×256. In experiments, values of Henon map parameters are $a = 1.4$ and $b = 0.3$. While the initial conditions $x(1) = 0.1354477 y(1) = 0.189463$ are considered. We have analyzed the performance of the proposed encryption scheme against the statistical attacks such as information entropy, histogram, 3-D plots. Also, the sensitivity of secret key, occlusion attack and noise attack is analyzed.

3.1 Information Entropy

Entropy is a statistical measure of randomness that is used to characterize the texture of the input image. For source m, information entropy $H(k)$ is defined as

$$H(k) = \sum P(m_k) \log_2 \frac{1}{P(m_k)}$$

where $P(m_k)$ stands for the probability of m_k. For a grayscale image estimate of entropy lies in the interval [0, 8]. For the proposed scheme entropy of grayscale input image of Cameraman has the value 7.0097 while the value of entropy of encrypted image is 7.9957. This high value of entropy shows that the encrypted image is quite random.

3.2 Histogram and 3-D Plots

For a good encryption scheme, histogram and 3-D plots of encrypted image should be different from that of the original input image. Histogram of an image provides us with information about the frequency distribution of its pixel. Figure 6 displays the histogram for grayscale image of Cameraman. It is clear from Fig. 6b that pixels in the encrypted image are uniformly distributed and hence do not provide any clue to the adversary. Figure 7 displays the 3-D plots of input image Cameraman. It is clear from Fig. 7 that 3-D plot of input and the recovered image is identically same whereas pixels of the encrypted image is uniformly distributed.

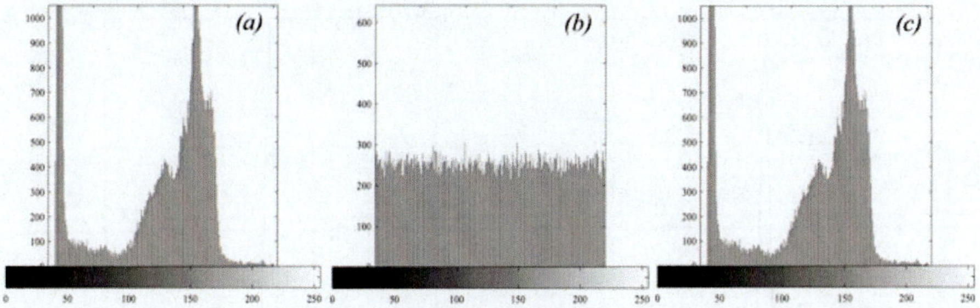

Fig. 6 Histogram plots of **a** input image, **b** encrypted image, **c** decrypted image of Cameraman

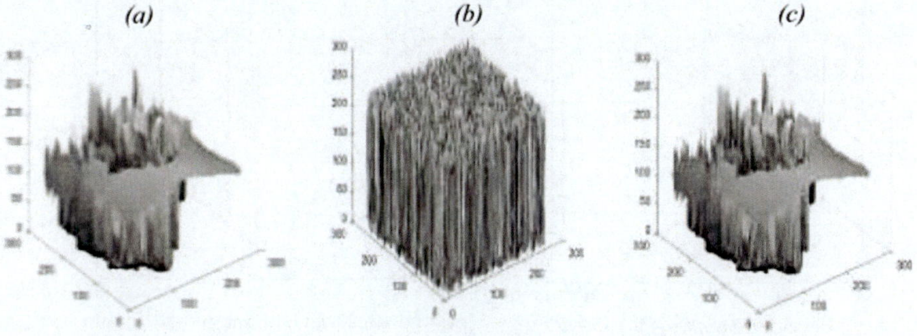

Fig. 7 3-D plots **a** input image, **b** encrypted image, **c** recovered image of Cameraman

3.3 *Key Sensitivity*

In the proposed scheme, Henon map parameters show very high sensitivity to its original value as shown in Fig. 8a, b. Herein, we have plotted key sensitivity plots in terms of deviation from the correct value for parameters a, b and initial conditions x_1 and y_1. Results show that parameters are sensitive to a change of 10^{-15} from its original values.

In the scheme, RPM2, RPM3 and parameters of Henon map serve as decrypted keys. Figure 9 shows the decrypted image when the wrong random phase mask (RPM4 in place of RPM2 and RPM3) is used. Figure 9a, b show the results of the recovered image when wrong random phase mask is used in place of RPM2 (with CC = 0.0026) and RPM3(with CC = 0.0032), respectively. Key sensitivity results show the strength of the scheme against the key sensitivity against brute-force attack.

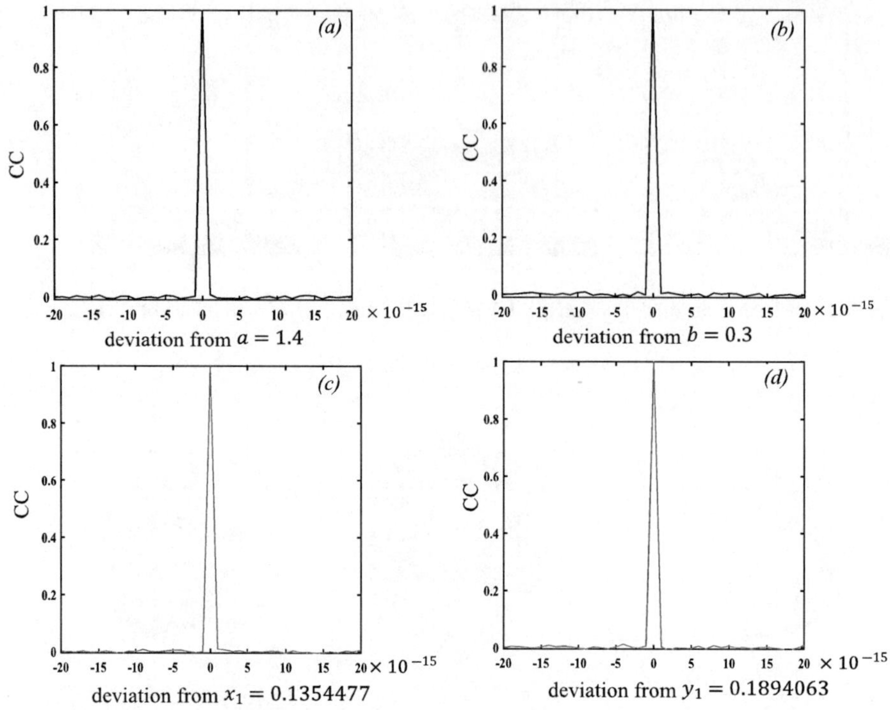

Fig. 8 Key sensitivity plots from **a** $a = 1.4$, **b** $b = 0.3$, **c** $x_1 = 0.1354477$, **d** $y_1 = 0.1894063$

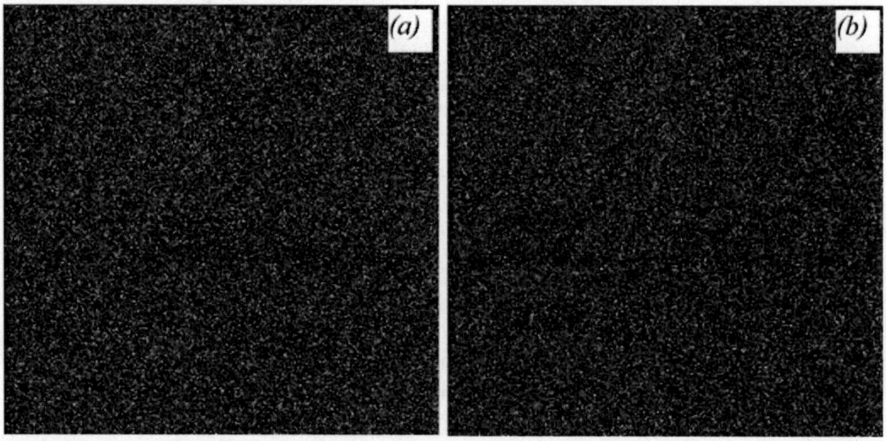

Fig. 9 decrypted image when wrong key is used in place of **a** RPM2, **b** RPM3

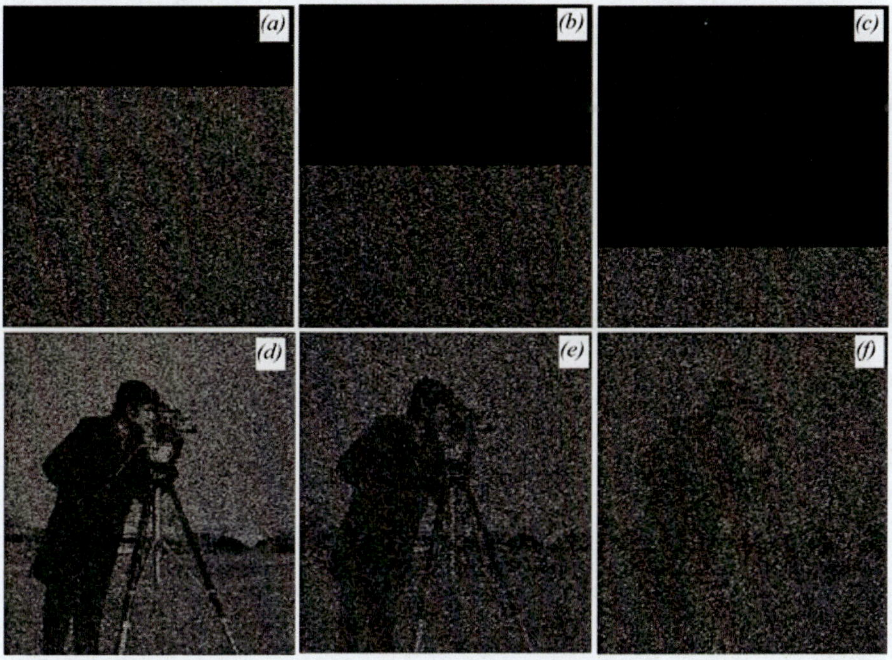

Fig. 10 Results of occlusion attack. Encrypted images with occlusion **a** 25%, **b** 50%, **c** 75%, **d–f** corresponding decrypted images of Cameraman, respectively

3.4 *Occlusion Attack Analysis*

The performance of the proposed scheme against occlusion attack has been investigated. Figure 10 displays the occluded encrypted images and its subsequent decrypted images. Figure 10a–c shown the encrypted image with the loss of data 25%, 50%, 75% and corresponding decrypted images shown in Fig. 10d, e, respectively. Results show that the quality of decrypted images is degrading with the increment of loss of data in encrypted image. It is clearly seen that the image is identifiable even when occluded data as high as 75%. Occlusion attack results show that the scheme is robust against the occlusion attack.

3.5 *Noise Attack Analysis*

The proposed scheme is also analyzed against noise attack. Salt and pepper noise is introduced in the encrypted image by using the relation

$$N = T(1 + kO),$$

Fig. 11 Recovered images when encrypted is polluted with noise strength **a** $k = 5$, **b** $k = 10$, **c** $k = 15$ of Cameraman image

where N is the noise-affected encrypted image, k represents noise strength coefficient, O is the salt and pepper noise with density 0.05. Figure 11a–c show the recovered image with noise strength coefficient 5, 10, 15, respectively. The CC between the original and the recovered image with noise strength $k = 15$ is around 0.7073. Results show that the scheme resists the noise attack.

4 Conclusion

In this paper, an image encryption scheme based on TRPE scheme with chaotic Henon map is proposed. Henon map is used to strengthen the image encryption scheme. Henon map has two parameters and two initial conditions which are highly sensitive to its original value. Experiments were carried out on grayscale images to validate the proposed scheme. In the simulation on Matlab, we get CC = 1, mean squared error(MSE) is $4.4160e^{-22}$ and peak signal-to-noise ratio(PSNR) is 602.7128 between the original and the recovered images. Statistical attacks such as information entropy, histogram analysis and 3-D plot analysis are successfully endured by the scheme. Performances against noise and occlusion attacks show the robustness of the scheme. The key sensitivity results indicate that the proposed scheme is highly secure.

References

1. Nishchal NK (2019) Optical cryptosystems. Institute of Physics (IOP) Publishing Ltd., Bristol, UK
2. Refregier P, Javidi B (1995) Optical image encryption based on input plane and Fourier plane random encoding. Opt Lett 20(7):767. https://doi.org/10.1364/OL.20.000767

3. Peng X, Wei H, Zhang P (2006) Chosen-plaintext attack on lensless double-random phase encoding in the Fresnel domain. Opt Lett 31(22):3261. https://doi.org/10.1364/OL.31.003261
4. Gopinathan U, Monaghan DS, Naughton TJ, Sheridan JT (2006) A known-plaintext heuristic attack on the Fourier plane encryption algorithm. Opt Express 14(8):3181. https://doi.org/10.1364/OE.14.003181
5. Liu X, Wu J, He W, Liao M, Zhang C, Peng X (2015) Vulnerability to ciphertext-only attack of optical encryption scheme based on double random phase encoding. Opt Express 23(15):18955. https://doi.org/10.1364/OE.23.018955
6. Carnicer A, Montes-Usategui M, Arcos S, Juvells I (2005) Vulnerability to chosen-cyphertext attacks of optical encryption schemes based on double random phase keys. Opt Lett 30(13):1644. https://doi.org/10.1364/OL.30.001644
7. Unnikrishnan G, Joseph J, Singh K (2000) Optical encryption by double-random phase encoding in the fractional Fourier domain. Opt Lett 25(12):887. https://doi.org/10.1364/OL.25.000887
8. Mehra I, Nishchal NK (2014) Image fusion using wavelet transform and its application to asymmetric cryptosystem and hiding. Opt Express 22(5):5474. https://doi.org/10.1364/OE.22.005474
9. Liu Z, Guo Q, Xu L, Ahmad MA, Liu S (2010) Double image encryption by using iterative random binary encoding in gyrator domains. Opt Express 18(11):12033. https://doi.org/10.1364/OE.18.012033
10. Vashisth S, Singh H, Yadav AK, Singh K (2014) Devil's vortex phase structure as frequency plane mask for image encryption using the fractional Mellin transform. International Journal of Optics 2014:1–9. https://doi.org/10.1155/2014/728056
11. Singh P, Yadav AK, Singh K, Saini I (2019) Asymmetric watermarking scheme in fractional Hartley domain using modified equal modulus decomposition 21:484–491
12. Rakheja P, Vig R, Singh P (2020) Double image encryption using 3D Lorenz chaotic system, 2D non-separable linear canonical transform and QR decomposition. Opt Quant Electron 52(2):103. https://doi.org/10.1007/s11082-020-2219-8
13. Zhu N, Wang Y, Liu J, Xie J, Zhang H (2009) Optical image encryption based on interference of polarized light. Opt Express 17(16):13418. https://doi.org/10.1364/OE.17.013418
14. Shi Y, Li T, Wang Y, Gao Q, Zhang S, Li H (2013) Optical image encryption via ptychography. Opt Lett 38(9):1425. https://doi.org/10.1364/OL.38.001425
15. Peng X, Cui Z, Tan T (2002) Information encryption with virtual-optics imaging system. Opt Commun 212(4–6):235–245. https://doi.org/10.1016/S0030-4018(02)02003-5
16. Chen W, Chen X, Sheppard CJR (2010) Optical image encryption based on diffractive imaging. Opt Lett 35(22):3817. https://doi.org/10.1364/OL.35.003817
17. Pérez-Cabré E, Cho M, Javidi B (2011) Information authentication using photon-counting double-random-phase encrypted images. Opt Lett 36(1):22. https://doi.org/10.1364/OL.36.000022
18. Javidi B, Nomura T (2000) Securing information by use of digital holography. Opt Lett 25(1):28. https://doi.org/10.1364/OL.25.000028
19. Clemente P, Durán V, Torres-Company V, Tajahuerce E, Lancis J (2010) Optical encryption based on computational ghost imaging. Opt Lett 35(14):2391. https://doi.org/10.1364/OL.35.002391
20. Ahouzi E, Zamrani W, Azami N, Lizana A, Campos J, Yzuel MJ (2017) Optical triple random-phase encryption. Opt Eng 56(11):1. https://doi.org/10.1117/1.OE.56.11.113114
21. Hai H, Pan S, Liao M, Lu D, He W, Peng X (2019) Cryptanalysis of random-phase-encoding-based optical cryptosystem via deep learning. Opt Express 27(15):21204. https://doi.org/10.1364/OE.27.021204
22. Elshamy AM et al (2013) Optical image encryption based on chaotic baker map and double random phase encoding. J Lightwave Technol 31(15):2533–2539. https://doi.org/10.1109/JLT.2013.2267891
23. Singh P, Yadav AK, Singh K (2017) Color image encryption using affine transform in fractional Hartley domain. Optica Applicata XLVII:421–433. https://doi.org/10.5277/oa170308

24. Elshamy AM et al (2016) Optical image cryptosystem using double random phase encoding and Arnold's Cat map. Opt Quant Electron 48(3):212. https://doi.org/10.1007/s11082-016-0461-x
25. Huang H, Yang S (2017) Colour image encryption based on logistic mapping and double random-phase encoding. IET Image Proc 11(4):211–216. https://doi.org/10.1049/iet-ipr.2016.0552
26. Singh M, Kumar A, Singh K (2008) Optical security system using jigsaw transforms of the second random phase mask and the encrypted image in a double random phase encoding system. Opt Lasers Eng 46(10):763–768. https://doi.org/10.1016/j.optlaseng.2008.04.021
27. Sharma N, Saini I, Yadav A, Singh P (2017) Phase-image encryption based on 3D-Lorenz chaotic system and double random phase encoding. 3D Res 8(4):39. https://doi.org/10.1007/s13319-017-0149-4
28. Mishra K, Saharan R (2019) A fast image encryption technique using Henon chaotic map. In: Pati B, Panigrahi CR, Misra S, Pujari AK, Bakshi S (eds) Progress in advanced computing and intelligent engineering, vol 713. Springer Singapore, Singapore, pp 329–339

Different Loading of Distributed Generation on IEEE 14-Bus Test System to Find Out the Optimum Size of DG to Allocation in Transmission Network

Rakesh Bhadani and K. C. Roy

Abstract Distributed generation (DG) units in Transmission Network have become more and more important in the present time. The aim of the optimum DG unit allocation in Transmission Network is to offer the best places in entire network. Here an endeavor has been made to control the most sensitive buses to voltage collapse in Transmission Network. Here a method for allocation of DG units in Transmission Network has been offered. The proposed method is based on the study of power flow continuation and determination of most sensitive buses to voltage collapse. The proposed method illustrations significant reduction in real and reactive power losses, improve voltage profile, it also may permit intensification in power transfer capacity, and voltage stability margin, and improves the loading capability of transmission network.

Keywords DG technologies · Placement algorithm · Effect of placement of DG units · Case study · Different DGs installed in transmission network · Conclusion

1 Introduction

An electric power system is a network of electrical tackles used to source, transmission and use of electric power. An electric power system mainly divided in two portions one is the generators that is source, its supply the power and second one is the transmission system that transfer the power from the generating station to the consumer and the distribution system that brought the power to neighboring homes and industries [1, 2].

Distributed generation comparatively limited- scale generators that produce some kilowatts to 10 MW of power and usually connected to the grid at the distribution

R. Bhadani (✉) · K. C. Roy
Rai University, Ahmedabad, India
e-mail: rksbhadani@gmail.com

K. C. Roy
e-mail: dr.kcroy@gmail.com

K. Ray et al. (eds.), *Proceedings of International Conference on Data Science and Applications*, Lecture Notes in Networks and Systems 148,
https://doi.org/10.1007/978-981-15-7561-7_6

or substation levels. Distributed generation units custom a wide range of generation technologies, including gas based turbines, diesel engine generators, solar photovoltaic generation, wind turbines, fuel cells, biomass energy and hydroelectric generators. Distributed generation, also called on-site generation, drive in generation, decentralized generation, dispersed energy and produce electricity from many trivial energy sources [3]. Distributed generation permits collection of energy from several sources and may provide lesser environmental effects and improved safety of source. As an import, the connection of DG to the network may impact the stability of the power structure, i.e., angle, frequency, and voltage stability [4].

The greatest and fast placement and transfer DG units are one of the main tasks in the design area and several approaches were used for discovering. Lagrange method, two degrees gradient method, and sensitivity analysis method have been used for that purpose [5, 6].

This research work presents a method for placement of DG units in transmission network; it is based on load flow study and determination of most sensitive buses to the voltage collapse. Then the DG units with assured capacity will be connected in these buses through an objective purpose and an iterative algorithm. In load flow study is used for determination of the voltage collapse bus or maximum loaded bus. This method will be performed on an IEEE 14 test system.

2 Dıstributed Generation (DG) Technologies

In the recent years the electrical power profits have proficient fast restructuring process universally. With a regulation, progress in technologies and uneasiness about the environmental impacts, competition is predominantly accepted in the generation side, thus permitting augmented interconnection of generating units to the utility networks. These types of generating sources are termed distributed generators (DG) and distinct as the plant which is directly connected with network and isn't centrally prearranged and transmitted. These are also entitled as a static or distributed generation units. DG systems are nothing but one type of a connection among the transmission system and the electricity users [7]. The size of the DG systems can lies between some kW to 100 MW. Several new forms of distributed generating systems like micro turbines, fuel cells, solar and wind power are generating new prospects for the addition of diverse DG systems to the utility. Interconnection of generators will offer a number of benefits such as enhanced reliability, power quality improvement, efficiency, development of system constraints with the environmental benefits [8].

The main features of DG units can be listed follows:

- Reduction of power loss in transmission line.
- Voltage profile improvement in transmission system.
- High reliability of electrical system.
- Power quality improvement in transmission system.
- Voltage stability improvement in transmission system.

3 Effect of DG on Voltage Stability

3.1 *Synchronous Generator*

Synchronous generators are proficient of both generating and absorbing reactive power in electrical system. Therefore the overexcited synchronous generators will allow on-site generation of reactive power in transmission system. The local generation of reactive power decreases its significance from the feeder, thus decreases the related losses, and improves the voltage profile [6, 9].

3.2 *Asynchronous Generator*

An asynchronous generator has a number of features that make it very appropriate for Distribution Generation unit.

Following are the features of Asynchronous generator:

- Relatively inexpensive prices,
- Low maintenance requirements,
- Robust,
- This type of DG consumes reactive power when directly connected to the transmission network. The reactive power consumption of asynchronous generators is generally reimbursed by the shunt capacitor banks.

This is a part solution to the voltage stability problem; meanwhile a voltage drop will decrease the amount of reactive power generated by the capacitor banks, whereas swelling the reactive power consumption of the asynchronous generator. Thus, there is a risk that instead of supportive the network at an under voltage situation, the asynchronous generator will further reduce the system voltage and it capacity cause a voltage stability [6, 9].

3.3 *Line-Commutated and Self-commutated Converter*

That is fact the conventional line-commutated converters continuously consume reactive power. The total consumed reactive power 30% high from the rated power of the converter. To reimburse the demand, capacitor banks are generally installed on the ac side of the converter. Therefore, below certain circumstances, the occurrence of such a converter can negatively disturb voltage stability. Furthermore, the capacities of DG are often quite small, which creates the utilization of unconventional power electronics devices economically beneficial. Consequently, it can be expected with certain degree of confidence that in the upcoming days most of the power electronics converters will be self-commutated.

The application of self-commutated converters for interconnection of DG units with the transmission network gives the permission for instant and unique control of the output voltage and angle. Thus, reactive power can be either produced or absorbed, depending on the control mode. Meanwhile normally the power factor of such a converter is close to unity, no reactive power is injected in the transmission network. Case studies presented [10] in report a important development of transient stability by a fuel cell power plant interfaced with power electronic converters [6, 9].

4 Flow Chart for DG Placement

For location of DG units, it is essential to describe objective function to solving this problem. As per the structure of placement algorithm in Fig. 1, the desire voltage profile will achieve.

5 Case Study (IEEE 14-Bus Test System)

The rating and nature of the test system play a main role in the study of the transient stability. A big system may increase the time and difficulty of the analysis whereas a small system may lead to ignoring essential factors. So a medium sized typical 14-bus test system has been select for the study. Figure 2 shows the single-line diagram of IEEE 14-bus test system as similar to typical system, and the characteristics of the test system are given in Table 1.

In this test system bus 1 is consider as a slack bus the buses 2, 3, 6, and 8 consider as a generator bus and also buses 4, 5, 7, 9, 10, 11, 12, 13, and 14 consider as a load bus.

The test system contains of a main source connected to bus no.1 and bus no. 2. There are three DGs connected to buses 14, 4, and 5. The DGs connected with buses 3, 6 and 8 are denoted by synchronous condenser which are shown in Fig. 3. In this test system buses 4, 5, 7, 9, 10, 11, 12, 13 and 14 are at a voltage level of 1.0 kV and only bus no. 8 is at higher voltage level of 1.09 kV.

Synchronous generator, transformer, loads, and induction generator have been taken with classic values in MATLAB. Here given test system study by Newton–Raphson Method for only load flow study and then we can compare different results obtain from respective DG installation.

This method was performed on an IEEE 14-bus test system and the outcomes presented the healthiness of this method for optimum and firm placement of DG units.

Now we can execute load flow analysis on IEEE 14-bus test system to find out the voltage collapse profile curve in normal state and three iterations of placement algorithm with different certain loading capacity of DGs.

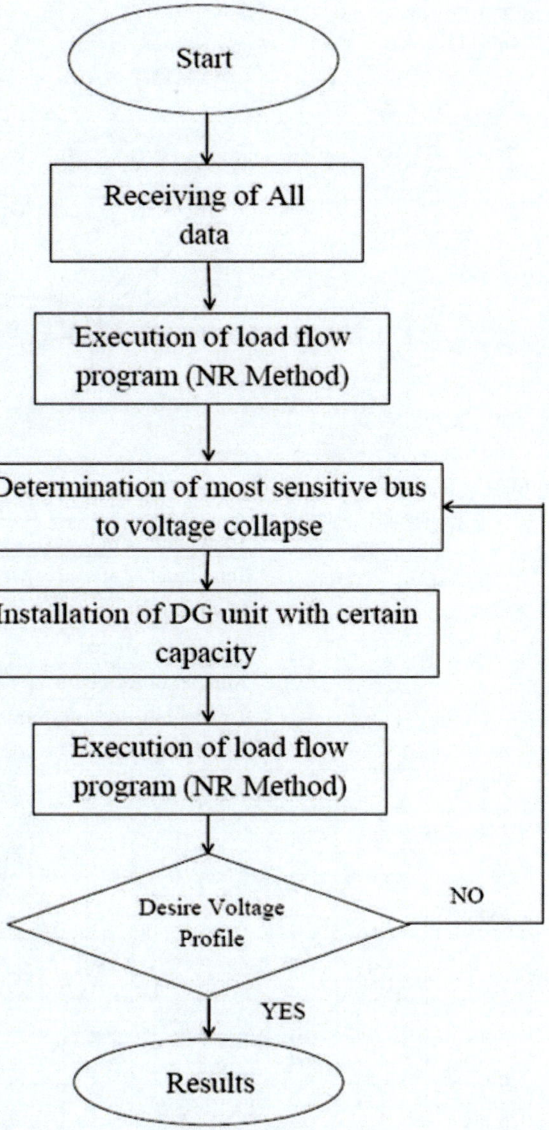

Fig. 1 Flowchart of DG placement method

Case: 1 With DG units (10 MW P_G) (Fig. 3; Table 2)

Case: 2 With DG units (10 MW P_G+ 10 Mar Q_G) (Fig. 4; Table 3)

Case: 3 With DG units (10 MVar Q_G) (Fig. 5; Table 4)

Case: 4 With DG units (20 MW P_G) (Fig. 6; Table 5)

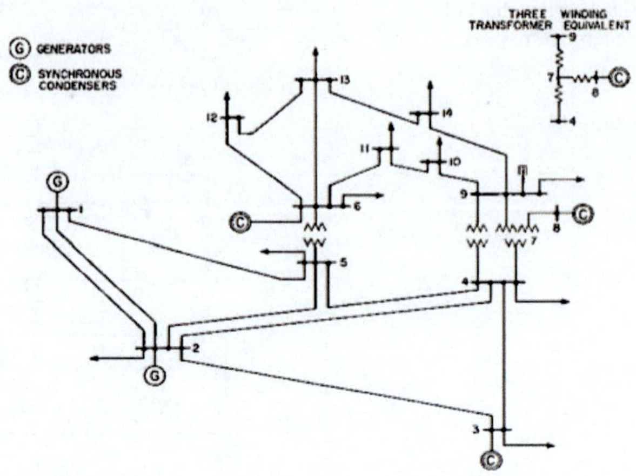

Fig. 2 IEEE 14-bus test system [11]

Table 1 Characteristics of the IEEE 14-bus test system

System characteristics	Value
Total number of buses	14
Total number of source	5
Main source	2
Distributed source	3
Number of load buses	9
Total number of transformers	3

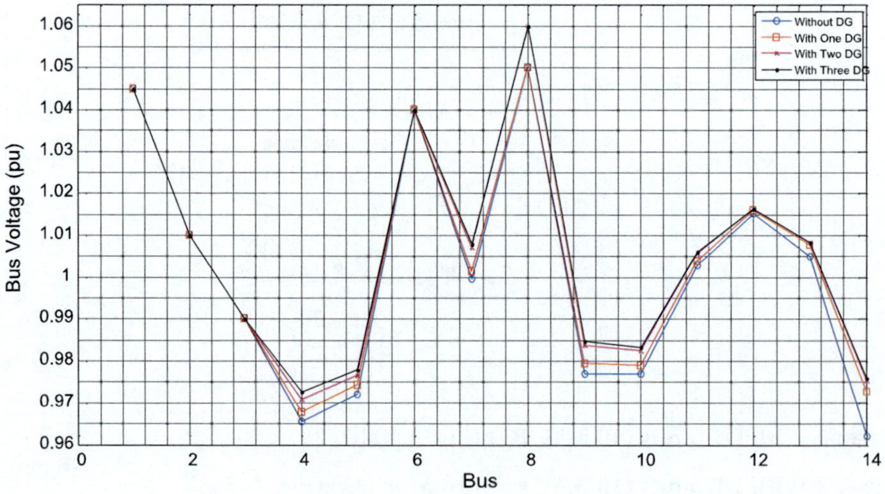

Fig. 3 Voltage profile curve for an IEEE 14-bus test network with DG units (10 MW P_G)

Table 2 Voltage profile for an IEEE 14-bus test network with DG units (10 MW P_G)

Bus no.	Without DG	With one DG	With two DG	With three DG
	V (pu)	V (pu)	V (pu)	V (pu)
1	1.045	1.045	1.045	1.045
2	1.01	1.01	1.01	1.01
3	0.99	0.99	0.99	0.99
4	0.9654	0.9678	0.9708	0.9727
5	0.972	0.9744	0.9766	0.9781
6	1.04	1.04	1.04	1.04
7	0.9995	1.0015	1.0071	1.008
8	1.05	1.05	1.06	1.06
9	0.977	0.9795	0.9837	0.9847
10	0.977	0.979	0.9826	0.9834
11	1.0028	1.0039	1.0057	1.0062
12	1.0151	1.0161	1.0164	1.0164
13	1.0049	1.0077	1.0083	1.0085
14	0.9619	0.9726	0.9753	0.976

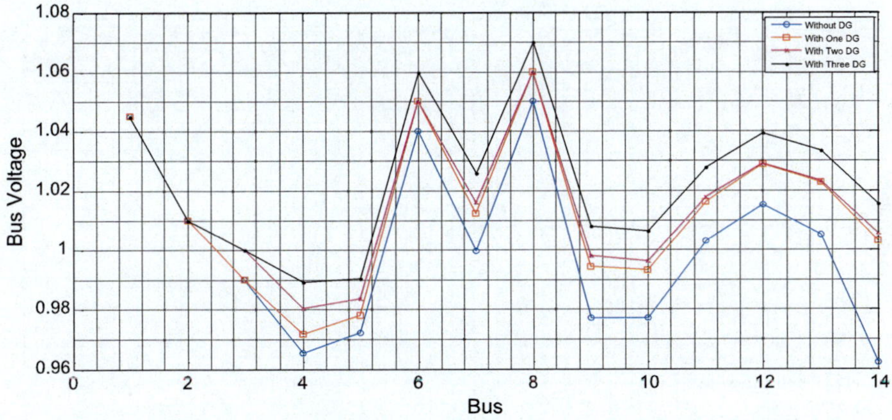

Fig. 4 Voltage profile curve for an IEEE 14-bus test network with DG units (10 MW P_G + 10 MVar Q_G)

Case: 5 Voltage profile curve for an IEEE 14-bus test network with DG units (20 MW P_G+ 20 MVar Q_G) (Fig. 7; Table 6)

Case: 6 With DG units (20 MVar Q_G) (Fig. 8; Table 7)

Case: 7 With DG units (25 MW P_G+ 20 MVar Q_G) (Fig. 9; Table 8)

Table 3 Voltage profile for an IEEE 14-bus test network with DG units (10 MW P_G + 10 MVar Q_G)

Bus no.	Without DG	With one DG	With two DG	With three DG
	V (pu)	V (pu)	V (pu)	V (pu)
1	1.045	1.045	1.045	1.045
2	1.01	1.01	1.01	1.01
3	0.99	0.99	1	1
4	0.9654	0.9717	0.9803	0.9892
5	0.972	0.978	0.9836	0.9905
6	1.04	1.05	1.05	1.06
7	0.9995	1.0122	1.0161	1.0259
8	1.05	1.06	1.06	1.07
9	0.977	0.9941	0.9979	1.008
10	0.977	0.993	0.9961	1.0064
11	1.0028	1.016	1.0176	1.0278
12	1.0151	1.0287	1.029	1.0392
13	1.0049	1.0225	1.0231	1.0333
14	0.9619	1.003	1.0054	1.0156

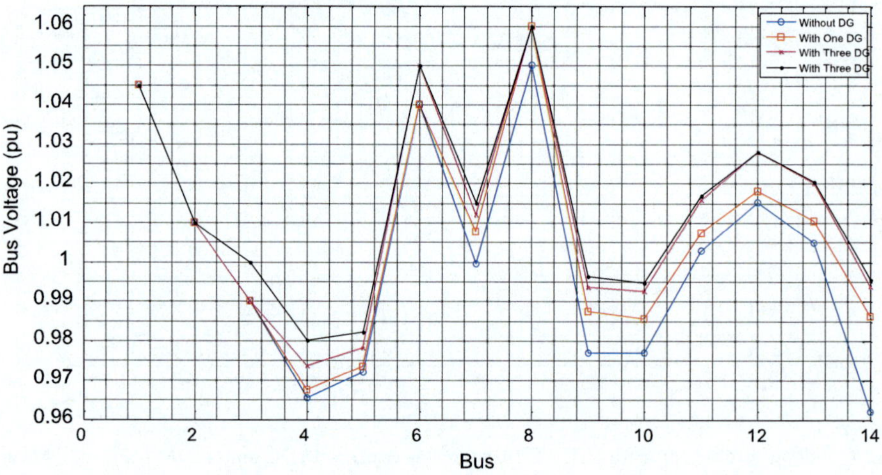

Fig. 5 Voltage profile curve for an IEEE 14-bus test network with DG units (10 MVar Q_G)

Table 4 Voltage profile for an IEEE 14-bus test network with DG units (10 MVar Q_G)

Bus no.	Without DG	With one DG	With two DG	With three DG
	V (pu)	V (pu)	V (pu)	V (pu)
1	1.045	1.045	1.045	1.045
2	1.01	1.01	1.01	1.01
3	0.99	0.99	0.99	1
4	0.9654	0.9676	0.9736	0.9803
5	0.972	0.9735	0.9783	0.9824
6	1.04	1.04	1.05	1.05
7	0.9995	1.0078	1.0122	1.0152
8	1.05	1.06	1.06	1.06
9	0.977	0.9875	0.9935	0.9964
10	0.977	0.9857	0.9925	0.9949
11	1.0028	1.0073	1.0157	1.0169
12	1.0151	1.018	1.0279	1.0281
13	1.0049	1.0104	1.0201	1.0205
14	0.9619	0.9861	0.9938	0.9957

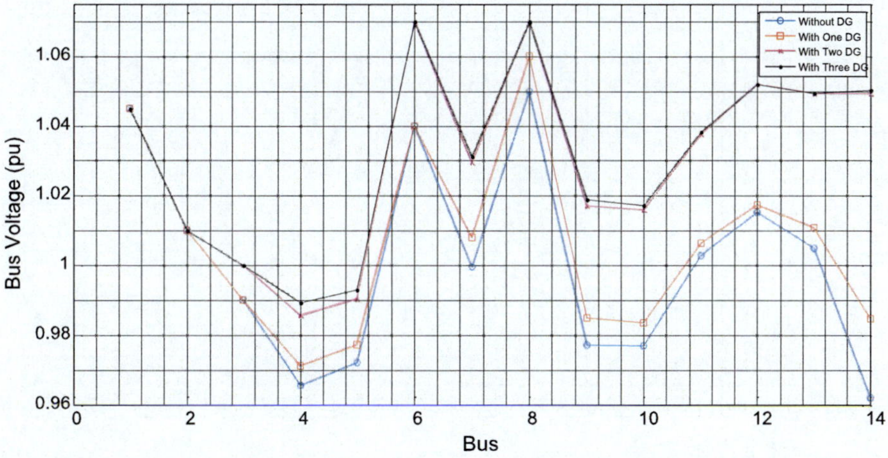

Fig. 6 Voltage profile curve for an IEEE 14-bus test network with DG units (20 MW P_G)

Table 5 Voltage profile for an IEEE 14-bus test network with DG units (20 MW P_G)

Bus no.	Without DG	With one DG	With two DG	With three DG
	V (pu)	V (pu)	V (pu)	V (pu)
1	1.045	1.045	1.045	1.045
2	1.01	1.01	1.01	1.01
3	0.99	0.99	1	1
4	0.9654	0.9711	0.9857	0.9893
5	0.972	0.9773	0.9902	0.993
6	1.04	1.04	1.07	1.07
7	0.9995	1.0079	1.0295	1.0312
8	1.05	1.06	1.07	1.07
9	0.977	0.9848	1.0171	1.0189
10	0.977	0.9835	1.0158	1.0173
11	1.0028	1.0062	1.0375	1.0383
12	1.0151	1.0172	1.0519	1.052
13	1.0049	1.0107	1.0492	1.0494
14	0.9619	0.9846	1.0493	1.0504

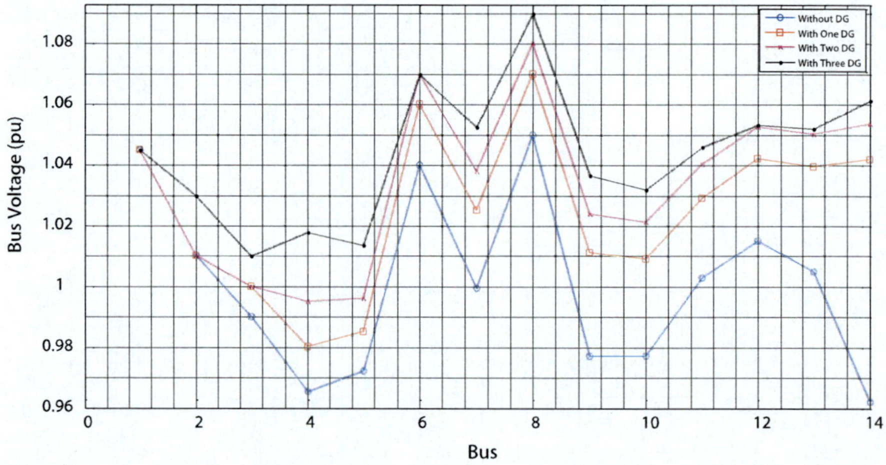

Fig. 7 Voltage profile curve for an IEEE 14-bus test network with DG units (20 MW P_G + 20 MVar Q_G)

Table 6 Voltage profile for an IEEE 14-bus test network with DG units (20 MW P_G + 20 MVar Q_G)

Bus no.	Without DG	With one DG	With two DG	With three DG
	V (pu)	V (pu)	V (pu)	V (pu)
1	1.045	1.045	1.045	1.045
2	1.01	1.01	1.01	1.03
3	0.99	1	1	1.01
4	0.9654	0.9802	0.995	1.0179
5	0.972	0.9852	0.996	1.0139
6	1.04	1.06	1.07	1.07
7	0.9995	1.0253	1.0379	1.0525
8	1.05	1.07	1.08	1.09
9	0.977	1.0112	1.0239	1.0367
10	0.977	1.009	1.0214	1.0321
11	1.0028	1.0291	1.0404	1.0459
12	1.0151	1.0421	1.0524	1.0534
13	1.0049	1.0396	1.0501	1.052
14	0.9619	1.0419	1.0535	1.0614

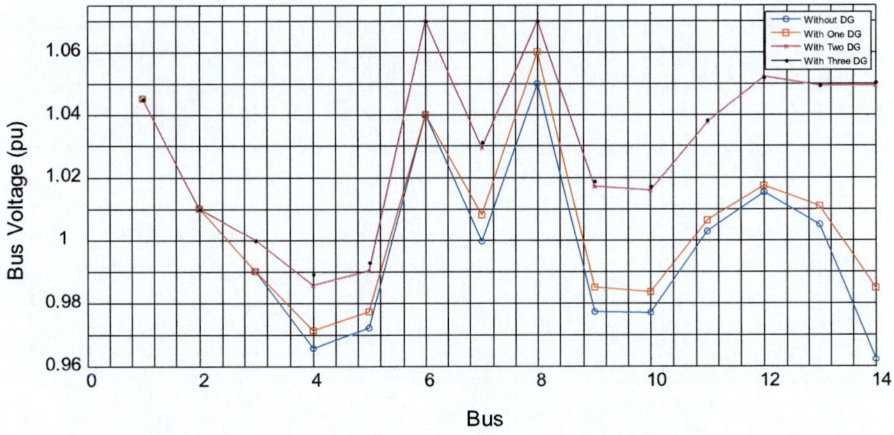

Fig. 8 Voltage profile curve for an IEEE 14-bus test network with DG units (20 MVar Q_G)

Table 7 Voltage profile for an IEEE 14-bus test network with DG units (20 MVar Q_G)

Bus No.	Without DG	With one DG	With two DG	With three DG
	V (pu)	V (pu)	V (pu)	V (pu)
1	1.045	1.045	1.045	1.045
2	1.01	1.01	1.01	1.01
3	0.99	0.99	1	1
4	0.9654	0.9722	0.9903	0.9952
5	0.972	0.9785	0.9924	0.9938
6	1.04	1.06	1.07	1.07
7	0.9995	1.0161	1.0315	1.0364
8	1.05	1.06	1.07	1.08
9	0.977	1.0026	1.0189	1.0215
10	0.977	1.0019	1.0172	1.0193
11	1.0028	1.0254	1.0382	1.0393
12	1.0151	1.04	1.0521	1.0509
13	1.0049	1.0339	1.0494	1.0453
14	0.9619	1.02	1.0504	1.0356

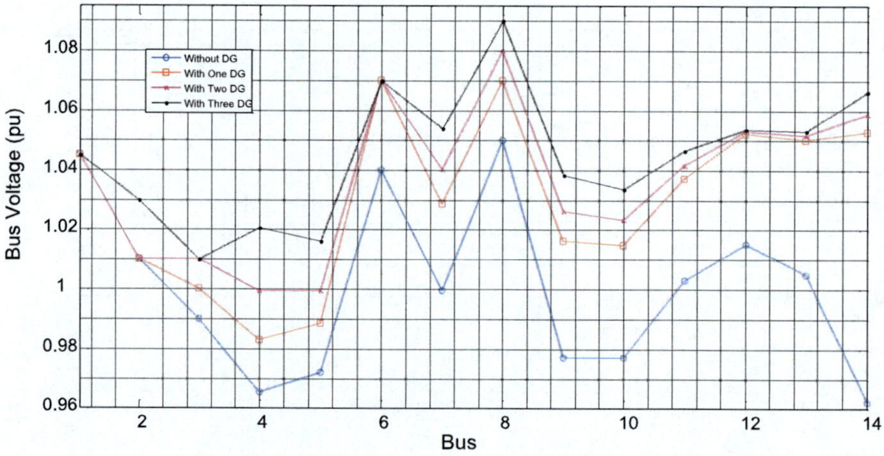

Fig. 9 Voltage profile curve for an IEEE 14-bus test network with DG units (25 MW P_G + 20 MVar Q_G)

Table 8 Voltage profile for an IEEE 14-bus test network with DG units (25 MW P_G + 20 MVar Q_G)

Bus no.	Without DG	With one DG	With two DG	With three DG
	V (pu)	V (pu)	V (pu)	V (pu)
1	1.045	1.045	1.045	1.045
2	1.01	1.01	1.01	1.03
3	0.99	1	1.01	1.01
4	0.9654	0.9829	0.9993	1.0205
5	0.972	0.9883	0.9991	1.016
6	1.04	1.07	1.07	1.07
7	0.9995	1.0284	1.0401	1.0539
8	1.05	1.07	1.08	1.09
9	0.977	1.0161	1.0262	1.0383
10	0.977	1.0149	1.0233	1.0334
11	1.0028	1.0371	1.0414	1.0466
12	1.0151	1.0522	1.0529	1.0538
13	1.0049	1.05	1.0515	1.0532
14	0.9619	1.0526	1.0587	1.0662

6 Conclusion

In this research we can carried out load flow study with the help of NR method on IEEE 14-bus test system and get the results as above. The different cases are described in this paper with 10–25% power value of their base MVA rating of this particular DG, Also get the result of active, reactive and active plus reactive power for better analysis and find out the optimum size of the DG. By comparing all the results we can conclude that approximate optimum size of the DG from given case study is 25 pu active and 20 pu reactive power for DG rated base 100 MVA. This method is simple and straightforward, because one has to simply detect the voltages at weak buses at the voltage collapse point, which are found at the end of NR method solution. The known week buses are selected for selecting the location of DG units. Based on survey this technique is effective for improvement of voltage profile, reduction of power losses and also an increase in maximum loading and voltage stability margin in transmission system.

References

1. Celli G, Pilo F (2001) Optimal distributed generation allocation in mv distribution networks. In: Proceedings 22nd IEEE PES International Conference on PICA 2001
2. http://en.wikipedia.org/wiki/Electric_power_system

3. Georgilakis PS (2013) Optimal distributed generation placement in power distribution networks: models, methods, and future research. IEEE Trans Power System 28(3)
4. Wan YH, Rau NS (1994) Optimum location of resources in distributed planning. Power System IEEE Trans 9:2014–2020
5. Griffin KT, Low A (2000) Placement of dispersed generations systems for reduced losses. In: 33rd Annual Hawaii International Conference on System Sciences (HICSS)
6. Reza M, Slootweg JG, Van der Sluis L (2003) Investigating impacts of distributed generation on transmission system stability. Power Syst IEEE PowerTech
7. Distribution generation impacts on transmission system http://en.wikipedia.org/wiki/Distributed_generation
8. David TA and Milanovic JV (2002) Stability of distribution networks with embedded generators and induction motors. Proc Winter Meet Power Eng Soc 2:1023–1028
9. Akbarimajd A, Hedayati H (2008) A method for placement of dg units in distribution networks. IEEE Trans Power Deliv 23
10. Mardanhe M (2004) Siting and sizing of DG units using GA and OPF based technique. IEEE
11. https://www.researchgate.net/figure/Single-line-diagram-of-the-IEEE-14-bus-system_fig2_233160893

Kidney Care: Artificial Intelligence-Based Mobile Application for Diagnosing Kidney Disease

Zarin Subah Shamma, Israt Jahan Rumman, Ali Mual Raji Saikot, S. M. Salim Reza, Md. Maynul Islam, Mufti Mahmud, and M. Shamim Kaiser

Abstract Prior identification is an important factor in controlling the chronic kidney disease (CKD). The clinical data and diagnostic results provide concealed facts which will help physicians to identify severity of CKD. In this paper, we propose a fuzzy analytical hierarchy process-based model for detecting CKD. In addition, a mobile app has been developed for collecting data from patient. The performance evaluation shows that relatively high accuracy can be achieved through the proposed method.

Keywords Fuzzy analytical hierarchy process · Kidney disease · Multicriteria decision making · Mobile App

Z. S. Shamma · I. J. Rumman · A. M. Raji Saikot · S. M. Salim Reza · Md. M. Islam
Department of Information and Communication Engineering,
Bangladesh University of Professionals, Dhaka, Bangladesh
e-mail: zarin.shamma46@gmail.com

I. J. Rumman
e-mail: isratrumman25@gmail.com

A. M. Raji Saikot
e-mail: amrsaikot@gmail.com

S. M. Salim Reza
e-mail: salim.reza@bup.edu.bd

Md. M. Islam
e-mail: maynul@bup.edu.bd

M. Mahmud
Computing & Technology, Nottingham Trent University, Nottingham, UK
e-mail: mufti.mahmud@ntu.ac.uk

M. Shamim Kaiser (✉)
Institute of Information Technology, Jahangirnagar University, Dhaka 1342, Bangladesh
e-mail: mskaiser@juniv.edu

K. Ray et al. (eds.), *Proceedings of International Conference on Data Science and Applications*, Lecture Notes in Networks and Systems 148,
https://doi.org/10.1007/978-981-15-7561-7_7

1 Introduction

Now-a-days, kidney disease is a common problem in a developing country. Kidney disease will have an effect on our body's ability to wash our blood and management of metallic element, salts, and acids. Once our kidney function is disturbed, waste merchandise and fluid will build up in our body. In that case, we can feel many abnormalities, that is called Kidney Disease. Acute and chronic are the two types of kidney disease. Acute kidney injury (AKI) is not a diagnosis, rather it describes the situation where there is a sudden and often reversible loss of renal function, which develops over days or weeks and is often accompanied by reduction in urine volume.

Chronic kidney disease (CKD) refers to irreversible deterioration of renal function that usually develops over a period of years. Initially, it manifests only biochemical test but in later stage, this disease can be diagnosed by physical examination and investigation. The social and economic consequences of CKD are not considerable in our country because long-term treatment is costly. Signs, symptoms, and some of risky diseases that affect kidney patients (Table 1)

The prevalence of CKD within the world is (5–7) %. Chronic kidney disease suggests that lasting harm to the kidneys that may aggravate over time. If the harm is incredibly dangerous, our kidneys are gone. This is often known as nephropathy, or end-stage urinary organ disease (ESRD). Quite 25 of the population aged over 75 years, usually reflects association in nursing increased vessel risk burden, as only a few can ever develop ESRD. Stages of CKD:

1. Kidney damage with normal or high GFR(greater than 90)—Normal function
2. Kidney damage and GFR (60–89)—Mild CKD
3. GFR (45–59)—Mild to moderate CKD, GFR (30–44)—Moderate to severe CKD
4. GFR (15–29)—Severe CKD
5. Low GFR(less than 15) or on dialysis—Kidney failure.

In kidney disease, number of CKD patients are more than the acute kidney injury (AKI) or acute renal failure patient. AKI may be a state of affairs wherever abrupt and obtained reversible loss of urinary organ performance that is developed over weeks. It always associates with reduction of excretion volume. In AKI, related to infection

Table 1 Causes of chronic kidney disease

Symptoms	Signs	Diseases
Increased frequency of micturition	Hypertension	Diabetes
Decreased frequency of micturition	Anemia	High BP
Anemia	Fever	Family with kidney disease
Lethargic	Nose bleeding	Being over 60yrs old
Swelling of the body	Rash	Acute GLumero Nephritis
Bone pain	Itching	Poly-cystic kidney
Poor appetite	Vomiting	Interstitial nephritis

and multiple organ failure, mortality is 50–70% and also the outcome is sometimes determined by the severity of the underlying disorder and different complications, instead of by urinary organ injury itself. That is why early detection of kidney disease is really needed. In this paper, we create an application for detecting kidney disease and kidney patient condition, because of immediate care, reduced risks, reduced diagnosis, decreasing frequent death, etc. and it is our main goal. For digital interaction, mobile applications are becoming the main medium and it is so helpful for client-side, that is why we choose it. To implement our goal, we use machine learning algorithm because machine learning algorithm is a rising field involved with the study of huge and multiple variable information. Advantage of machine learning (ML) is that it uses mathematical models, heuristic learning, knowledge acquisitions, and call trees for deciding. In this paper, we used fuzzy analytical hierarchy process (FAHP) that is employed for composition and examining sophisticated issues by mathematical calculations. We get such attributes for detecting the condition of kidney and calculating the weights of multiple attributes using AHP and thus it is easier to detect the exact problem of a kidney patient. The proposed model answers the following questions:

1. What is the current stage of chronic kidney disease patient?
2. Are there any precautions when complications are not available?
3. What is the predictive result for kidney disease?

2 Literature Review

A model has been presented using different data mining techniques like SVM, Naive Bayes, decision tree etc., to predict the chronic kidney disease in [1]. In this paper, the accuracy of predicting CKD is analyzed by using different data mining techniques. After analyzing, it is stated that no single classifier produces the best result on predicting CKD. An existing model is compared in [2] using KNN and SVM. A new predictive decision support system is designed by comparing the performance of SVM and KNN to predict chronic kidney disease. In this article, it is realized that KNN is more precise and gives more specific decision than SVM. Best classification algorithm is chosen in [3] to predict CKD based on the accuracy and execution time. For SVM, the accuracy value is 76.32 and precision value is 0.820. For Naïve Bayes, the accuracy value is 70.96 and precision value is 0.809. The prediction of CKD is done and the classification techniques are compared in the bioinformatics field in [4]. The paper emphasizes on prediction of precision for CKD. For this, six classifiers are used here and studied their performances on different parameters. The precision of Naïve Bays is 95%, multilayer perception is 99.75%, SVM is 62%, J48 is 99%, conjunctive rule is 94.75%, and decision tree is 99%. CKD is predicted using SVM and KNN and then their performance is compared according to accuracy and execution time in [5]. The execution time of SVM is 3.22 and for ANN, it is 7.26. The accuracy of SVM is 76.32 and for ANN, it is 87.70. So, it is stated that

SVM is more accurate and has more execution value than ANN. The risk of kidney stones is classified in [6] by using decision tree, multilayer perception, and genetic algorithm. This paper states that multilayer perception has the best redaction. UCI machine learning repository is used in [7] to detect CKD from by using six machine learning algorithms. With sensitivity, specificity, and accuracy, the RF executes better than other classifiers. The sensitivity, specificity, and accuracy values are 1.00, 1.00, and 1.00, respectively. Different data mining methods are compared and the prediction of CKD is classified based on the accuracy to measure the performance of the classifications in [8]. Several research papers are reviewed using different data mining classifiers. As per the result, it is stated that SVM, KNN, random forest, and multilayer perceptron give the highest accuracy to predict CKD. Fuzzy analytical hierarchy is used for a software called Symptom Checking that can detect various diseases of elderly people by using multicriteria decision-making process in [9] taking different symptoms as input. An app is developed for early diagnosis of elderly illness. The app provides some predictive results that is defined based on the relative weights of the symptoms. The healthcare motive toward patients of different hospitals and healthcare centers are evaluated by a predictive model proposed in [10] using fuzzy analytical hierarchy process. One from the five types of care is chosen using FAHP and each of the care will be dependent on five other criteria. The decision is taken according to the relative fuzzy weights. A multicriteria decision-making problem like supply selection of the supply chain management is solved in [11] by using fuzzy analytical hierarchy. The fuzzy weights are calculated here using extent analysis. In this algorithm, the company with highest final weights will be selected as it will have the highest priority. Following graphs shows the research interest in kidney disease detection field (Figs. 1 and 2).

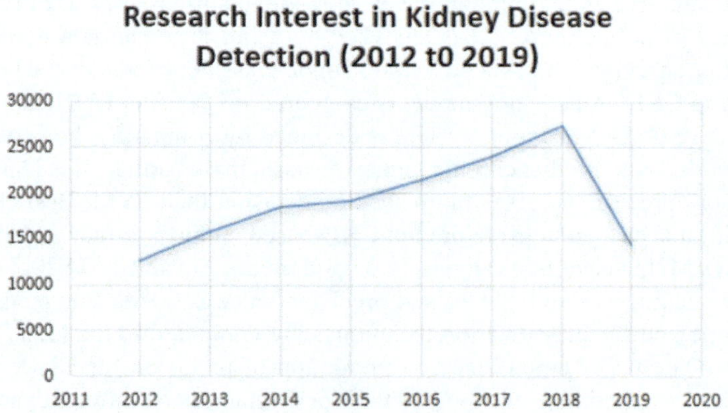

Fig. 1 Research interest in kidney disease detection

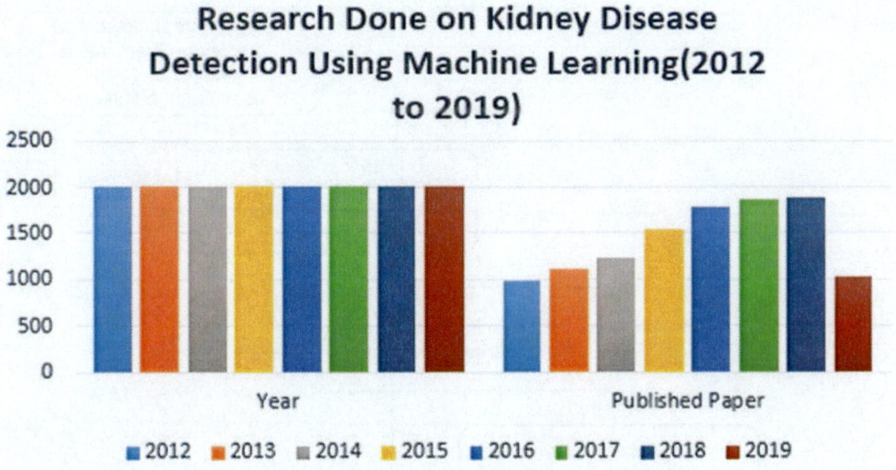

Fig. 2 Research done on kidney disease detection using machine learning

3 Methods

Fuzzy analytic hierarchy system (AHP) proves to be a very useful technique for decision making in fuzzy settings of various parameters, which has found significant applications in recent years. Multicriteria decision-making approaches may help decision-makers cope skillfully with conflict states with these predominant problems [12, 13]. In this analysis, we used the fuzzy analytical hierarchy model's approach combined with a fuzzy theory in the analytical hierarchy system [14–17]. Saty has developed AHP which is a hierarchical method, through which maximum set of goals in multicriteria decision-making issues can be found. The fuzzy analytical hierarchy process is done using steps of Fig. 3.

3.1 Weight Determination

Nine linguistic parameter scales are identified to describe the apprehensive acknowledgement of the kidney care with fuzzy AHP where there will be triangular fuzzy scales in order to remark the relevant choice designed to meet the needs of validity and dependability of the evaluation on the basis of the importance of differentiating several elements of the matter [18, 19] (Table 2).

There are many factors that need to be addressed when discussing issues with kidney care problems [20]. Nine units are considered to explain the intensity of similarity among the symptoms, S = high BP, metallic taste, shortness of breath, fatigue, trouble sleeping, frequent urination at night, swelling, skin itching, loss of

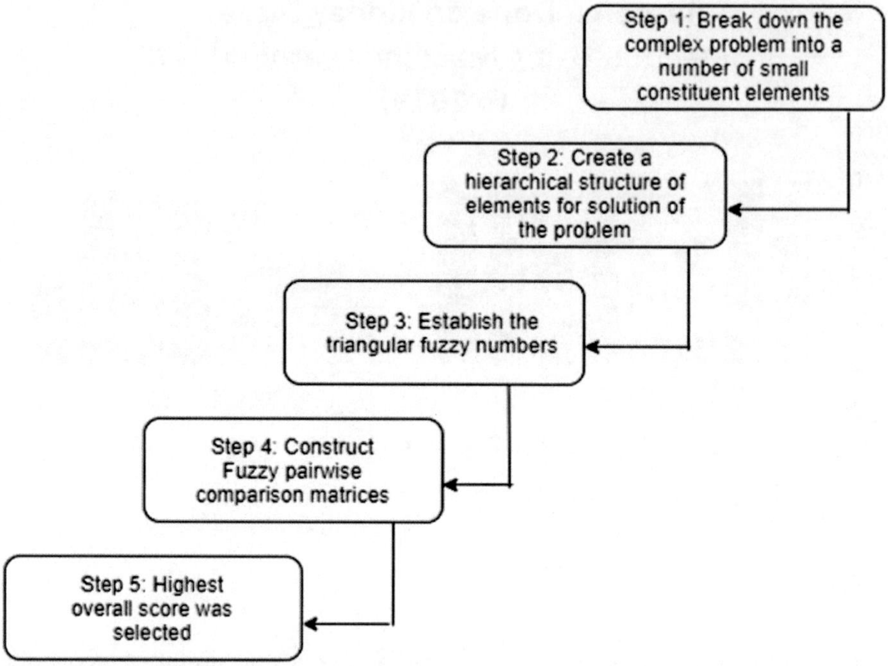

Fig. 3 Steps of fuzzy analytical hierarchy

Table 2 Categorizations of AHP scale

Linguistic variable	Scale	Triangular scale	Fuzzy scale
Equally important	1	(0.03, 0.19, 0.90)	(0.03, 0.19, 0.90)
Weakly important	2	(0.08, 0.33, 0.80)	(0.33, 0.08, 0.80)
Moderately important	3	(0.23, 0.38, 0.70)	(0.38, 0.24, 0.70)
Moderately plus	4	(0.33, 0.39, 0.60)	(0.39, 0.33, 0.60)
Strongly important	5	(0.44, 0.38, 0.50)	(0.38, 0.44, 0.50)
Strongly plus	6	(0.58, 0.34, 0.40)	(0.34, 0.58, 0.40)
Very strong	7	(0.73, 0.29, 0.30)	(0.29, 0.73, 0.30)
Very very strong	8	(0.90, 0.20, 0.20)	(0.20, 0.90, 0.20)
Extremely important	9	(2, 0, 0)	(0, 2, 0)

appetite on the decision criteria, and the decisions, D = good kidney, precautions needed, kidney problem, chronic kidney disease as alternatives (Fig. 4).

Each feasible connection and dependency among several choices of users' symptoms are estimated using the framework in order to compute the performance analysis. Enough particularities of complex weights are accumulated by the structure in order to separate and give feasible symptom recognizes. The kidney care's design operation is shown in the following manner.

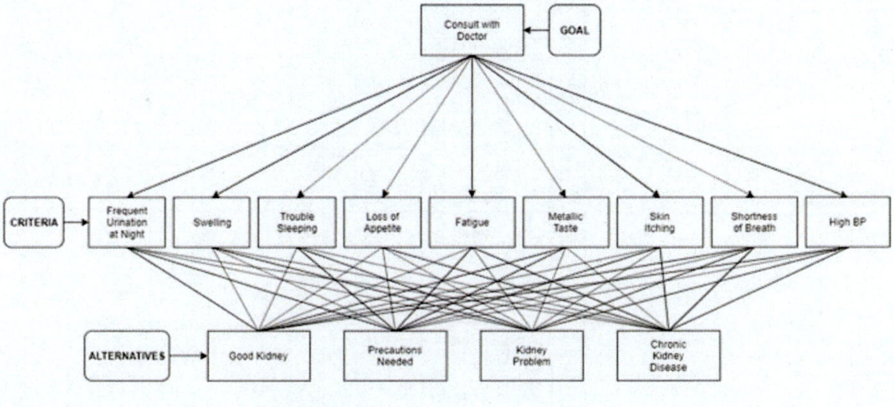

Fig. 4 Structural depiction of kidney care

Since the case study had nine sets of symptom criteria, there could be nine correlations based on the criteria, generating a ratio of 9 × 9 (Table 3).

After complete observations of the matrix, a generalized matrix is designed using the equation:

$$k_{ij} = \frac{b_{ij}}{\sum_{p}^{m} b_{pj}} \quad (1)$$

Ai was accomplished with regard to ith parameters, where C represents the norms and n tends to the number of norm, m states the number of decision, and rows and columns are represented by i and j, respectively.

$$A_j = \sum_{j}^{m} M_{C_i}^{j} \left[\sum_{i}^{n} x \sum_{j}^{m} M_{C_i}^{j} \right]^{-1} \quad (2)$$

The relative weights are determined as follows after obtaining the weights of the various symptom parameters correlated with the fluffy artificial degree of TFNs (Table 4):

$$w_i = \frac{w_i}{\sum_{j}^{m} w_j} \quad (3)$$

An identical weight estimation methodology was used to calculate the weights provided for the alternatives in regard to the symptoms. With respect to the symptoms, the risk of different diseases with final priority weight is calculated as follows (Fig. 5; Table 5):

$$\text{Weight} = \sum p_{ij} * w_i \quad (4)$$

At first, an account needs to be created by users with their personal information to use kidney care, and then, they will be contacted to their registered mobile number

Table 3 Pairwise comparison matrix of criteria symptoms

	BP	T	Br	F	Sl	Ur	Sw	I	A
BP	(1, 1, 1)	(0.03, 0.29, 0.9)	(0.08, 0.34, 0.8)	(0.24, 0.38, 0.7)	(0.24, 0.38, 0.7)	(0.23, 0.38, 0.7)	(0.08, 0.34, 0.8)	(0.44, 0.38, 0.5)	(0.33, 0.39, 0.6)
T	(2 0.4, 6 0.7, 6)	(1, 1, 1)	(0.33, 0.39, 0.6)	(0.08, 0.34, 0.8)	(0.03, 0.29, 0.9)	(0.08, 0.34, 0.8)	(0.24, 0.38, 0.7)	(0.9, 0.2, 0.2)	(0.08, 0.34, 0.8)
Br	(2.5, 5.5, 15.4)	(3, 4.7, 5.6)	(1, 1, 1)	(0.03, 0.29, 0.9)	(0.08, 0.34, 0.8)	(0.24, 0.38, 0.7)	(0.08, 0.34, 0.8)	(2, 0, 0)	(0.58, 0.34, 0.4)
F	(2.8, 4.8, 8.8)	(2.5, 5.5, 15.4)	(1.4, 6.7, 60)	(1, 1, 1)	(0.03, 0.29, 0.9)	(0.03, 0.29, 0.9)	(0.03, 0.29, 0.3)	(0.08, 0.34, 0.8)	(0.03, 0.29, 0.9)
Sl	(2.8, 4.8, 8.8)	(2.4, 6.7, 60)	(2.5, 5.5, 15.4)	(1.4, 6.7, 60)	(1, 1, 1)	(0.03, 0.29, 0.9)	(0.73, 0.29, 0.3)	(0.73, 0.29, 0.3)	(0.33, 0.39, 0.6)
Ur	(2.5, 5.5, 15.4)	(2.5, 5.5, 15.4)	(2.8, 4.8, 8.8)	(2.4, 6.7, 60)	(2.3, 6.7, 60)	(1, 1, 1)	(0.03, 0.29, 0.90)	(0.73, 0.29, 0.3)	(0.33, 0.38, 0.6)
Sw	(2.4, 6.7, 60)	(2.8, 4.8, 8.8)	(2.5, 5.5, 15.4)	(2.4, 6.7, 60)	(6, 6.7, 2.8)	(20, 20, 0.24)	(1, 1, 1)	(0.33, 0.39, 0.60)	(0.90, 0.2, 0.2)
I	(3.6, 4.8, 4)	(20, 20, 20.4)	(0, 0, 2)	(2.5, 5.5, 15.4)	(6, 6.6, 2.8)	(3.6, 4.9, 4.5)	(3, 4, 7 ,5.7)	(1, 1, 1)	(0.24, 0.38, 0.7)
A	(3, 4.7, 5.7)	(2.5, 5.5, 15.4)	(4.4, 5.5, 3.2)	(2.4, 6.7, 60)	(3.0, 4.7, 5.6)	(2.5, 5.4, 15.4)	(20, 20, 2.4)	(2.8, 4.8, 8.9)	(1, 1, 1)

Legend T–Taste; Br–Breath; F–Fatigue; Sl–Sleeping; Ur–Urination; Sw–Swelling; I–Itching; A–Appetite

Table 4 Relative weights of symptoms

	Sum of the row	Synthetic index	Sum of the fuzzy numbers	Weight
BP	(2.9, 3.9, 7.2)	(0.004, 0.03, 0.07)	0.09	0.02
Taste	(4.7, 9.0, 66.2)	(0.02, 0.05, 0.63)	0.68	0.08
Breath	(7.2, 20.0, 33.9)	(0.02, 0.06, 0.33)	0.39	0.05
Fatigue	(6.6, 26.6, 87.9)	(0.02, 0.09, 0.84)	0.93	0.22
Sleeping	(9.0, 30.9, 135.8)	(0.02, 0.22, 2.29)	2.4	0.28
Urination	(9.3, 36.2, 149.6)	(0.02, 0.24, 2.42)	2.56	0.29
Swelling	(33.7, 47.2, 137.5)	(0.04, 0.29, 2.29)	2.52	0.29
Itching	(35.7, 43.2, 40.5)	(0.05, 0.28, 0.39)	0.60	0.07
Appetite	(35.2, 50.6, 99.8)	(0.05, 0.32, 0.95)	2.29	0.27

along with a one-time password. After the creation of the profile, users can choose several symptoms according to their sicknesses. The program can create potential decisions after the symptom collection is completed and recommend health services that can help users to find effective recovery strategies.

4 Results and Discussion

For each parameter, the final assessment of each option is tabulated on the basis of the calculation (Tables 6 and 7).

Now, using these weights, the predictive decisions will be provided. We can see that among the symptoms, two symptoms weights are the highest; one is urination at night and another one is swelling. Then comes trouble sleeping. In this way, weights will define the severity of the disease according to the symptoms given by the users. Then comes the decision part. In these decisions, CKD has the greatest weights, then comes kidney disease. With these weights, decisions will be predicted and will be given to the users.

5 Conclusion

In kidney care, it is the first concern to identify the kidney disease as quick as possible along with proper analysis in order to diminish health risks. A fuzzified predictive kidney treatment in order to allocate the most convenient prerequisites and danger to

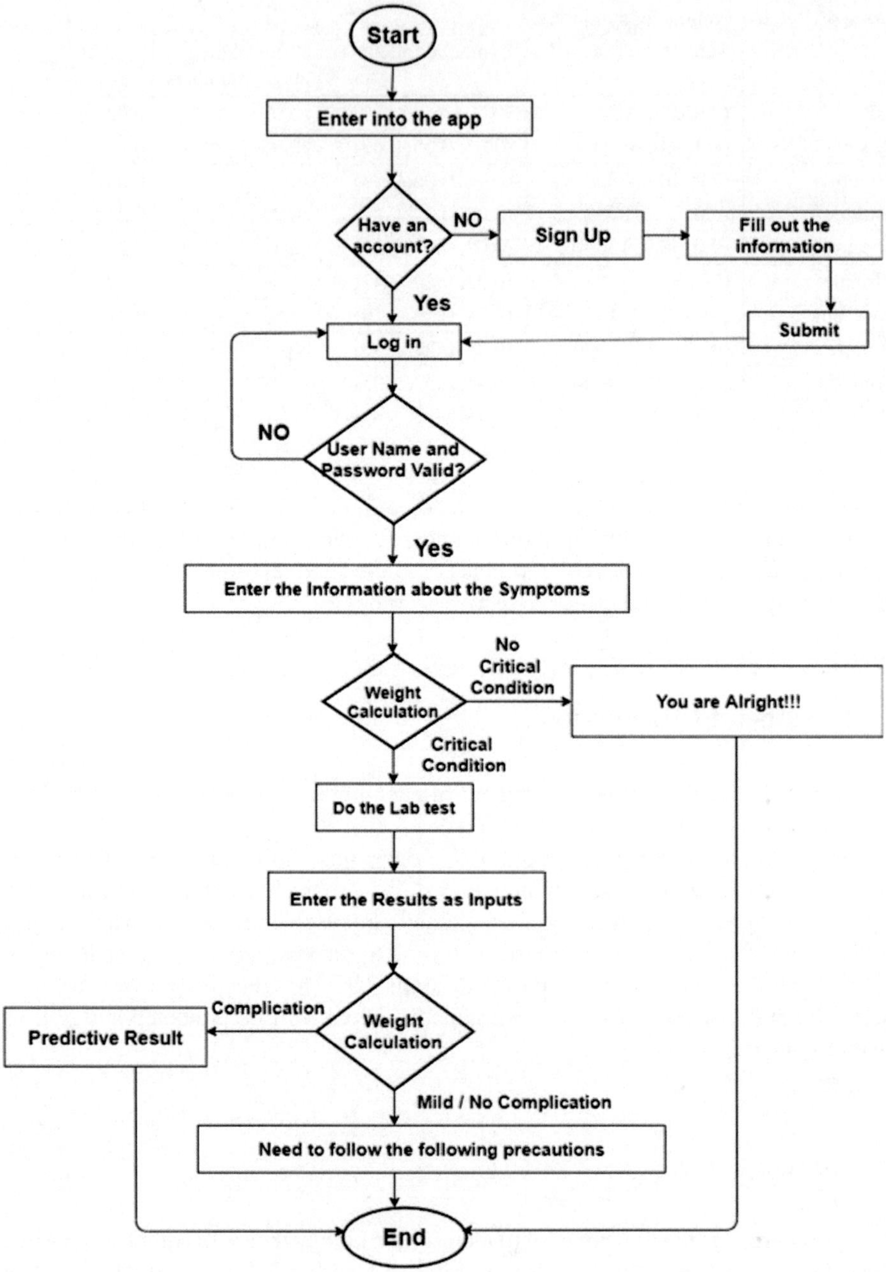

Fig. 5 Flow chart of kidney care

Table 5 Calculation of the symptom frequent urination with decision

U	HK	PN	CKD	KP	Weight
HK	(1, 1, 1)	(0.20, 0.40, 0.70)	(0.03, 0.29, 0.9)	(0.03, 0.29, 0.9)	0.22
PN	(2.8, 4.4, 20.0)	(1, 1, 1)	(0.03, 0.29, 0.90)	(.03, 0.29, 0.90)	0.26
CKD	(2.4, 6.7, 60.0)	(2.4, 6.7, 60.0)	(1, 1, 1)	(1, 1, 1)	0.39
KP	(2.4, 6.7, 60.0)	(2.4, 6.7, 60.0)	(2.4, 6.7, 60.0)	(2.4, 6.7, 60.0)	0.43

Legend U–URINATION; HK–Healthy Kidney; PN–Precautions Needed; KD–Kidney Disease

Table 6 Final Weights of alternating decisions

	BP	Taste	Breath	Fatigue	Sleeping	Urination	Swelling	Itching	Appetite	Weights
HK	0.273	0.098	0.292	0.238	0.233	0.207	0.053	0.074	0.074	0.209
PN	0.093	0.226	0.298	0.339	0.293	0.256	0.309	0.257	0.344	0.282
CKD	0.405	0.393	0.338	0.357	0.395	0.396	0.458	0.495	0.345	0.293
KD	0.324	0.437	0.450	0.450	0.467	0.432	0.468	0.443	0.509	0.298

Table 7 Full weights and total target rating with respect to the target

Decisions	Final weights	Overall ranking
HK	0.209	4
PN	0.282	3
CKD	0.293	1
KD	0.298	2

the kidneys with complex effects. The development method often incorporates early diagnosis quantitative and qualitative information which helps users select the right practitioner for any kidney problems.

References

1. Patil PM (2016) Review on prediction of chronic kidney disease using data mining techniques. Int J Comput Sci Mobile Comput 5:135–141
2. Parul Khare Sinha PS Comparative study of chronic kidney disease prediction using knn and svm
3. Vijayarani S, Dhayanand SS (2015) Data mining classification algorithms for kidney disease prediction. Int J Cybernetics & Inform 4:13–25
4. Jena L, Kamila NK (2015) Distributed data mining classification algorithms for prediction of chronic-kidney-disease. J Emergency Manage 9359(11):110–118
5. Vijayarani DS, Dhayanand S (2015) Kidney disease prediction using svm and ann algorithms. Int J Comput Bus Res (2015)
6. Oladeji FA, Idowu P, Egejuru N, Faluyi S, Balogun J (2019) Model for predicting the risk of kidney stone using data mining techniques. Int J Comput Appl 182:36–56
7. Kumar M (2016) Prediction of chronic kidney disease using random forest machine learning algorithm. Int J Comput Sci Mob Comput 5:24–33

8. Huang YP, Basanta H, Kuo HC, Huang A (2018) Health symptom checking system for elderly people using fuzzy analytic hierarchy process. Appl Syst Innovation 1(2). https://www.mdpi.com/2571-5577/1/2/10
9. Savarimuthu SJ, Raj AR Applying fuzzy analytic hierarchy process to evaluate the motive of healthcare towards patients. Int J Math Appl 4:229–238
10. Aktepe A, Ersoz S (2011) A fuzzy analytic hierarchy process model for supplier selection and a case study. Int J Eng Res Develop 3:33–36
11. Chen Z et al (2018) Development of a personalized diagnostic model for kidney stone disease tailored to acute care by integrating large clinical, demographics and laboratory data: The diagnostic acute care algorithm—kidney stones (daca-ks). BMC Medical Informatics and Decision Making, vol 18
12. Iqbal MA, Zahin A, Islam ZS, Kaiser MS (2012) Neuro-fuzzy based adaptive traffic flow control system. In: Proceeding of 2012 CODIS, pp 349–352
13. Roy S, Rahman A, Helal M, Kaiser MS, Chowdhury ZI (2016) low cost rf based online patient monitoring using web and mobile applications. In: Proceeding 2016 ICIEV. pp 869–874
14. Heilpern S (1992) The expected value of a fuzzy number. Fuzzy Sets and Syst 47(1):81–86
15. Chutia R, Mahanta S, Datta D (2011) Arithmetic of triangular fuzzy variable from credibility theory. Int J Energ Inf Commun 2
16. Ruan J, Shi Y (2014) Situation-based allocation of medical supplies in unconventional disasters with fuzzy triangular values. In: 2014 IEEE international conference on fuzzy systems (FUZZ-IEEE), pp 1178–1182
17. Zhou J, Yang F, Wang K (2016) Fuzzy arithmetic on lr fuzzy numbers with applications to fuzzy programming. J Intell Fuzzy Syst 30:71–87
18. Kaiser MS et al (2018) Advances in crowd analysis for urban applications through urban event detection. IEEE Trans Intell Trans Syst 19(10):3092–3112
19. Mahmud M et al (2018) A brain-inspired trust management model to assure security in a cloud based IoT framework for neuroscience applications. Cognitive Comput 10(5):864–873
20. Biswas S, Anisuzzaman Akhter T, Kaiser MS, Mamun SA (2014) Cloud based healthcare application architecture and electronic medical record mining: an integrated approach to improve healthcare system. In: Proceeding of 2014 ICCIT, pp 286–291

A Framework to Evaluate and Classify the Clinical-Level EEG Signals with Epilepsy

Linkon Chowdhury, Bristy Roy Chowdhury, V. Rajinikanth, and Nilanjan Dey

Abstract Brain health examination attracted the research community due to its clinical significance. The abnormality in brain will lead to death and temporary/permanent disability. The proposed study considers the evaluation of the clinical-level EEG signals with epilepsy. Recently, a number of procedures are implemented to evaluate the epilepsy, to estimate the regions, which initiates the abnormal discharge of brain signal. During the treatment planning stage, the clinical-level assessment of single-/multi-channel electroencephalogram (EEG) plays a vital role, and its frequency bands such as delta, theta, alpha, and beta are a reliable index. Disappearance of the healthy index, such as a chaotic tendency in the normal spontaneous EEGs, indicates an initiation of the illness, and therefore, EEG rhythms can be used to predict the state of the patient. Recently, metadata description and ontology are highlighted as successful tools to distribute expert knowledge. OdML started as a part of the INCF data sharing program and have been developed in related electrophysiological communities. Exploring the possibility to extend existing scheme to EEG, clinical knowledge sharing is discussed in this work. EEG metadata clarifies properties on epilepsy with signal amplitude, frequency, areas of occurrence, and

L. Chowdhury
Department of Computer Science & Engineering, East Delta University, Chittagong 4209, Bangladesh
e-mail: linkoncuet@gmail.com

B. R. Chowdhury
Department of Computer Science& Engineering, BGC Trust University, Chandanaish, Bangladesh
e-mail: c.bristy18@gmail.com

V. Rajinikanth (✉)
Department of Electronics and Instrumentation, St. Joseph's College of Engineering, Chennai, Tamil Nadu 600119, India
e-mail: v.rajinikanth@ieee.org

N. Dey
Department of Information Technology, Techno India College of Technology, Kolkata, West Bengal 740000, India
e-mail: neelanjan.dey@gmail.com

K. Ray et al. (eds.), *Proceedings of International Conference on Data Science and Applications*, Lecture Notes in Networks and Systems 148,
https://doi.org/10.1007/978-981-15-7561-7_8

estimated causality. The OdML platform allows data models to store fully annotated and reviewed by professionals.

Keywords Epileptic seizure · Hippocampal region · EEG waveforms · DWT · OdML

1 Introduction

Epilepsy is neurological disorder with several responsible human brain lobe. Fornatal lobe epilepsy (FLT) and temporal lobe epilepsy (TLE) are more frequent type epilepsy. Most of FLT originates from frontal lobe during the sleep or awake time [1]. FLE also known as focal epilesy, in which neuro imaging process used for its investigation [2]. The investigation process differs from lobe origins including perirolandic, supplementary sensorimotor, anterior front polar, and cingulate; eventhough it is difficult to understand the cognitive functionalites of human frontal lobe [3]. In several studies, it is significantly noticeable for cognitive dysfunction and behavioral disorder in FLE patients in their daily activities [4].

Though increasing the modern technology, epilespical seizures are complexed for investigation of brain signals. Several regions such as sensory regions and motor-related regions are involved in motor modulation, sensory functionalities, and other input signals of related frontal lobe parts. In brain functionalities, basal ganglia (BG) is important part for sensorimotor modulation and motor functionalities. FLT patients disorder was accomplished from in sensori motor parts to BG part.

Basically, BG part consists subthalamic nucleus and substantia nigra to making linked between the distinct functional regions in cerebal cortex [5, 6]. In epilepsy investigation, the structural and functional examine of BG region is important [7, 8]. The BG activities are crucially observed for the medullation process of epilepsical seizures. The brain signal pathway from frontal lobe through BG regions to cerebellum involve for altering the functional seizures in FLE [9].

Basically, epilepsies are two types: localization-related epilepsy and generalized epilepsy. Idiopathic (known as partial seizure) and symptomatic (known as complex partial seizure) are the localization-related epilepsy (Fig. 1). Idiopathic is genetic, and children are affected more. Benign childhood epilepsy with centrotemporal spikes, childhood epilepsy with paroxysms, and panayiotopoulos syndrome are the syndromes of idiopathic epilepsy. Similarly, brain damage from prenatal or perinatal injuries, congenital abnormalities, and an infection of the brain such as meningitis, encephalitis, and neurocysticercosis are the syndromes of symptomatic epilepsy. Adult are affected by symptomatic epilesy and it in non-genetic. Generalized epilepsy have two groups one is genetic (known as idiopathic or myoclonic seizure) and other one is non-genetic (known as cryptogenic or tonic seizure). Understanding the brain lobe functionalities and regions is important for detecting these epilepses.

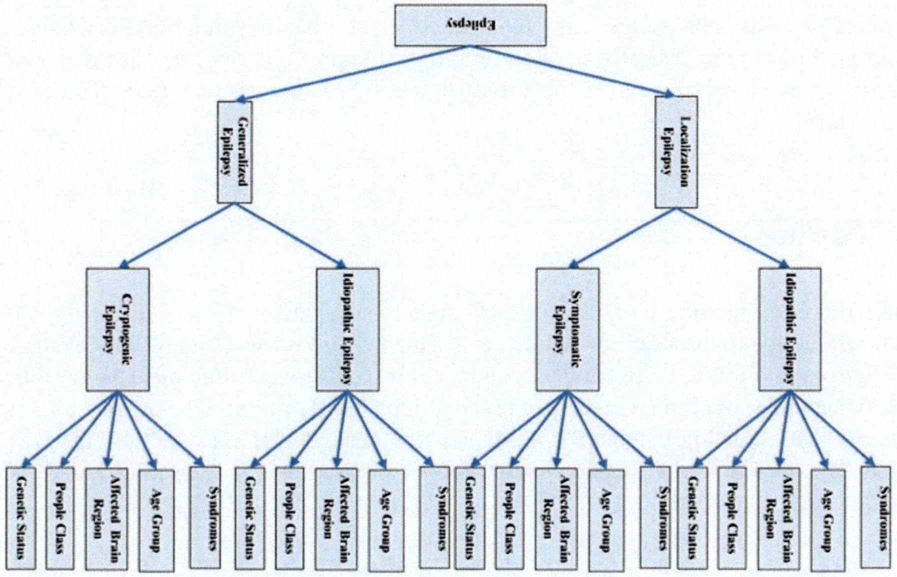

Fig. 1 Hierarchy and classification of different types of epilepsy

Recently, different types of approaches have been used for finding brain disorder, and voxel-based morphometry (VBM) is one of them. VBM defined the subtle structure of brain and it voxel-by-voxel or seed-to-voxel process applied for neuropsychiatric disorders. And, gray matter volume (GMV) is another method that spreading the gray matter from epileptic active area to seizures areas by measuring the changes of volume of cortical and subcortical nuclei [10, 11]. In addition, this GMV-based studies are performed on FLE patients based on increasing or decreasing volume of gray matter [12–14].

A significant numbers of statistical connectivity methods between the unique time series of brain and demographic data that help the researchers to investigate patterns of circuitry functions of brain [15]. Statistically analysis also explores the another time series which can be used to predict another through Granger causality analysis (GCA). GCA is also predicted with multivariate auto-regression, which functionality performed on effective connectivity concepts [16]. Some researchers also consider fMRI signals to analyze dynamic VBM and GCA causality [17]. Another non-prior knowledge is also applied on childhood epilepsy to observe on brain disorder regions [18]. Recently, EEG and MRI signals have applied to find the connectivity between the seizures mechanism [19]. Another time frequency domain graphs are applied for health control of GCA patients [20]. We apply EEGs signals and DFT conversion into DC and AC for investigate epileptically disorder. We classified these signals based on stimulus or sine wave's duration based on amplitude, frequency, and mean intensity. We generate metadata connectivity among the responsible parameters. We also make an ontological connectivity between EEG signal parameters with patient clinical records such as age, age on seizure one set, histology, interictal discharge, and ictal discharge. We also applied ontological connectivity approaches between the parameters to find out the connectivity between EEG signals features and clinical data link.

2 Method

The proposed method operates on EEG signals and patient clinical data (Fig. 2). The EEG signals process for feature or segmented by using Discrete Wavelength Transformation (DWT). In parallel, patient clinical data defines into meta form. Both EEG features and clinical meta data process into ODML engine. The organizational design and modeling language (ODML) engine generates eplipesy patient features tree. The detection of epilepsy patient process is based on IF–Then rules based on node value.

2.1 EEG Datasets

The sample datasets are collected from Dhaka Medical college hospital. About twenty patients datasets are simulated for this experiments. The primary focus of the simulation is to investigate and understand the seizures and epilepsy patterns among the various EEG records and symptoms of the patients. There are thirteen male and seven female. Among male, eight patients are age over sixty five and rest are fifty five years old. Currently used approaches reflect a few limitaions. To overcome the

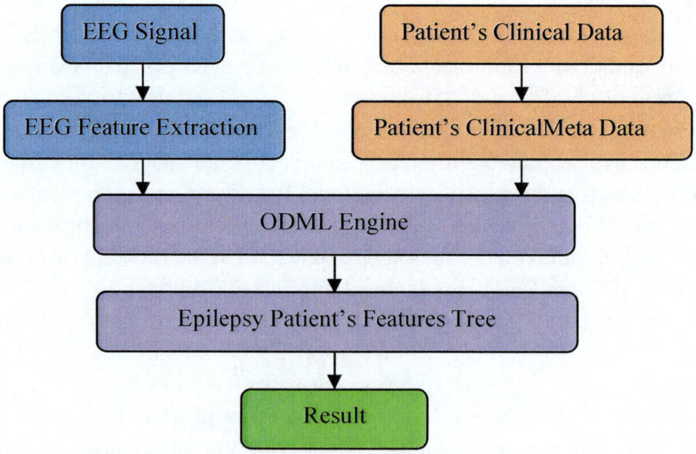

Fig. 2 An integrated approach for detecting epilepsy patient

issues, NIX-odML centric integrated framework allows for testing EEG reliability. Significant number of researchers have been widely used for their algorithm [21–23]. We transform the EEG signals data by using discrete wavelength transformation. Supervised training signal is used to check the outcomes. Experimental environment is composed of MATLAB 2016 with 2.4 GHz processor.

2.1.1 Discrete Wavelet Transformation

To understand the orientations of the EEG spikes and variations, accurate transformations are essential. From a in-depth study, digitized wavelet transformation identified the targeted features (Fig. 3). A wavelet consists both localized function in frequency and time domains. This wavelet transformation decomposed the signal and scaled into $f(x, y(t)$ to single function $f(t)$, which is called main process of wavelet transformation.

$$f_{x,y}(t) = \frac{1}{\sqrt{|x|}} f\left(\frac{t-y}{x}\right) \quad (1)$$

where x and y are the scaling and translation parameters, respectively, with satisfying condition, $x, y \in R$ and $x \neq 0$. DWT [24] was obtained by two parameters x and y which are discretizing parameters. Another dyddic sampling approach for frequency division with parameter $x = 2^i$ and $y = k2^i$, where $i, k \in Z$. By substitution the values x and y, we obtained the dyadic wavelets from Eq. 2.

$$f_{x,y}(t) = 2^{-\frac{i}{2}} f\left(2^{-j}t - k\right) \quad (2)$$

DWT also can be written as

$$C_{i,j} = \int_{-\infty}^{+\infty} f_1(t) 2^{-\frac{i}{2}} f\left(2^{-\frac{i}{2}} f\left(2^{-j}t - k\right)\right) \mathrm{d}t = \left(f_1(t), f_{i,k}(t)\right) \quad (3)$$

where $C_{i,j}$ are known as wavelet coefficients at level i and location k [25]. These coefficients basically used to find feature factors for each EEG signals.

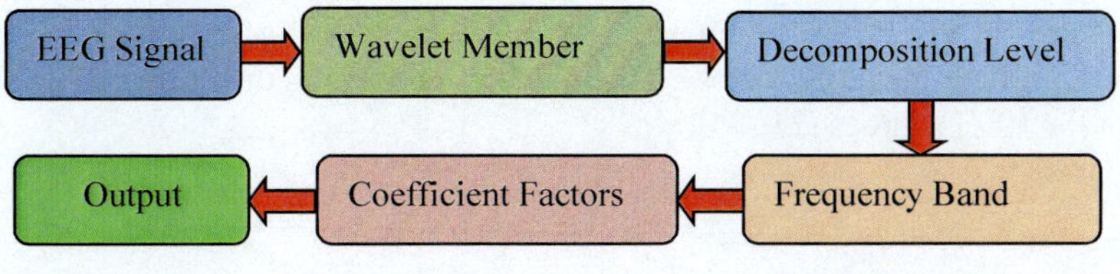

Fig. 3 DWT decomposition process for two phases: frequency band and features Vector of EEG signals

Here, we demonstrate the process of wavelet transformation how to extract features vector from EEG signals to frequency domain to leads coefficient factors.

2.1.2 Decomposition Level

Decomposition level is an significant parameter of DWT process. Each level consists specific frequency band in DWT process. Collection of several level provide more information of EEG signals. Though, considering all level of signal produces the features that increase complexity and reduce accuracy. The maximum level L of decomposition level is calculated from the signal and main wavelet transformation condition:

$$L < \log_2 \frac{S_N}{F_N - 1} + 1 \tag{4}$$

where S_N is the signal size, and F_N is the filter size [26]. Our datasets have 4021 samples in EEG sagments.

2.1.3 Frequency Band

In DWT, each signal contains the decomposition level of frequency band for each EEG signal. Supposing that input EEG signal has frequency band (x, y), according to study [27–30], with decomposition variations of N, finally, overall changes are:

$$\left(x, x + \frac{y - x}{2^N}\right) \tag{5}$$

The details form of Eq. 6 is:

$$\left(x + \frac{y - x}{2^N}, x + \frac{y - x}{2^{N-1}}\right) \tag{6}$$

In clinical process, EEG explained by rythmic activity which is DWT-based EEG analysis. From the EEG frequency with certain EEG rhythm, two classes are identified such as presence of seizures and its absence. EEG segments are classified by their features bands. But, sometimes same frequency band made confusion for classification, and we abandoned those. This reduction improves the frequencies in both sides of the EEG sources [31–34].

Table 1 Features of EEG signals considered

Features name of	Description
Max	The maximum coefficient
Min	The minimum coefficient
Mean	The mean of coefficients
STD	The standard deviation of coefficients
Energy	The squared sum of all coefficients

2.1.4 Coefficient Factors

It is important to select suitable features from EEG characteristics signals for EEG classification. DWT coefficient factors derived from EEG frequency band segmentation. These EEG coefficient factors are given in Table 1.

2.2 ODML Engine for Epilepsy Tree

The ODML engine performs on EEG segments and patient clinical records and designed an ontological tree. The formal definition of ODML specification φ is given bellow:

$$\varphi = \{N, H,\}$$

$$N = \{N_0, N_1, \ldots, N_n\} \tag{7}$$

$$N_i = \{t, p, I, H\}$$

The large scale of the ODML engine specification is made up of the set N of *nodes*, and it has logical relationships among the parameters of the inputs. For example, in our method, there would be nodes corresponding to age, EEG frequency, and frequency mean and made the relationships between them. Each N_i contains a number of elements, defined below:

t: The node's type. This label must be unique within the set of epilepsy nodes that make up the patient tree

$$N.t = \{\text{symbol}\}$$

p: This is local signal parameter.

$$N.p = \{(\text{symbol}, \text{type}), \ldots\}$$

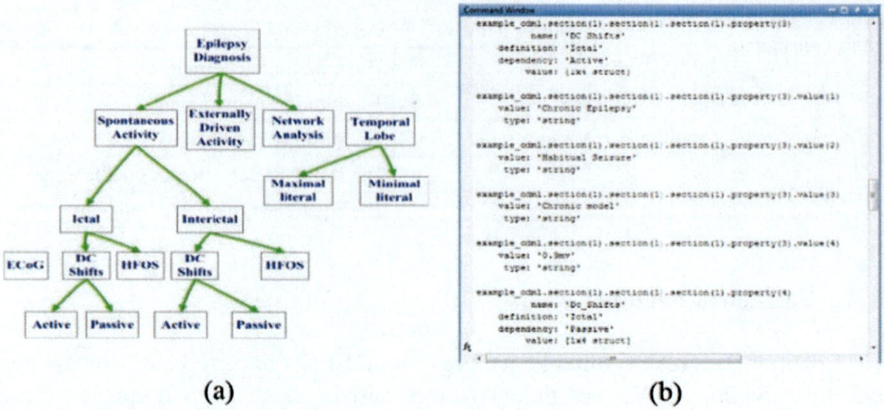

Fig. 4 ODML structure of epilesy. **a** Conceptual tree **b** implementation structure using MATLAB tool

I: This reflects about the node with a specific relations. If we assume that a patient have certain feature $I = \{\text{age, ictal spike}\}$, then

$$N.I = \{(\text{type}), \ldots.\}$$

H: The magnitude of the instance that have a value. For example, the frequency of EEG signal is 5 Hz.

$$N.H = \{(\text{symbol, type, value}), \ldots\}$$

The ODML engine designs the epilepsy tree (Fig. 4a, b). This tree made an decision using IF-Then condition for detecting epilepsy patient. The epilepsy patient is detected from the value of epilepsy tree node.

3 Result

The primary outcomes locate the frontal epilepsy in collected datasets. Especially, ten patients data show particular spikes for the seizures. The OdML framework holds both key-value seizures that helps essay representations of the epilepsy (Table 2).

The accuracy measurement based on several EEG datasets by using DWT process predicted is about more than 88%. DWT process predicts more accurately which is about more than 90% in coif series. We also measure the accuracy of generalization epilepsy and localization epilepsy using ODML process and without ODML process (Fig. 5).

Table 2 Attained accuracy with respect to the decomposition level

Wavelet	Maximum decomposition level	Accuracy (%)
bior2.4	10	88.25
bior2.6	8	87.32
bior2.8	8	87.45
bior3.1	9	89.23
bior3.3	10	89.22
coif1	10	90.12
coif3	9	90.32

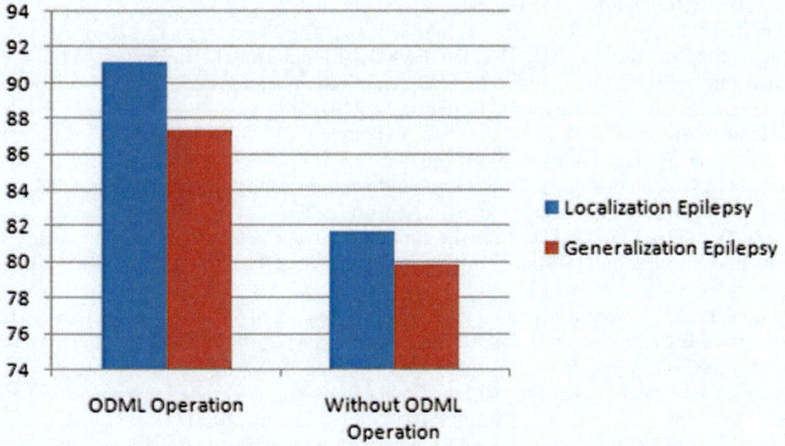

Fig. 5 Accuracy measurement of localization and generalized epilepsy using ODML and without ODML operation

The ODML operations performed better in both localization and generalized epilepsy prediction, though odml more accurately predict localized epilepsy than generalized pediction. In odml operation, the accuracy of localization epilepsy is more than 3% generalized epilepsy. But, Odml operates more 10% accurate than non-odml operation in both types of epilesy.

4 Conclusion

Because of less prior knowledge and adaptive resolution, NIX-odML structure support core identifications of the epilepsy regions. The digital transform helps classify the datasets with accurate formats. NIX-odML grabs these formats in the light of key-value pairs and associates the locations of the epilepsy with high volume of frequent seizures. However, the current datasets only deals with small amount of datasets. The datasets will be verified with deep-feature-based analysis in near future.

References

1. Haut S (2009) Frontal lobe epilepsy. Medscope
2. Doelken MT, Mennecke A, Huppertz HJ (2012) Multimodality approach incryptogenic epilepsy with focus on morphometric 3T MRI. J Neuroradiol 39(2):87–96 (2012). https://doi.org/10.1016/j.neurad.2011.04.004
3. Bagla R, Skidmore CT (2011) Frontal lobe seizures. The Neurologist 17:125–135
4. Braakman HM, Vaessen MJ, Hofman PA, Debeij-van Hall MH, Backes WH, Vles JS, Aldenkamp AP (2011) Cognitive and behavioral complications of frontal lobe epilepsy in children: a review of the literature. Epilepsia 52:849–856
5. McHaffie J, Stanford T, Stein B, Coizet V, Redgrave P (2005) Subcortical loops through the basal ganglia. Trends Neurosci 28:401–407
6. Selemon L, Goldman-Rakic P (1985) Longitudinal topography and interdigitation of corticostriatal projections in the rhesus monkey. J Neurosci 5:776–794
7. Li Q et al (2009) EEG-fMRI study on the interictal and ictal generalized spike-wave discharges inpatients with childhood absence epilepsy. Epilepsy Res 87(1–2):160–168. https://doi.org/10.1016/j.eplepsyres.2009.08.018
8. Luo C et al (2015) Altered structural and functional feature of Striato-cortical circuit in benign epilepsy with Centrotemporal spikes. Int J Neural Syst 25(06):1550027. https://doi.org/10.1142/S0129065715500276
9. Braakman HM et al (2013) Frontal lobe connectivity and cognitive impairment in pediatric frontal lobe epilepsy. Epilepsia 54(3):446–454. https://doi.org/10.1111/epi.12044
10. Keller SS et al (2002) Voxel based morphometry of grey matter abnormalities in patients with medically intractable temporal lobe epilepsy: effects of side of seizure onset and epilepsy duration. J Neurol Neurosurg Psychiatry 73(6):648–655
11. Bernasconi N, Duchesne S, Janke A, Lerch J, Collins D, Bernasconi A (2004) Whole-brain voxel-based statistical analysis of gray matter and white matter in temporal lobe epilepsy. Neuroimage 23(2):717–723
12. Lawson J, Cook M, Vogrin S, Litewka L, Strong D, Bleasel A, Bye A (2002) Clinical, EEG, and quantitative MRI differences in pediatric frontal and temporal lobe epilepsy. Neurology 58(5):723–729
13. Shah S, Kumar A, Kumar R, Dey N (2019) A robust framework for optimum feature extraction and recognition of P300 from raw EEG. U-Healthcare Monit Syst 1:15–35. https://doi.org/10.1016/B978-0-12-815370-3.00002-5
14. Chen Y et al (2019) A distance regularized level-set evolution model based MRI dataset segmentation of brain's caudate nucleus. IEEE Access, 7:124128–124140. https://doi.org/10.1109/access.2019.2937964
15. Dey N et al (2019) Social-group-optimization based tumor evaluation tool for clinical brain MRI of Flair/diffusion-weighted modality. Biocybernetics Biomed. Eng. 39(3):843–856. https://doi.org/10.1016/j.bbe.2019.07.005
16. Acharya UR et al (2019) Automated detection of Alzheimer's disease using brain mri images– a study with various feature extraction techniques. J Med Syst 43(9):302. https://doi.org/10.1007/s10916-019-1428-9
17. Liao W et al (2010) Evaluating the effective connectivity of resting state networks using conditional Granger causality. Biol Cybern 102(1):57–69. https://doi.org/10.1007/s00422-009-0350-5

18. Luo C et al (2016) Altered functional and effective connectivity in anticorrelated intrinsic networks in children with benign childhood epilepsy with centrotemporal spikes. Med (Baltimore) 95(24):e3831. https://doi.org/10.1097/md.0000000000003831
19. Ji G, Zhang H, Wang J, Liu D, Zang Y (2013) Disrupted causal connectivity in mesial temporal lobe epilepsy. PLoS One 8(5):e63183. https://doi.org/10.1371/journal.pone.0063183
20. Wei H et al (2016) Altered effective connectivity among core neurocognitive networks in idiopathic generalized epilepsy: an fMRI evidence. Front Hum Neurosci 10:447. https://doi.org/10.3389/fnhum.2016.00447
21. Fotiadis DI (2016) Handbook of research on trends in the diagnosis and treatment of hronic conditions. Medical Information Science Reference (an imprint of IGI Global), Hershey PA, USA. https://doi.org/10.4018/978-1-4666-8828-5
22. Shoeb AH (2009) Application of machine learning to epileptic seizure onset detection and treatment. Massachusetts Institute of Technology
23. Andrzejak RG, Lehnertz K, Mormann F, Rieke C, David P, Elger CE (2001) Indications of nonlinear deterministic and finite-dimensional structures in time series of brain electrical activity: dependence on recording region and brain state. Phys Rev E. 64(6):1–8. https://doi.org/10.1103/physreve.64.061907
24. Mallat SG (1989) A theory for multiresolution signal decomposition: the wavelet epresentation. IEEE Trans Pattern Anal Mach Intell 11(7):674–693. https://doi.org/10.1109/34.192463
25. Magosso E, Ursino M, Zaniboni A, Gardella E (2009) A wavelet-based energetic approach for the analysis of biomedical signals: application to the electroencephalogram and electro-oculogram. Appl Math Comput 207(1):42–62. https://doi.org/10.1016/j.amc.2007.10.069
26. Wu YL, Agrawal D, Abbadi AE (2000) A comparison of DFT and DWT based similarity search in time-series databases. In: Proceedings of the ninth international conference on information and knowledge management. ACM, pp 488–495. https://doi.org/10.1145/354756.354857
27. Mallat S, Zhong S (1992) Characterization of signals from multiscale edges. IEEE Trans Pattern Anal Mach Intell 14(7):710–732. https://doi.org/10.1109/34.142909
28. Ali MNY, Sarowar MG, Rahman ML, Chaki J, Dey N, Ravares JMRS (2019) Adam deep learning with SOM for human sentiment classification. Int J Ambient Comput Intell (IJACI) 10(3):92–116. https://doi.org/10.4018/ijaci.2019070106
29. Lakehal A, Alti A, Laborie S, Roose P (2020) semantic agile approach for reconfigurable distributed applications in pervasive environments. Int J Ambient Comput Intell (IJACI) 11(2):48–67. https://doi.org/10.4018/ijaci.2020040103
30. Chandrakar P (2019) A secure remote user authentication protocol for healthcare monitoring using wireless medical sensor networks. Int J Ambient Comput Intell (IJACI) 10(1):96–116. https://doi.org/10.4018/ijaci.2019010106
31. Kamal MdS et al (2018) Big DNA datasets analysis under push down automata. J Intell Fuzzy Syst 35(2):1555–1565. https://doi.org/10.3233/jifs-169695
32. Kamal MS et al (2017) Self-organizing mapping based swarm intelligence for secondary and tertiary proteins classification. Int J Mach Learn Cybernet 10(2):229–252. https://doi.org/10.1007/s13042-017-0710-8
33. Kamal S et al (2016) Evolutionary framework for coding area selection from cancer data. Neural Comput Appl 29(4):1015–1037. https://doi.org/10.1007/s00521-016-2513-3
34. Kamal S et al (2016) A mapreduce approach to diminish imbalance parameters for big deoxyribonucleic acid dataset. Comput Methods Programs Biomed 131:191–206. https://doi.org/10.1016/j.cmpb.2016.04.005

Designing of UWB Monopole Antenna with Triple Band Notch Characteristics at WiMAX/C-Band/WLAN

S. K. Vijay, M. R. Ahmad, B. H. Ahmad, S. Rawat, P. Singh, K. Ray, and A. Bandyopadhyay

Abstract In this paper, the design of ultra-wideband (UWB) monopole with triple band notch characteristics is obtained. The antenna has overall dimension of 28 × 30 mm^2 which is compatible with wireless devices. The proposed antenna comprises of elliptical radiating patch with partial ground to enhance bandwidth which includes the UWB frequency range from 2.5 to 15.0 GHz. To obtain triple notch characteristics, two elliptical complementary split ring resonators (ECSRRs) and symmetrical slits are used. In this paper, an elliptical monopole UWB antenna is suggested with triple band-notched features for WiMAX-band (3.3–3.7 GHz), downlink C-band (3.8–4.2 GHz), and WLAN-band (5.2–5.8 GHz) using ECSRR slots and symmetrical slits on patch, respectively. The effects of each individual unit over band notch characteristics are also investigated. All simulation works have been done using electromagnetic software Ansoft High Frequency Structure Simulator (HFSS).

S. K. Vijay
Department of Electronics & Communication Engineering, Amity University, Jaipur, Rajasthan, India

M. R. Ahmad
CETRI, University Teknikal Malaysia Melaka (UTeM), 76100 Durian Tunggal, Melaka, Malaysia

B. H. Ahmad
FECE, University Teknikal Malaysia Melaka, Melaka, Malaysia

S. Rawat
Department of Electronics & Communication Engineering, Manipal University, Jaipur, Rajasthan, India

P. Singh · K. Ray (✉)
Department of Physics, Amity University Rajasthan, Jaipur 303007, India
e-mail: kray@jpr.amity.edu

A. Bandyopadhyay
International Center for Materials and Nanoarchitectronics (MANA), Research Center for Advanced Measurement and Characterization (RCAMC), National Institute for Materials Science, 1-2-1 Sengen, Tsukuba, Ibaraki 3050047, Japan

K. Ray et al. (eds.), *Proceedings of International Conference on Data Science and Applications*, Lecture Notes in Networks and Systems 148,
https://doi.org/10.1007/978-981-15-7561-7_9

Keywords UWB antenna · CSRR antenna · Microstrip antenna · Planar antenna · Triple band notched antenna

1 Introduction

As the development popularity and massive demand of wireless devices have enforced makers and scholars to create the cheap, small size, low visibility as well as broadband planar antennas for telecommunication system. Ultra-wide band (UWB) technology had attracted many researchers due to un-licensed use in commercial application after declaration by the US Federal Communication Commission (FCC) in 2002 [1]. To design the UWB (frequency ranges from 3.10 to 10.60 GHz) antennas, printed microstrip antennas have been exceedingly comprehend in recent years by cause of its small size, low visibility, light weight and cheapness, but the major downside of the UWB communication is that many narrow band wireless applications coexist in UWB spectrum like WiMAX-band, C-band, WLAN-band, and X-range for satellite communications causing electromagnetic interferences [2]. It is essential to model UWB antennas which have band filtering characteristics [3]. Several researches have been done for implementing band notch characteristics in the UWB antenna since last few years [3–10]. These include various kinds of slots on ground or patch, use of split ring resonators, EBG structures asymmetrical resonators, and defected ground planes [3–10]. Slots etching can be done as, etching of L & C-shaped slot [4], elliptical complementary split ring resonator [5], L-shaped slot with asymmetrical resonator [6], slot-loaded EBG structure [7], U-shaped slot [8], S-shaped slot [9], and a Π-shped slot in patch [10].

In this paper, an elliptical monopole UWB antenna is suggested with triple band-notched features for WiMAX-band (3.3–3.7 GHz), downlink C-band (3.8–4.2 GHz), and WLAN-band (5.2–5.8 GHz) using ECSRR slots and symmetrical slits on patch, respectively. The overall antenna dimension is 28×30 mm^2. The whole work is systematized as follows; Sect. 2 defines the antenna design, synthesis, and analysis. Section 3 refers to results as well as validation, then Sect. 4 designates recommended antenna's applications in big data systems, and finally, conclusion of the work describes in Sect. 5. All the simulation works have executed by applying Ansys electromagnetic software HFSSv15.

2 Design and Analysis

The planning of band-notched ultra-wideband is initiated with the consideration of elliptical microstrip antenna structure with partial ground, because it is simply matched over the complete UWB bandwidth beginning from 2.5 to 15 GHz and that also represented in Fig. 1a. The optimized antenna is fed with microstrip feed line to accomplish 50 Ω characteristics impedance matching between patch and feed line.

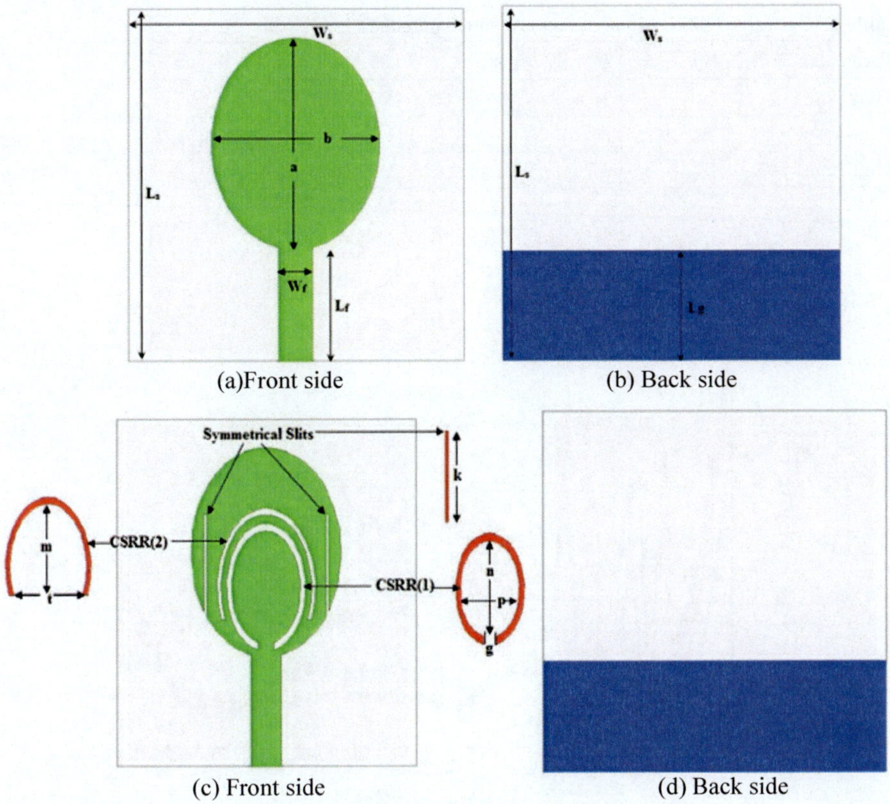

Fig. 1 **a** and **b** Primary UWB antenna **c** and **d** Band-notched UWB antenna

The radiating component coupled intensely with the conducting partial ground plane and also the aimed antenna is proficient of validating multiple resonances prominent to the wide operating band.

Triple band notching is produced when ECSRR slots and symmetrical slits cut in the patch. This innovative antenna using ECSRR-1 in patch provides the first notch at 3.25–3.65 GHz band intended for WiMAX-band applications, an another notch with ECSRR-2 in patch has notch at 3.7–4.2 GHz band for C-band applications and finally symmetrical slits cut in the patch to produce notch at 5.2–5.8 GHz band for WLAN-band.

This type of arrangement has been displayed in Fig. 1c and d. The entire dimensions that described in Fig. 1 are listed in Table 1. The desired triple band notch characteristics have been accomplished using the described approach for recommended antenna. The considered UWB antenna provides the VSWR range below 2 except the notch bands as displayed in Fig. 2.

Table 1 Optimized dimensions of recommended antenna

Parameter	W_S	L_S	W_f	L_g	a	b	L_f
Unit (mm)	28	30	3	9.3	9	7	9.7
Parameter	m	n	p	k	g	t	
Unit (mm)	3	3.7	2.45	9.3	1.4	3	

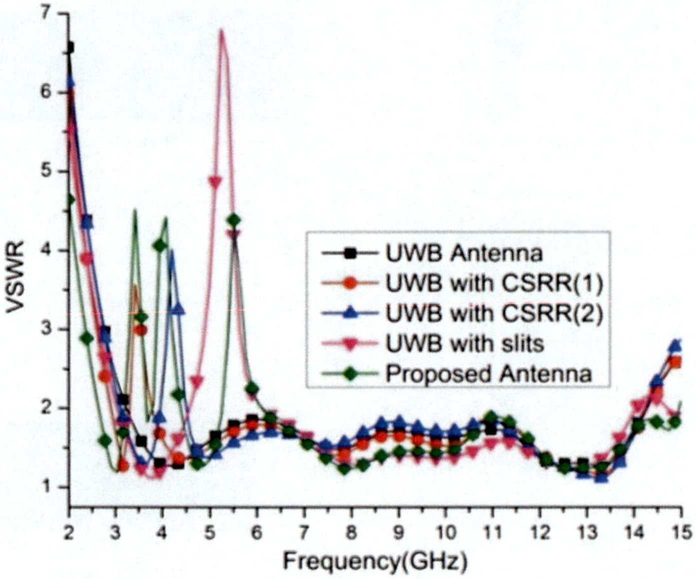

Fig. 2 Compared VSWR of suggested antenna

3 Result and Discussion

The proposed antenna produces wideband from 2.5 to 15 GHz having bandwidth of 12.5 GHz. The suggested design shows very good band-notched characteristics at WiMAX/WLAN/C-band in Fig. 3. The value of VSWR at notched band is more than 4, which is good sign of band notching.

The significance of surface current above the suggested antenna configuration has been demonstrated in Fig. 4. On desired notching frequencies of 3.5, 4, and 5.5 GHz, the disseminations of vector current are discontinuous and are concentrated at slot edges as illustrated in Fig. 4b–d, although in Fig. 4a and e, the current distribution is identical at 3 and 8 GHz frequency which is like pass band for antenna structure.

These all features can also be proved from radiation and gain characteristics. The radiation patterns of the presented antenna illustrated in Fig. 6 proclaim the suitable

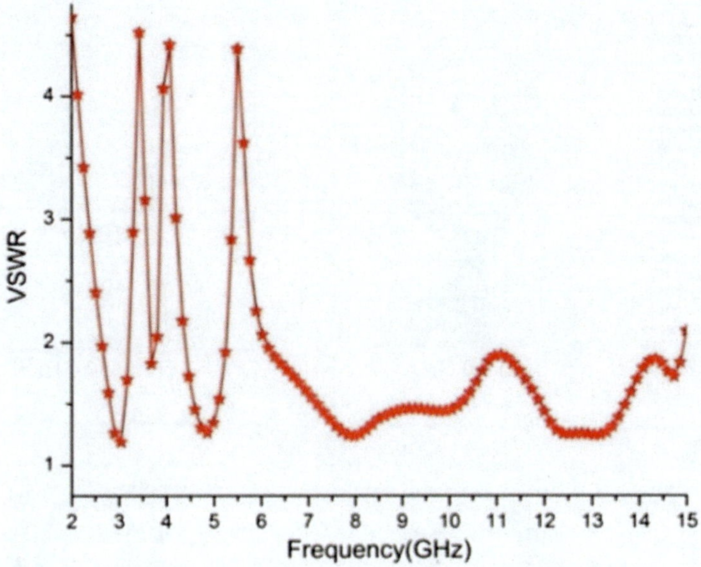

Fig. 3 Simulated VSWR of Suggested antenna

matching results with pass band characteristics. Figure 5 shows the antenna radiation efficiency and peak realized gain. It can be observed that gain is approximate below 0.25 dBi for notched frequency band whereas 1.75 dBi for other frequencies, similarly efficiency is approximately 10% for notched bands which replicate that antenna has effectively stopped the desired bands. Moreover for entire UWB bandd efficiency is around 60–70%.

4 Proposed Antenna for Big Data Application

Over past years, plenty of organizations have performed extensive and profound investigations upon the through-wall detection of human beings by means of UWB radar [11, 12]. In paper [11], authors proposed a principle meant for through-wall detection of human being based upon wavelet packet transform (WPT) and also statistical process control (SPC). The authors uses the UWB radar system for anomaly detection. Another application of UWB radar system is for moving object detection, where object detection through human can be achieved by 2D double stage detector with tracking filter and also by evaluating the frequency element of an echo signal [12].

Another application of UWB antennas in big data perspective is in the wireless body area network (WBAN). The UWB communication prototype is comprehensively applicable in WBAN. UWB RF technique supports the resilient and

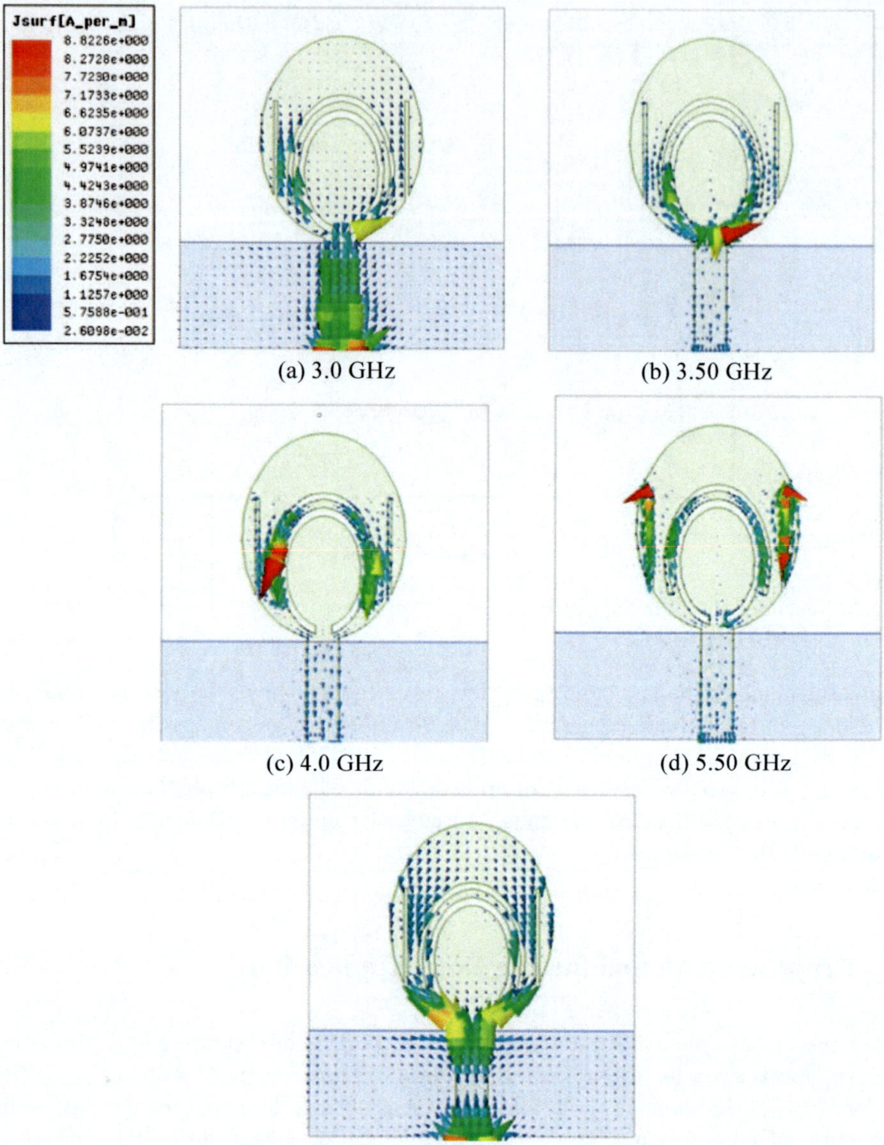

(a) 3.0 GHz

(b) 3.50 GHz

(c) 4.0 GHz

(d) 5.50 GHz

(e) 8.0 GHz

Fig. 4 Current distribution of proposed antenna

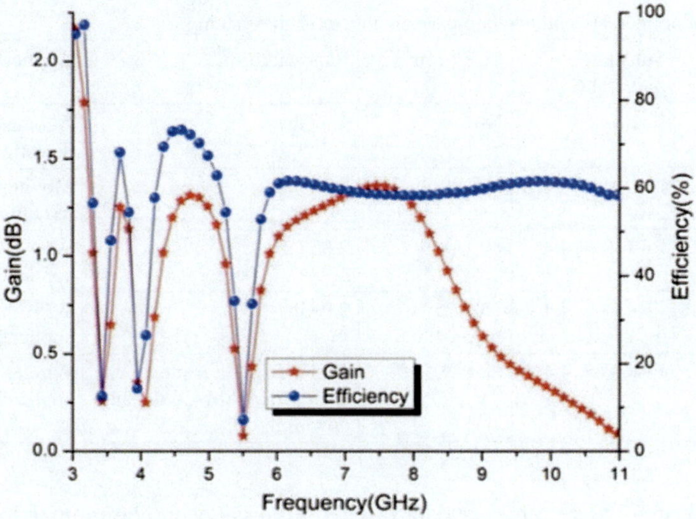

Fig. 5 Realized peak gain and antenna efficiency of recommended antenna

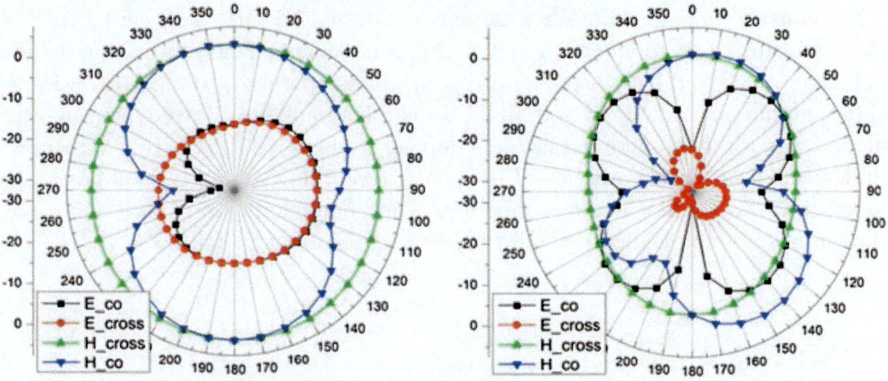

(a) 3 GHz (b) 8 GHz

Fig. 6 Radiation pattern of recommended antenna

energy-effective transmission of data and signals over wireless systems [13]. A low power level of UWB antenna suits for WBAN application with high wireless data transmission/reception rate.

However, the UWB antenna is also employed for Wireless Personal Area Network (WPAN). WPAN also employed for low power high data transmission. Since the required power is much low for the devices, the UWB antenna is the best candidate for WPAN application [14].

Table 2 Recommended antenna comparison with existent system

Ref. No.	Substrate material (ε_r)	Size (mm^2)	Bandwidth	Application
[11]	NA	NA	2.3 Ghz	Anomaly detection
[12]	NA	NA	2.25 Ghz	Moving target tracking
[13]	FR4 $\varepsilon_r = 4.4$	80 × 80	22.5 Ghz	WBAN application
[14]	FR4 $\varepsilon_r = 4.4$	25 * 25	2.9 Ghz	WPAN application
Proposed design	FR4 $\varepsilon_r = 4.4$	28 * 30	12.5 GHz with triple notch at WLAN/WiMAX/C-band downlink	Wireless application

Another area where a higher data rate required is ocean communication. UWB antenna with omni-directional radiation pattern and higher efficiency leads in underwater communication [15]. Various parameters to be considered are underwater communication conductivity, attenuation constant, permittivity, and permeability.

The considered antenna has a wide bandwidth of 12.5 GHz well suited for big data application. Due to low power level, proposed antenna could be used in indoor application. Due to the short pulses of the proposed antenna system, it is easier to achieve higher data rate up to 500 Mbps. The higher data rate allowed the proposed antenna to work in PC peripherals like wireless printers, wireless LANs, webcams, wireless monitors, keyboards, etc. Low power level allows the proposed design to recognized WPAN and WBAN technologies. The recommended antenna is compared with various literatures mentioned for big data applications in Table 2.

5 Conclusion

Proposed antenna encompasses UWB band varying from 2.5 to 15 GHz. Band-stop filtering features of ECSRR slots have been utilized to reduce the intervention from WiMAX, C-band, and WLAN-band operations. Initially, an elliptical monopole microstrip is designed for wireless application, then the size of ground is reduced to broaden the bandwidth of antenna to accomplish UWB spectrum. Finally, band notching structures are embedding in patch to provide security for WLAN, WiMAX, and C-band signals. Finally, the application of proposed antenna is discussed with by comparing various existing literatures. Recommended antenna has compact 28 × 30 mm 2 size and simple structure. All results of considered antenna specify that prospective antenna perchance a strong applicant for the band-notched application in UWB spectrum.

References

1. Kim YM (2003) Ultra wide band (UWB) technology and application. NEST group, The Ohio State University, July 10, 2003
2. Schantz H (2005) The art and science of ultra wideband antennas. Artech House Inc., Norwood, MA
3. Lin CC, Jin P, Ziolkowski RW (2012) Single, dual and tri-band notched ultrawideband antennas using capacitively loaded loop resonators. IEEE Trans Antennas Propag 60(1):102–109
4. Peng Gao LX, Dai J, He S, Zheng Y (2013) Compact printed wide slot UWB antenna with 3.4/5.5 GHz dual band-notched characteristics. IEEE Antennas Wireless Propag Lett 12:983–986
5. Sarkar D, Srivastava KV, Saurav KA (2014) Compact microstrip-fed triple band notched UWB monopole antenna. IEEE Antennas and Wireless Propag Lett 13
6. Wang Z, Liu J, Yin Y (2016) Triple band-notched UWB antenna using novel asymmetrical resonators. Int J Electron Commun. https://doi.org/10.1016/j.aeue.2016.10.001
7. Jaglan N, Kanaujia BK, Gupta SD, Srivastava S (2016) Triple band notched UWB antenna design using electromagnetic band gap structures. Progress In Electromagnetics Res C 66:139–147
8. Toshniwal S, Sharma S, Rawat S, Singh P, Ray K (2015) Compact design of rectangular patch antenna with symmetrical U slots on partial ground for UWB applications. In: 6th international conference on innovations in bio-inspired computing applications. AISC Series, Springer, vol 424, pp 535–542
9. Yadav A, Agrawal S, Yadav RP (2017) SRR and S-Shape slot loaded triple band notched UWB antenna. Int J of Electron Commun https://doi.org/10.1016/j.aeue.2017.06.003
10. Arora M, Sharma A, Ray K (2013) A Π- slot microstrip antenna with band rejection characteristics for ultra wideband applications Shodhganga Conference, Jaipur, 2013, pp 1135–1147
11. Wang W, Zhou X, Zhang B, Mu J (2015) Anomaly detection in big data from UWB radars. Security Comm Networks 8:2469–2475. https://doi.org/10.1002/sec.745
12. Kocur D et al (2010) Imaging method: an efficient algorithm for moving target tracking by UWB radar. Acta Polytechnica Hungarica 7(3):5–24
13. Yang D, Hu J, Liu S (2018) A low profile UWB antenna for WBAN applications. IEEE Access 6:25214–25219. https://doi.org/10.1109/ACCESS.2018.2819163
14. Ashtankar PS, Dethe CG (2012) Design and modification of circular monopole UWB antenna for WPAN application. Int J Electron Comput Sci Eng 3:960–970
15. George N, Ganesan R, Dinakardas CN (2014) Design of wide band antenna for ocean communication: review. Int J Adv Comput Res 4(1):258–265

The Dynamic Performance of Gaze Movement, Using Spectral Decomposition and Phasor Representation

Sergio Mejia-Romero, J. Eduardo Lugo, Delphine Bernardin, and Jocelyn Faubert

Abstract Motion analysis is widely used to monitor the dynamics of the gaze movement signal. The performance analysis in the spectral domain shows a variation of the characteristics due to the development of changes in dynamic behavior. Each dynamic reaps characteristics in its spectral domain. After examining the spectrum of gaze movement, one can provide valuable information about defects that develop with visible artifacts. The evaluation of the gaze movement is based on the presence of characteristic frequencies and its harmonics in the acquired movement signal. In this work, a signal processing analysis of the movement of the gaze in the domain of time and frequency was performed. The empirical mode decomposition has been proposed to plot the five first intrinsic mode functions in the polar form to detect the alteration of the gaze movement dynamics. The results show that the evaluation of the performance of the gaze movement dynamics can be done by observing the polar representation only.

Keywords Gaze movement · Dynamic performance · Empirical mode decomposition · Phasor representation

1 Introduction

Many movements and physiological movements are analyzed as a complex dynamic signal, characterized by irregular and nonlinear properties systems [1]. The evaluation of these hidden and essential characteristics is a complicated task, and

S. Mejia-Romero · J. Eduardo Lugo (✉) · D. Bernardin · J. Faubert
FaubertLab, School of Optometry, Université de Montréal, Montréal, QC, Canada
e-mail: eduardo.lugo@gmail.com

S. Mejia-Romero
e-mail: sergio.mejia.romero@umontreal.ca

D. Bernardin
Essilor Canadá Ltd., Montréal, QC, Canada

K. Ray et al. (eds.), *Proceedings of International Conference on Data Science and Applications*, Lecture Notes in Networks and Systems 148,
https://doi.org/10.1007/978-981-15-7561-7_10

several methodologies and algorithms for the analysis provide different relationships between the phenomenon and the variables to study.

This evaluation can be obtained through observations of the system that produce a biological time series, the analysis of which requires appropriate techniques. Very often, nonlinear time series analysis methods are chosen, which allow both the structure of a system and its dynamics to be evaluated. For example, different techniques are applied for the analysis of an electroencephalogram (EEG) signal [2] or to look for characteristic patterns of electrocardiogram (ECG) signals [3].

When we study the characteristics of the movement of the gaze direction, one can find variations of the measurements obtained for a specific subject or group of subjects despite maintaining identical experimental conditions as possible. This variation may result from the health status of the study subject or noise introduced by the experimental measurement. However, it should also be considering that the signals contain the intrinsic variability of the biological systems caused, for example, from experience, reaction time, etc.

During everyday life, a person visually explores their environment constantly to detect, anticipate, observe, and follow any information relevant to their behavior. For visually exploring the environment, the person's gaze is prone to perform various elementary movements such as fixations, saccades, and smooth pursuit [4].

The strategy of visual exploration depends on the context, that is, if the person drives, walks, etc. Besides, the visual exploration strategy is different from one person to another and may vary according to multiple factors, such as age, ametropia, motor, sensory, or cognitive abilities.

Therefore, visual exploration involves a significant number of coordinated movements between the eyes and head. Taking the eye-tracking as an example: the subject moves its eyes in direction to see an object. As the object is captured in the fovea, the head begins to rotate in the same direction as the eye, but because of this head movement, the eyes are making the rotation movement in the opposite direction. This dynamic continues as long as the subject does not take his eyes off from the object.

It takes approximately 15° to see a target only by eye movement [5], and when the amount of eye movement exceeds 20°, almost 80% of people move their heads [6]. Based on these factors, the movement of the head and the movement of the eyes combine to capture the visual information effectively. This mechanism is called gaze direction.

During a driving task, the gaze pattern is modeled using 3D-directional vectors points in space, which contains the information of the dynamics of the visual exploration following dynamic objects of interest, considering objects inside and outside the car, such as other vehicles, pedestrians, traffic lights, speedometer, side mirrors, etc.

Gaze movements are a series of movements caused by eye muscles considered as a biological system. It is represented as a time series, which varies over time, and contains a set of ocular positions consisting of two main events of eye movement: fixation and saccade [7].

In several studies, where the performance of the visual exploration is evaluated, the movement of the gaze is considered as a three-dimensional signal varying in time, whereby the number of fixations and saccade movements can be extracted to derive the evaluation of the driver and estimate the level of attention behavior to predict fatigue, loss attention, etc.

It is widely accepted that deficiencies in visual attention are responsible for a large proportion of road traffic accidents [8]. An understanding of the visual search strategies of drivers is thus fundamental, and that is why much research has been conducted in this area. Gaze movement recording and analysis provide essential techniques to understand the nature of the driving task and are essential for developing effective driving strategies and accident countermeasures [9].

The objective of this work is to present a tool for assessing gaze movement using an empirical model decomposition and polar representation of the gaze movement signal. We propose as an efficiency evaluator of the visual exploration strategy, the spectral frequencies distribution analysis of the gaze movement. In addition to this, using empirical model decomposition, we first graph each decomposed signal, and second, we plotted them in polar coordinates. From the results, we observe the signals of movements of the gaze that have an ordered structure, their distribution of the spectrum is narrow, and in the polar representation, each fundamental frequency is described with its amplitude varying in time. Otherwise, when the gaze that has spontaneous and random movements, the distribution of the spectrum is extensive, and the polar graph shows different amplitudes with values close to the origin (zero amplitude).

2 Methods

To obtain the direction of the gaze, we must consider two things: the movement of the eyes and the movement of the head; gaze estimation has been a problem in computer vision; most of the existing work follows a model-based approach to the estimation of gaze that a 3D eye model assumes, where the center of the eye is the origin of the gaze vector.

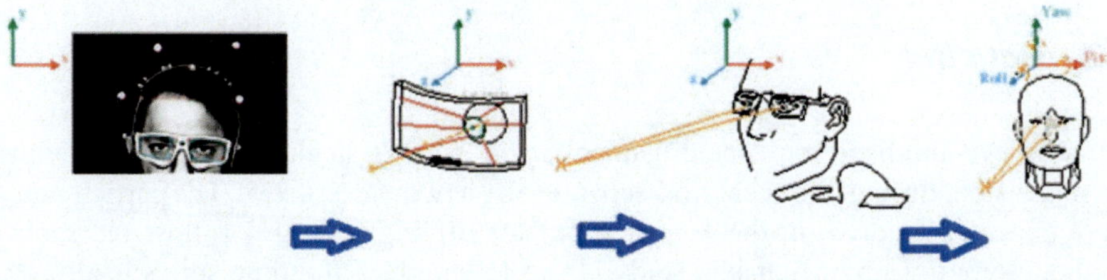

Fig. 1 Diagram of the gaze estimation method for the mapping between camera coordinates and gaze orientation

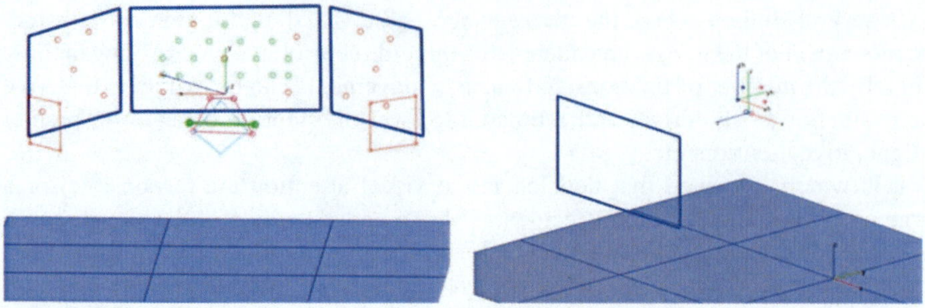

Fig. 2 Coordinate system transformations. Illustration of the different reference frames involved in the transformations: Left, gaze reference frame configuration with respect to the eye reference frame; Right, world reference frame, and eye centering with respect to the helmet reference frame

In this work, we use a similar model (see Fig. 1). The gaze estimation method was developed using the SMI eye-tracking system [10] to record the direction of the eyes and OptiTrack [11] to know the orientation of the head in a fixed reference frame to the ground.

The algorithm steps are displayed in Fig. 1, following the arrow, the coordinates (x, y) of the subject's position in the focal plane of the camera measured by a central rotation system are transformed into three-dimensional coordinates with respect to the reference frame of the four head marks [12]. A second coordinate system provided by the eye tracker is transformed, assuming that the head does not move. From the eye tracker, we can get the eye direction of all the points inside the calibration plane (Fig. 2a), the pupil center reflection in the eye-tracking coordinates is taken as an arbitrary reference, if the camera does not move, any change in the position of the pupil will represent a rotation of the eye and in the direction of the gaze. Finally, the resulting gaze direction vector in the calibration area is transformed into a ground-based coordinate system using head position and orientation information (see Fig. 2b).

To carry out the estimation of the direction of the gaze, we have developed a tool in MATLAB that performs the coordinates transformation of the eye-tracking systems and the head tracking data into the gaze direction coordinates, and all signals are synchronized with the time clock of the computer.

2.1 Data Set

We used eye and head movement data collected from 20 healthy participants, with normal vision, during two scenarios, separated by a two-week break. The participants were expected to drive in the Virage simulator driving task [13] follow the same methodology used by Michael's study [14], the length of the time series is almost 6 min per session. We only use a subset from these data to show our proposal.

Before running any test, the participant performed a calibration step that required the subject to look at three points at various locations on the screen, thus obtaining accurate data from the gaze. After this calibration test, the participant followed eighteen points on the screen (Fig. 2a).

Eye movements were registered with a 120 Hz sampling rate SMI eye tracker, and the head movement was registered with a 120 Hz sampling rate Opti Track, and the signal was processed by a toolbox developed on MATLAB resulting in the gaze position and gaze rotation.

2.2 Empirical Model Decomposition (EMD)

This algorithm has advantages in biomedical applications [15]. It is a nonparametric and self-adaptive method which decomposes a signal to a get a finite number of functions called intrinsically mode functions (IMF) [16]. This method does not require any predefined base function to represent the signal, which is the advantage of EMD over Fourier and wavelet analysis. The decomposition is based on the assumptions:

- the signal has at least two extrema: one maximum and one minimum;
- the time-lapse between the extrema defines the characteristic time scale, and
- if the data were devoid of extrema but contained only inflection points, then it can be differentiated once or more times to reveal the extrema.

The integration of the components can obtain the final results. The advantage of EMD over the Fourier analysis is that every IMF is considered as a band signal showing different temporal scale data [16].

The process is briefly presented [17, 18]. For a given signal $x(t)$, the local highest and lowest points are found as the first step of EMD. A cubic spline curve is used to connect all the local highest and lowest points (extrema), giving upper envelope $x_u(t)$ and low envelope $x_l(t)$. Further, to observe the values at every point of the envelopes, the mean value curve is calculated. The mean value $m_1(t)$ is defined as follows for two envelopes:

$$m_l(t) = [x_u(t) + x_l(t)]/2 \tag{1}$$

In this way, the value of the first IMF $h_1(t)$ can be calculated as follows:

$$h_1(t) = x(t) - m_1(t) \tag{2}$$

The process of obtaining the IMF is generally known as a shifting process. This process is used to cut off the riding waves and to ensure the symmetry of wave-profiles. It is a recurring process. For this, during the next shifting process, $h_1(t)$ is considered as original data and the notation for second IMF is as follows:

$$h_{11}(t) = h_1(t) - m_{11}(t) \tag{3}$$

The shifting process is repeatable for k times. It is continued till $h_{1k}(t)$ is considered as an IMF.

$$h_{1k}(t) = h_{1/k-1}(t) - m_{1k}(t) \tag{4}$$

The relation between first IMF$C_1(t)$ and $h_{1k}(t)$ can be defined as follows:

$$C_1(t) = h_{1k}(t) \tag{5}$$

Furthermore, it is mandatory to use some stopping criteria to end the shifting process so that it can ensure the continuity of sufficient physical sense of both amplitude and frequency modulations of IMF elements. The size of standard deviation (SD) is the stopping criteria and can be designated as:

$$\mathrm{SD} = \sum_{t=0}^{T} \left[|h_{k-1}(t) - h_k(t)|^2 \right] / h_{k-1}^2 \tag{6}$$

The first IMF $C_1(t)$ is achieved for the smaller value of SD compared to the threshold value. The value of residue is computed using the following equation:

$$r_n(t) = x(t) - \sum n_{i=1} C_i(t) \tag{7}$$

2.3 The Phasor

A phasor is a complex number representing a sinusoidal function whose amplitude (A), angular frequency (w), and initial phase θ are time-invariant. It is related to a more general concept called analytic representation. Euler's formula indicates that sinusoids signals can be represented mathematically as the sum of two complex-valued functions:

$$A\cos(wt + \theta) = A\frac{e^{i(wt+\theta)} + e^{-i(wt+\theta)}}{2}, \tag{8}$$

or as the real part of one of the functions:

$$A\cos(wt + \theta) = \mathrm{Re}\{Ae^{i(wt+\theta)}\} = \mathrm{Re}\{Ae^{i\theta}e^{iwt}\} \tag{9}$$

The function $Ae^{i(wt+\theta)}$ is called the analytic representation of $A\cos(wt + \theta)$; it is sometimes convenient to refer to the entire function as a phasor [19].

2.4 The Applied Method

In this work, the gaze signal was processed by the same noise filtering method. The choice of filtering methods and their parameters was based on the review of the literature. We apply the Savitzky-Golay (SG) filter [20], third-grade filter, with a window length of 15. The polynomial of the given order approximates the underlying signal within the window.

Subsequently, each time series, which represents the gaze movement obtained by applying the standard procedure, was subjected to an additional analysis that consisted of the following methodology:

- signals filtering and gaze estimation
- calculation of the power spectrum density (PSD by Fourier transform)
- intrinsically mode functions (IMF) calculation using the empirical model decomposition (EMD),
- graph each IMF in a polar graph,
- Shannon entropy calculation to each data series
- compare the first and second sessions using mean square error (MSE).

As a complement, we have calculated Shannon's entropy at each gaze direction signal and the mean square error between the signals to be compared.

3 Results

Below, we present as an example, the results obtained for a subject during two scenarios of the task.

As seen in Fig. 3, there is a change in the spatial distribution of the direction of the gaze from one stage to another, indicating a change in the dynamics of the movement. In Table 1, we show the values of the classical analysis for the spatial distribution of the gaze direction.

From Table 1, we observed a change in the number of fixations, changed the fixations duration, saccade amplitude, reflecting a less dispersed distribution in the

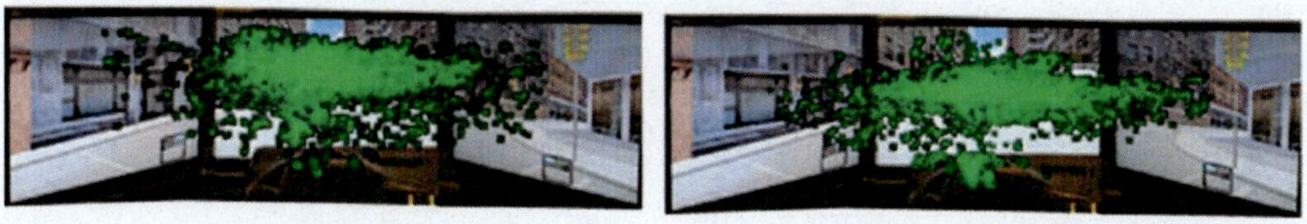

Fig. 3 We show the movement of the gaze on the driving simulator during the task. The left panel corresponds to the visual exploration during the first session, and on the right, it is during the second session

Table 1 Classical measures for visual exploration

Heading level	Mean ± SD	Mean ± SD
Saccade amplitude (°)	19.16 ± .705	17.46 ± 1.05
Fixation duration (ms)	261.38 ± 108.02	461.38 ± 78.02
Fixation rate (count/m)	96.91 ± 16.50	76.91 ± 16.50
Dominant frequency (Hz)	3.91	2.1
Gaze area (m^2)	82.76	78.76

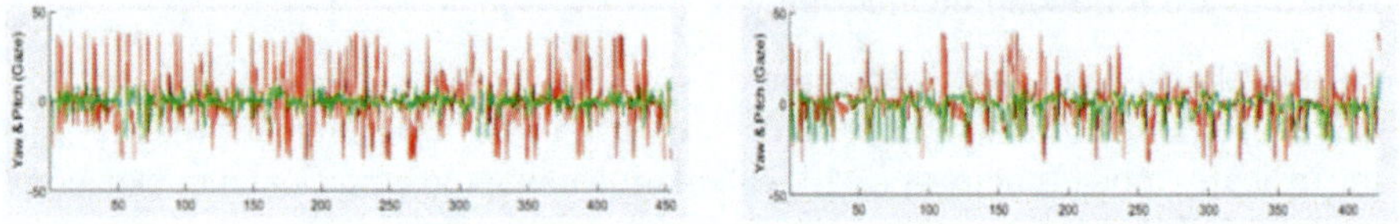

Fig. 4 The 2D coordinate position of gaze movement; the red corresponds to *X* position and green to *Y*

movement of the gaze that implies a less random visual scan pattern. Figure 4 shows the time series of gaze movement in coordinates (x, y) during the 450 s of test duration.

The power spectra of the gaze signal of the first and second sessions are shown in Fig. 5. The band frequency ranges from 0.08 to 10 Hz, where saccade and fixation movements are present, and microsaccades and tremors are discarded for this analysis due to the limitations of the sampling frequency.

Figure 6 shows the results of decomposition by EMD. From these figures, it can be seen that the distribution of the frequency components is separated from the original signals satisfactorily by EMD.

In Fig. 7, the polar graph of the original and its first five IMF signals are plotted; the first graph is the polar representation of the original signal, and consequently, their corresponding IMFs.

The IMFs polar graphs cover significant physiological frequency ranges corresponding to different eye movements. For example, the IMFs 2a-3a polar graphs(from Fig. 7) cover the saccade frequency, and the IMFs 4a 5a are the regression frequency

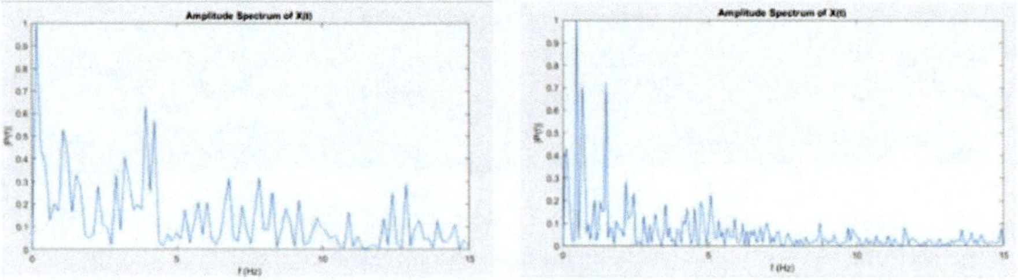

Fig. 5 Power spectrum of original signals (see Fig. 4)

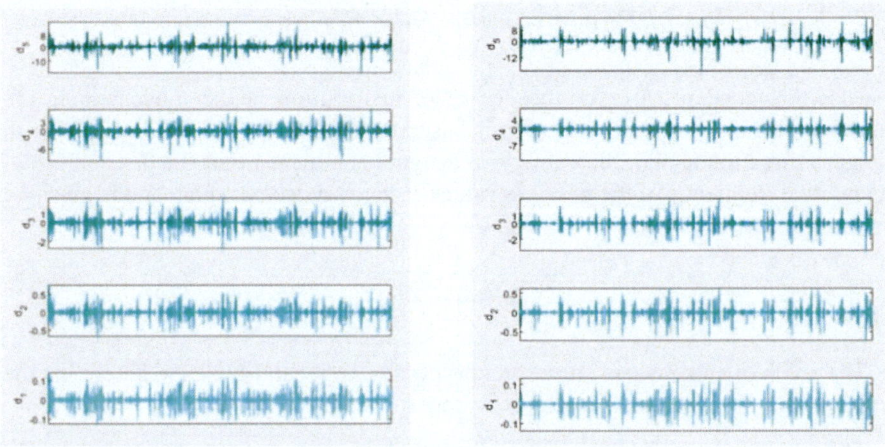

Fig. 6 EMD of gaze movement signal, the left panel corresponds to the first scenario, and the right is the second scenario, only the first five IMFs were calculated

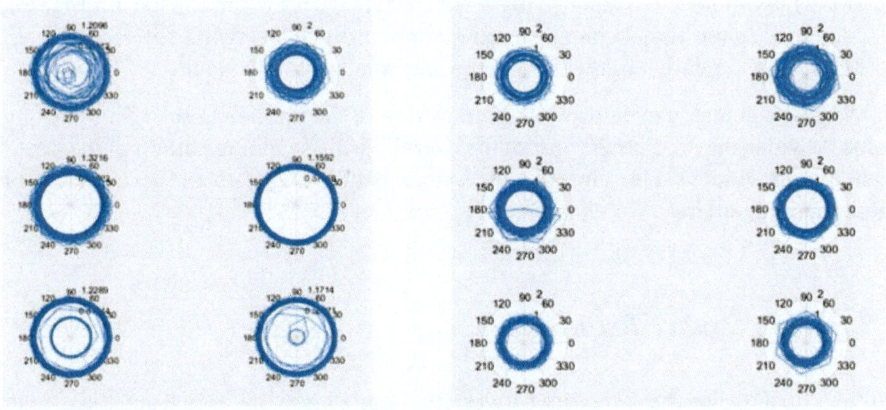

Fig. 7 Polar representation of the intrinsic mode function (IMFs) of the first session (left) and second session (right)

or microsaccades. Thus, we can say that visual exploration has changed. Time-domain MFIs can be further processed to extract features to distinguish different gaze activities.

As expected, the amplitude of the gaze movements, the range of fixation duration, and the density of frequencies present during the driving test changes between the different scenarios corresponding to the different workload presented by the tested scenario. While the eye movement increased, the fixation number decreased, suggesting an increase in the amount of time in the exploration and less cognitive load, as has been reported.

3.1 Evaluation of Entropy in the Time Series

We use this concept to know the degree of distribution of the movement in the direction of gaze, and this evaluation is based on the entropy value associated with the gaze direction signal; the entropy of a signal is a measure of the distribution of the relative frequency of the possible values P in the detected signal and is given by

$$H = -\sum_{j=0}^{N} P_j \log P_j \tag{10}$$

The value of entropy has interesting properties, so it can be used to determine the performance of a detected signal; such properties are:

- $H = 0$, if and only if all the values P_j are equal to one or zero. The entropy value is minimal. That is, the information contained is predictable; in other words, the scan has a small random component in its movement; in this case, the relative frequency is concentrated in few positions only.
- H is a maximum and is equal to $\log P_j$, if the entropy is maximum; the information contained in the signal contains high randomness (i.e., a significant variance). In this case, the relative frequency covers almost all possible values.

We also calculate the entropy at the transition of the fixation points. This entropy value provides the predictive value of the overall visual scanning pattern of the series or in a time window. This entropy value when is high suggests a less structured or more random pattern

3.2 Mean Square Error (MSE)

It is used to describe the similarity between two signals, the first signal and the second signal.

$$\text{MSE} = \frac{1}{N}\sum_{i-1}^{N}[s1_i(t) - s2_i(t)]^2 \tag{11}$$

where $s1$ is for the first session signal, $s2$ is the second session signal, and N for the signal length or the number of samples (Table 2).

Table 2 Entropy and MSE results

Heading level	Mean
Entropy signal 1	3.76
Entropy signal 2	2.23
MSE signal 1 versus signal 2	5.5516

4 Conclusions

In this paper, we have presented a method for efficiency evaluation of gaze exploration using the decomposition analysis model and polar diagrams. The proposed method allows us to evaluate the scan performance objectively; Besides, it was found that efficiency is linked to lower randomness in the movement.

Experimental results have shown that the polar representation of the signal that describes the direction of movement of the gaze, which contains saccadic movements and fixations, can distinguish subjects with different levels of exploration efficiency.

To achieve a complete characterization of the dynamics of the gaze direction data for each subject, we also estimated the randomness within each gaze signal using information entropy. Our result is similar to the one obtained by standard measurements used to evaluate visual exploration.

Acknowledgements We want to extend our gratitude to Jesse Michaels for providing the raw data collected during his Ph.D. research project and all the participants that were involved in this study, as well as Amandine Debieuvre of the Vision Sciences department, Essilor international R&D, Paris, and Essilor Canada Group Inc. for providing software analysis toolbox. This research was partly funded by an NSERC Discovery grant and Essilor Industrial Research Chair (IRCPJ 305729-13), Research and development cooperative NSERC—Essilor Grant (CRDPJ 533187-2018), Prompt.

Author Contributions M-R.S. designed and implemented the research method and conducted the data analysis. M-R.S was involved in preparing and carrying experiments. All authors took part in the paper preparation and edition.

Conflicts of Interest The authors declare no conflict of interest. The founding sponsors had no role in the design of the study; in the collection, analyses, or interpretation of data; in the writing of the manuscript, and in the decision to publish the results.

References

1. Guckenheimer J, Holmes P (2016) Nonlinear oscillations, dynamical systems, and bifurcations of vector fields. Springer-Verlag. Conference 2016, LNCS, vol 9999, Springer, Heidelberg, pp 1–13
2. Zhang X, Roy RJ (2001) EEG complexity as a measure of depth of anesthesia for patients. IEEE Trans Biomed Eng 48(12):1424–1433
3. X Zhang, Zhu Y, Thakor NV, Wang Z (1999) Detecting ventricular tachycardia and fibrillation by complexity measure. IEEE Trans Biomed Eng 46:548–555
4. Wood JM, Owsley C (2016) Vision and driving: a look at the research road ahead. Clin Exp Optom 99:393–394. https://doi.org/10.1111/cxo.12450

5. Bahill AT, Adler D, Stark L (1975) Most naturally occurring human saccades have magnitudes of 15 degrees or less. Invest Ophthalmol 14:468–469
6. Gresty MA (1974) Coordination of head and eye movements to fixate continuous and intermittent targets. Vision Res 14:395–403
7. Holmqvist K, Nyström M, Mulvey F (2012) Eye tracker data quality: what it is and how to measure it. In: Proceedings of the symposium on eye tracking research and applications, Santa Barbara, CA, USA, 28–30 March 2012; ACM: New York, NY, USA, pp 45–52
8. Insurance Institute for Highway Safety (2014). Statuts report: fit for the road 49:1. http://www.iihs.org/iihs/sr/statusreport/article/49/1/
9. Owsley C, McGwin G (2010) Vision and driving. Vision Res 50:2348–2361. https://doi.org/10.1016/j.visres.2010.05.0213
10. Senso Motoric Instruments and Noldus Information Technology combine eye tracking and video analysis. Noldus. Retrieved April 2, 2014
11. NaturalPoint (2017) Motion capture systems—OptiTrack webpage. [Online]. Available: optitrack.com. Accessed 09 Jan 2017
12. Cesqui B, van de Langenberg R, Lacquaniti F, d'Avella A (2013) A novel method for measuring gaze orientation in space in unrestrained head conditions. J Vision 13(8):28, 1–22
13. VS500M car driving simulator used in this study (Virage Simulation Inc®)
14. Michaels J, Chaumillon R et al (2017) Driving simulator scenarios and measures to faithfully evaluate risk driving behavior: a comparative study of different driver age groups. PLoS ONE 12(10):e0185909
15. Salisbury JI, Sun Y (2005) Assessment of chaotic parameters in nonstationary electrocardiograms by use of empirical mode decomposition. Ann Biomed Eng 32(10):1348–1354
16. Huang NE, Shen Z, Long SR, Wu MC, Shih HH, Zheng Q, Liu HH (1998) The empirical mode decomposition and the Hilbert spectrum for nonlinear and non-stationary time series analysis. Proc R Soc Lond A: Mathe Phys Eng Sci 454(1971):903–995
17. Zhang DX, Wu XP, Guo XJ (2008) The EEG signal preprocessing based on empirical mode decomposition. In: The 2nd international conference on IEEE bioinformatics and biomedical engineering, ICBBE 2008, pp 2131–2134
18. Alam ME, Samanta B (2017) Empirical mode decomposition of EEG signals for brain computer interface. Southeast Con 2017, Charlotte, NC, 2017, pp 1–6
19. Phadke AG, Thorp JS (2008) Synchronized phasor measurements and their applications. Springer, New York, NY, USA, pp 29–48
20. Schafer RW (2011) What is a savitzky-golay filter. IEEE Sig Process Mag 28:111–117

Novel Hairpin Band-Pass Filter Using Tuning Stub

Sonu Jain, Taniya Singh, Ajay Yadav, and M. D. Sharma

Abstract This paper presents the design and simulation of band-pass filter of pass band range from 1.83 to 2.45 GHz frequencies. Design consists of five hairpin resonator with *E*, *U*-shape stub and via. This design is presented in the compact size of 25.5 × 28 mm and a thickness of 1.6 mm and the center frequency of 2.26 GHz. The compact size gives extreme parallel coupling which enhances the performance of the hairpin filter.

Keywords Hairpin filter · Resonator · Band-pass filter · Wireless local area network (WLAN) · Radio frequency (RF)

1 Introduction

Filters are the most useful contribution in today's leading world where the introduction of the RF and millimeter wave gives a new and wide scope and also became the field of commercial interest. In the world of GSM (global area network), WLAN (wireless local area network), and much more, it is very important and difficult to cope with different networks without any interference and interruption with another band of frequencies and so become the challenging area. These crucial issues can only be solved by the use of the filter. The major attraction in RF field is achieved by the microstrip band-pass filter which includes combline filter, parallel coupled filter, planar microstrip filter, and hairpin filter. These filters have high transmission, steep, and well-defined edges and outstanding blocking between pass bands. Parallel coupled filter does not gain much attention because of the drawback of weak coupling which could be overcome by using half wave long resonator and admittance [1], whereas combline filter is popular because of small size but has high losses

S. Jain (✉) · T. Singh · A. Yadav · M. D. Sharma
Department of ECE, Global Institute of Technology, Jaipur, India
e-mail: jainsonu14@gmail.com

M. D. Sharma
Department of ECE, Malaviya National Institute of Technology, Jaipur, India

K. Ray et al. (eds.), *Proceedings of International Conference on Data Science and Applications*, Lecture Notes in Networks and Systems 148,
https://doi.org/10.1007/978-981-15-7561-7_11

in inner conductor which decrease the Q factor; this problem can be minimized by periodic N-disk loaded to inner conductor and so increase Q factor [2].

Hairpin filter attains a rapid rise because of its striking feature of optimal size utilization, easy to manufacture and generally needs no ground connection for the resonator. Many reports are proposed in this field; from them, some of the researchers achieve great result [3–8].

Filters are extremely used in applications such as in wireless communication and global positioning. It plays a major role in the field of communication. It is widely used at the receiver and the transmitter, and at the transmitter it ensures the limit of the signal bandwidth in the allocated band; at receiver, it allows signal of selected range of frequency. The use of split-ring resonators and receiver front ends assimilate have many remunerations for relevance to mobile radio equipment [9]. Band-pass filters are defined by their center frequency F0 along with their 3 dB bandwidth by using detected ground structure (DGS); also, superior bandwidth characteristics can be accomplished [10]. The hairpin filter is easy to manufacture. The design is derived from the edge-coupled filter. Parallel coupling present in the hairpin filter between the U-shaped resonator could also decrease the ripple by microstrip hairpin resonator and have application to develop cross-coupling microstrip resonator [11]. Many filters like CQ (cascade quadruplet) and end-coupled filter are proposed for the reduction of ripple.

This paper focuses on the design of a novel hairpin filter which is modified by the use of shorting pins which are nothing but a cylindrical hole engraved into the substrate it is also recognized by via also and E- and U-shaped stub loaded on the substrate. The implementation of these proposed factors is responsible for getting wide band rejection of discarded frequencies and only passes the desired range of frequency (1.83–2.45 GHz). The design of hairpin filter is made by the array of U-shape resonator cascaded with one another where $n = 5$ units are considered which significantly improve the quality factor (Q) simultaneously. As we increase the number of section n, we have a stepper transition from the pass band to stop band while maintaining pass band characteristic and cut off frequency. This filter is designed on a low-cost FR4-epoxy dielectric medium with a dielectric of 4.4 and height 1.6 mm. FR4 epoxy is so chosen because it cost low in fabrication and it works very well at the low frequencies and also it is easily available. Several filters mentioned in literatures have been compared in Table 1.

Table 1 Comparative literature review details

References	Size (mm ^{2})	Substrate	Frequency range (GHz)
[4]	30 × 68	Taconic RF35	2.1–2.7
[6]	40.04 × 32.43	FR4	2.55–2.61
[7]	33 × 80	Al_2O_3	0.025–0.085
[8]	29.9 × 14.3	Roger R04003	2.11–2.17
Presented work	25.5 × 28	FR4	1.8–2.26

2 Design and Analysis

The design and structure of this filter have been simulated and using ANSOFT HFSS software. The presented design introduces the ground plane and the dielectric substrate with rectangular radiating patch, feed line, stub and shorting pins loaded on it. Figure 1 shows the initial design of the hairpin filter. All the optimized dimensions of the basic hairpin filter are listed in Table 2.

The basic design of the hairpin filter is modified by reducing the width of each resonator which resulted to increase the frequency at the desired output of the hairpin structure, the gap between each of the resonators is providing the capacitance in the designing structure so distance between each of the resonators are kept small so as to achieve low capacitance which resulted to increase the coupling both of these factors contribute to tunability of the proposed hairpin filter but with wide band of frequency in which the chances of interruption of other frequencies of undesired range may be possible. Simulation of primary design works from 2.27 to 1.35 GHz with a center frequency of 0.92 GHz.

Designing equation for the length 'l' of individual element have shown in Fig. 2 are:

Fig. 1 Basic Hairpin filter

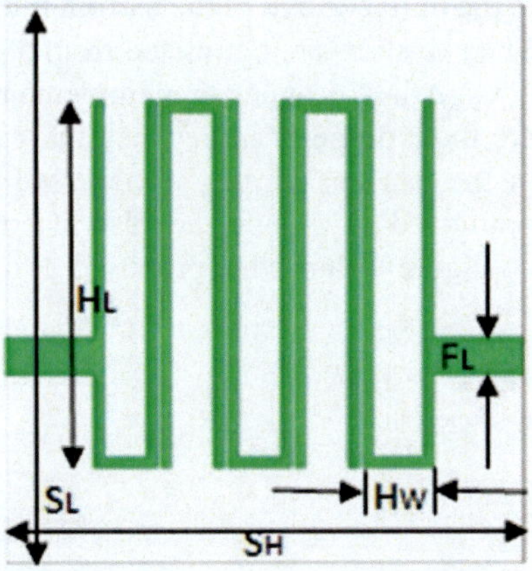

Table 2 Parameters of filter design

Parameters	H_L	H_W	S_L	S_H	F_L
Size (mm)	18.4	3.5	28	25.5	1.85
Parameters	F_H	U_H	E_H	E_W	U_W
Size (mm)	4.4	18.5	18.5	14.9	8.1

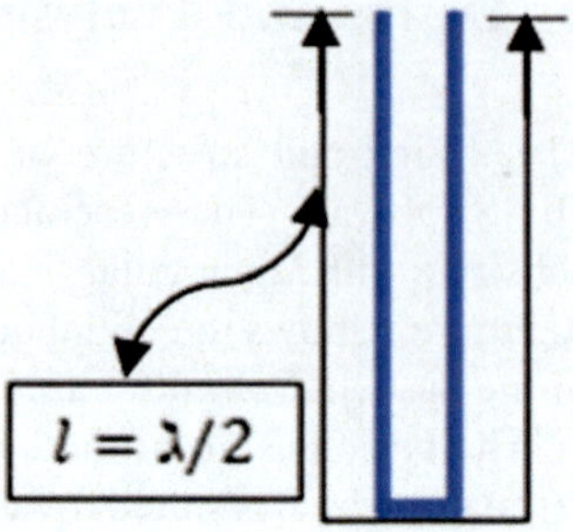

Fig. 2 Individual hairpin element '*l*'

$$l = \frac{C}{2f\sqrt{\mathcal{E}r}} \tag{1}$$

where

f Resonating Frequency in Hz
ε_r Dielectric Constant of the Substrate.

Figure 3 shows the design of hairpin filter with *E* and *U* stub loaded on it. Here the use of this stub is to decrease the capacitance up to a greater extent as compared to the initial design of the hairpin filter by decreasing the gap between the resonators. The two stubs are optimized from the top of a rectangular patch of 2.3 and reduce to 0.3 mm thickness and then implemented to the design. The arms of both *E* and *U* stub are fixed properly in between the gap of the resonators by doing so the frequency of the band-pass filter is increased and optimized result can be obtained. All the optimized dimensions are also listed in Table 2.

Figure 4 shows the proposed design of the hairpin filter with the shorting pins of

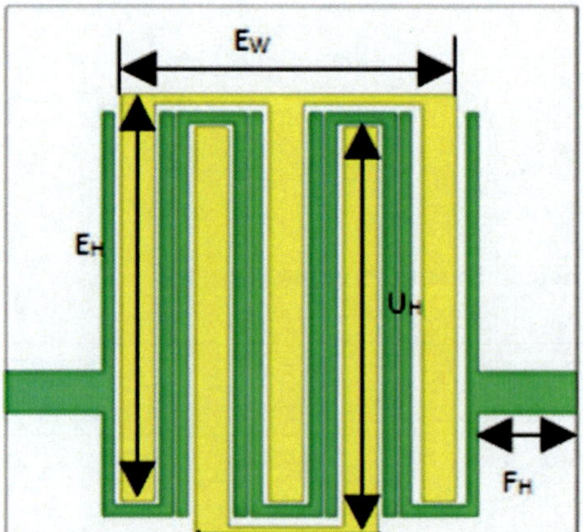

Fig. 3 Design with *U*- and *E*-shaped stub

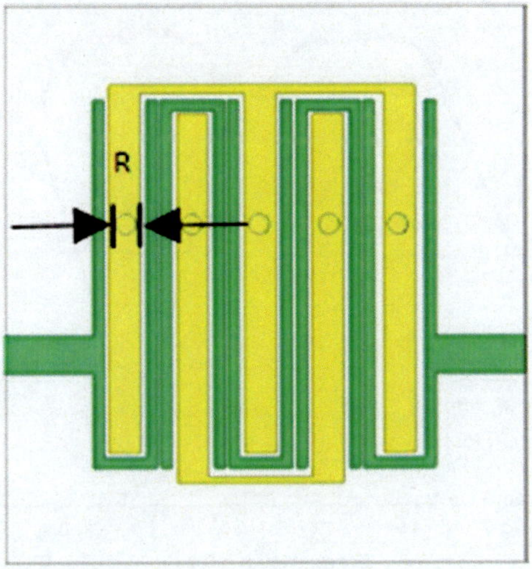

Fig. 4 Proposed design of hairpin filter

the diameter of 1 mm, here in the design shorting pins is providing the inductance in the hairpin filter which is responsible for getting a narrow band which satisfies the trait of the band-pass filter. Finally, the combination of the shorting pins and stubs contribute to the great result in the hairpin filter which works from 1.83 to 2.45 GHz and reject the other discarded frequencies with the high tunability and so the problem of interference of different bands of frequencies can be solved.

3 Results and Discussion

Figure 5 displays the simulated S_{11} and S_{21} graph of the primary design. We can see from the simulated result of the proposed hairpin filter. Now it is clearly visible that this filter has cut off frequency of 2 GHz with wider bandwidth, which does not show good affiliation and also does not assist with the good characteristics of the filter.

So from Figs. 6 and 7, it is clearly visible that by introducing stubs and shorting pins in the same design we can stop the involvement of undesired frequencies up to great extent and achieve filter trait. It can be seen that current density is highly concentrated at 2 GHz that is a positive sign to create band-pass filtering characteristics at these frequencies and can be seen from Fig. 8. The comparison of all designs also discussed in Table 3.

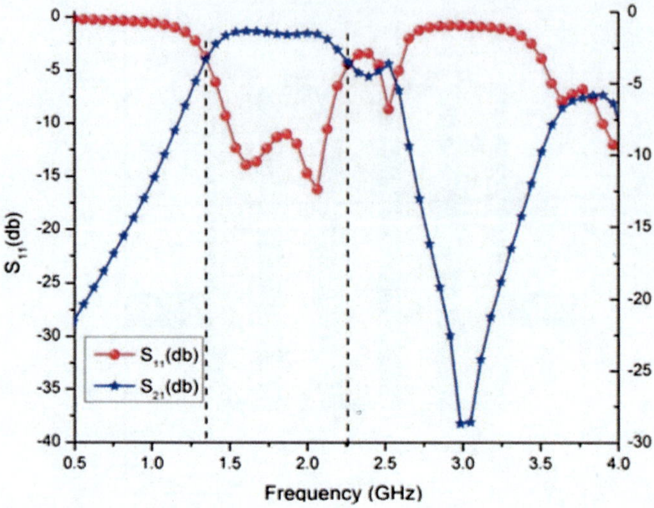

Fig. 5 Simulated S_{11} and S_{21} of primary design

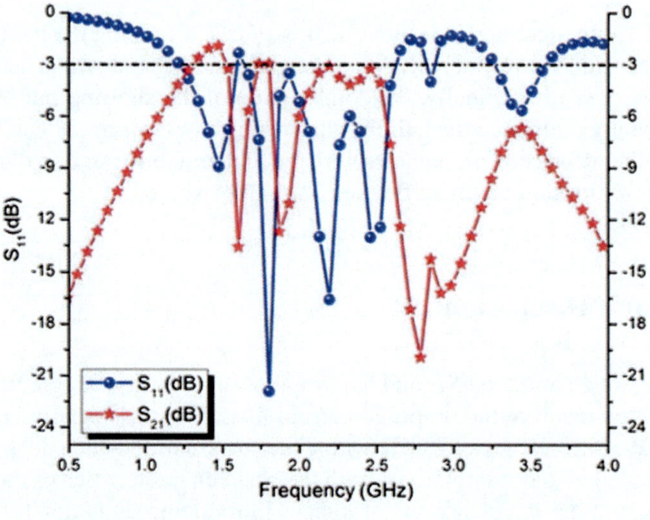

Fig. 6 Simulated S_{11} and S_{21} of primary design with E and U stub

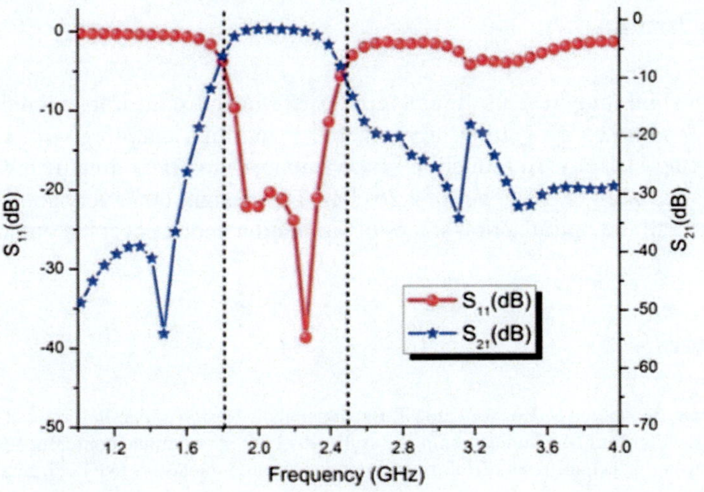

Fig. 7 Simulated S_{11} and S_{21} of proposed design

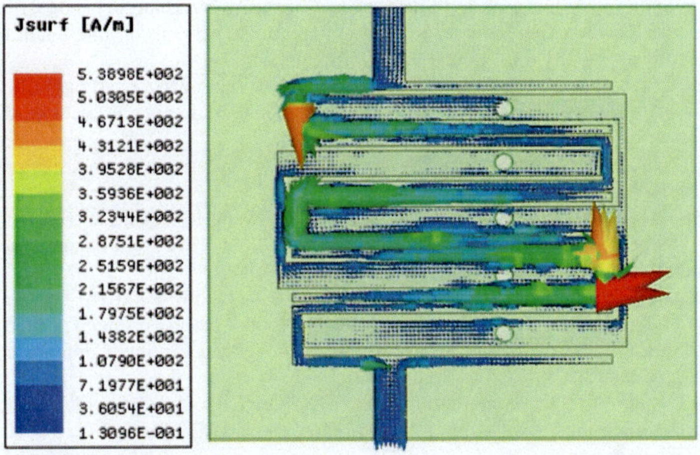

Fig. 8 Current distribution at 2 GHz

Table 3 Optimization stages of proposed HAIRPIN filter

Parameters	Resonant frequency (GHz)	3 dB bandwidth
Primary design	2.07	0.92
Design with E and U stub	1.80	1.02
Proposed design	2.26	1.26

4 Conclusion

This worthwhile study reveals the enhanced performance of the implemented hairpin filter which is based on getting high tunability and high coupling with extremely high blocking capacity. Introduction of the stub and via offers great result at 1.83–2.45 GHz with a center frequency of 2.26 GHz. This design shows very good dexterity between the all the optimization stages of simulation and at accepted outcomes.

References

1. Seghier S, Benahmed N, Bendimerad FT, Benabdallah N, Design of parallel coupled microstrip bandpass filter for FM Wireless applications. In: 2012 6th international conference on sciences of electronics, technologies of information and telecommunications (SETIT), Sousse, 2012, pp 207–211
2. Shen G, Budimir D Novel resonator structures for combline filter applications. In: 2002 32nd European microwave conference, Milan, Italy, 2002, pp 1–3
3. Darwis F, Setiawan A, Daud P (2016) Performance of narrowband hairpin bandpass filter square resonator with folded coupled line. In: 2016 international seminar on intelligent technology and its applications (ISITIA), Lombok, pp 291–294
4. Srisathit K, Tangjit J, Kumpontorn W (2010) Miniature microwave band pass filter based on modified hairpin technology In: 2010 IEEE international conference of electron devices and solid-state circuits (EDSSC)
5. Wei F, Chen L, Shi X (2012) Compact lowpass filter based on coupled-line hairpin unit. Electron Lett 48(7):379–381
6. Mokhtar MH, Jusoh MH, Sulaiman AA, Baba NH, Awang RA, Ain MF (2010) Multilayer hairpin bandpass filter for digital broadcasting. In: 2010 IEEE symposium on industrial electronics and applications (ISIEA), Penang, pp 541–544
7. Schuster C, Wiens A, Schüßler M, Kohler C, Binder J, Jakoby R (2016) Hairpin bandpass filter with tunable center frequency and tunable bandwidth based on screen printed ferroelectric varactors. In: 2016 11th European Microwave Integrated Circuits Conference (EuMIC), London, pp 496–499
8. Djaiz A, Denidni A (2006) A new compact microstrip two-layer bandpass filter using aperture-coupled SIR-hairpin resonators with transmission zeros. IEEE Trans Microw Theory Tech 54(5):1929–1936
9. Sagawa M, Takahashi K, Makimoto M (1989) Miniaturized hairpin resonator filters and their application to receiver front-end MICs. IEEE Trans Microw Theory Tech 37(12):1991–1997
10. Othman MA, Mohd Zaid NF, Abd Aziz MZA, Sulaiman HA (2014) 3 GHz hairpin filter with defected ground structure (DGS) for microwave imaging application. In: 2014 international conference on computer, communications, and control technology (I4CT), Langkawi, pp 411–414
11. Hong J-S, Lancaster MJ (1998) Cross-coupled microstrip hairpin-resonator filters. IEEE Trans Microw Theory Tech 46(1):118–122

Recognition of Faults in Wind-Park-Included Series-Compensated Three-Phase Transmission Line Using Hilbert–Huang Transform

Gaurav Kapoor

Abstract This work presents an HHT (Hilbert–Huang transform)-based fault recognition technique for wind-park-included series-compensated three-phase transmission line (WPISCTPTL). The single-side captured fault currents of the WPISCTPTL are used to evaluate the amplitudes of HHT outputs. The fault factors of WPISCTPTL are varied. It is investigated that the HHT well detects every type of fault. It is also explored that the HHT is robust to the modification in the fault factors of WPISCTPTL.

Keywords Fault recognition · Hilbert–Huang transform · Wind-park-included series-compensated three-phase transmission line protection

1 Introduction

The series-compensated transmission lines (SCTLs) carry enormous capacity of electrical power. The fault occurrence and consecutive tripping of SCTLs would outcome in extensive interruption of electrical power. Thus, an accurate recognition of fault in a SCTL turns out to be very crucial for reducing the loss of profit and providing fast preservation. In the latest years, lots of investigations have been devoted for the protection of TLs. In [1], multi-location LLG fault detection/classification technique based on WT (wavelet transform) has been presented for twelve-phase TL with series compensation. In [2–4], for TLs, a protection technique-based on MMF (mathematical morphological filters) has been proposed. Reference [5, 6] showed that WT can be utilized as an effective fault recognition tool for TLs with series compensation. In [7], an HHT (Hilbert–Huang transform)-based protection technique has been reported for DSTATCOM-compensated TL. In [8], MODWT (maximum overlap discrete WT)-based protection method has been presented. Extreme learning machine in [9] and

G. Kapoor (✉)
Department of Electrical Engineering, Modi Institute of Technology, Kota, India
e-mail: gaurav.kapoor019@gmail.com

K. Ray et al. (eds.), *Proceedings of International Conference on Data Science and Applications*, Lecture Notes in Networks and Systems 148,
https://doi.org/10.1007/978-981-15-7561-7_12

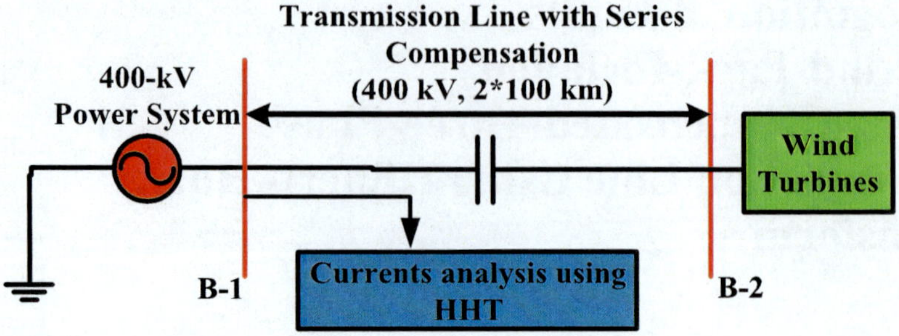

Fig. 1 The graphic of WPISCTPTL

a combination of ANN (artificial-neural network) and phasor data in [10] have been used for micro and smart grids protection.

In this work, the HHT is executed for the recognition of faults in the WPISCTPTL. Such type of investigation has not been depicted so far to the best of the information of the author. The results show that the HHT competently detects the faults, and the uniformity of the HHT is not susceptible to the deviations in fault factors.

This chapter is structured as: Sect. 2 presents the specifications of WPISCTPTL. Section 3 describes the process of HHT. Section 4 presents the performance appraisal of the investigations which are carried out in this study. Section 5 concludes the chapter.

2 The Specifications of WPISCTPTL

The simulation model (Fig. 1) consists of WPISCTPTL of 400-kV, divided into two parts each of length 100-km. The WPISCTPTL is fed from a 400-kV source at the B-1 and integrated with two wind-turbines at B-2. The current transformers are installed at the B-1 for fault current measurements.

3 The Flowchart for HHT-Technique

Figure 2 illustrates the procedure for HHT. The steps are shown below.

1. Simulate the WPISCTPTL and generate fault currents.
2. Analyze the currents using HHT for characteristics retrieval.
3. The phase will be declared as the faulted phase if its HHT output crosses the predefined threshold else repeat the process.

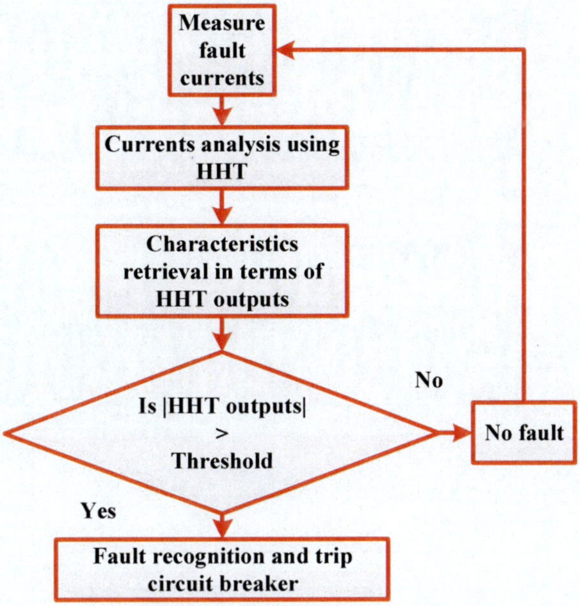

Fig. 2 The procedure for HHT

4 Performance Appraisal

The HHT has been tested for various situations such as: no-fault, converting faults, varying V_S and F_S, faults around capacitor bank, different faults, F_R-variation and varying wind-turbine units. The results are presented below.

4.1 The Effectiveness of HHT for No-Fault

Figure 3 exemplifies the currents and voltages for no-fault. Figure 4 shows the HHT outputs for no-fault. Table 1 presents the outcomes of HHT for no-fault situation.

4.2 The Effectiveness of HHT for Converting Faults

The exploration of HHT has been done for the converting faults. Figure 5 depicts the plot when the A-G fault (with $F_R = 3.5\ \Omega$ and $G_R = 4.5\ \Omega$) is switched at 50% TL length at 0.07 s and this fault after 12 cycles delay is converted into the AB-G fault at 50% at 0.3 s. Figure 6 exemplifies the HHT outputs when the A-G fault is converted into the AB-G fault. All the faults are simulated at 50% of transmission

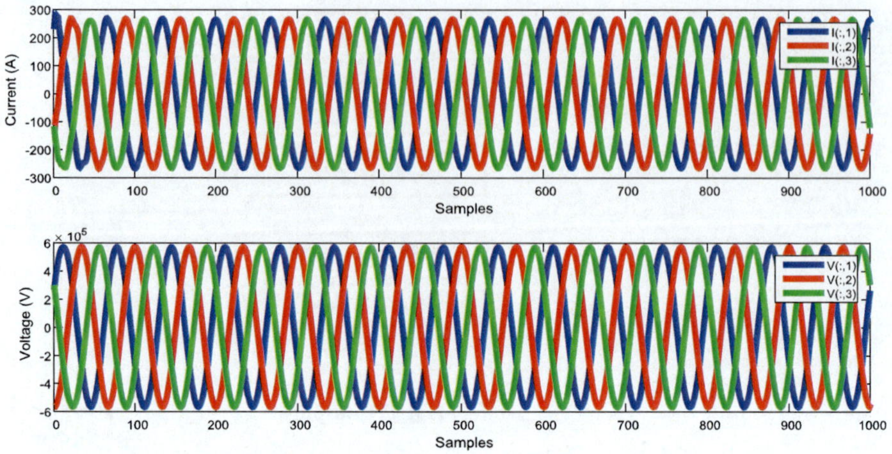

Fig. 3 The currents and voltages for no-fault situation

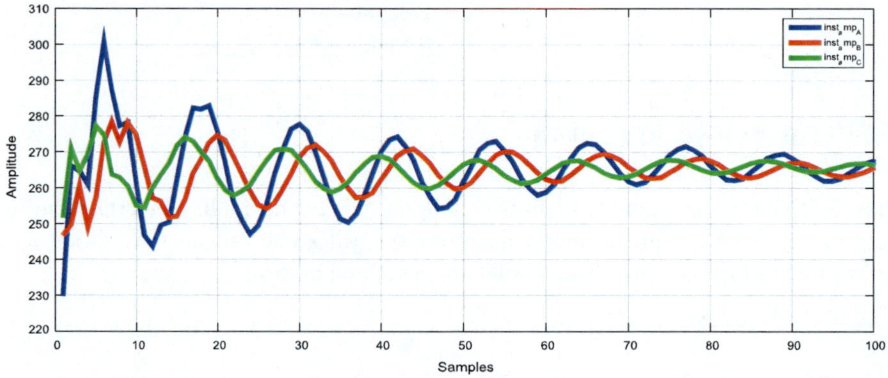

Fig. 4 HHT outputs of currents for no-fault situation

Table 1 Outcomes of HHT for no-fault

	HHT outputs of phases	
A	B	C
300.7661	278.2735	277.1177

line length. Table 2 reports the outcomes for different types of converting faults. It is evident from Table 2 that the HHT is insensitive to the converting faults.

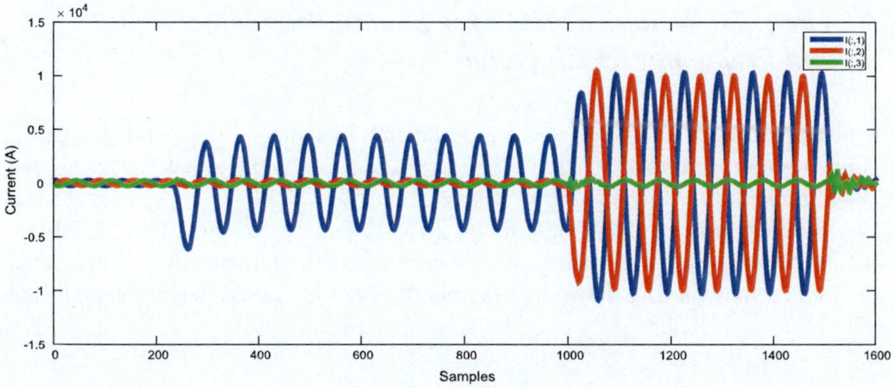

Fig. 5 The waveform when A-G fault at 50% length at 0.07 s is converted into AB-G fault at 50% length at 0.3 s

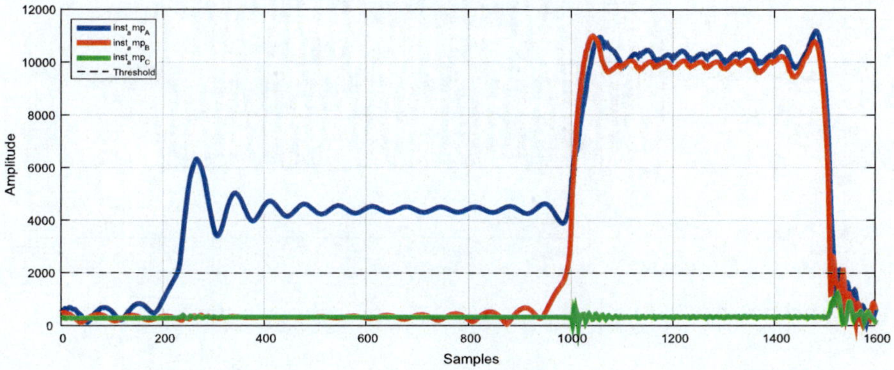

Fig. 6 HHT outputs when A-G fault at 50% length at 0.07 s is converted into AB-G fault at 50% length at 0.3 s

Table 2 Outcomes of HHT for converting faults

Fault-1	F_L (%)	F_R (Ω)	G_R (Ω)	Fault-2	HHT outputs of phases		
					A	B	C
AG (0.07 s)	50	3.5	4.5	ABG (0.3 s)	1.1161×10^4	1.0970×10^4	1.2127×10^3
ABG (0.08 s)	50	3.5	4.5	BG (0.2 s)	1.3223×10^4	1.1869×10^4	1.0893×10^3
CG (0.09 s)	50	3.5	4.5	ABCG (0.31 s)	1.6123×10^4	1.2993×10^4	1.1231×10^4
BG (0.06 s)	50	3.5	4.5	AG (0.26 s)	7.6281×10^3	6.0280×10^3	908.2038
CG (0.08 s)	50	3.5	4.5	BG (0.3 s)	1.1133×10^3	7.0938×10^3	7.5877×10^3

4.3 The Effectiveness of HHT for Source Voltage and Frequency Modification

The investigation of HHT is done for modification in source voltage and frequency. Figure 7 represents the ABC-G fault (among $F_R = 5.5\ \Omega$ and $G_R = 4.5\ \Omega$) triggered at 50% transmission line length at 0.08 s among V_S = 385-kV, F_S = 49.90-Hz. Figure 8 exemplifies the HHT outputs for ABC-G fault at 50% line length. Table 3 tabularizes the results for modification in source voltage and frequency. It is inspected from Table 3 that varying the source factors does not affect the performance of the HHT.

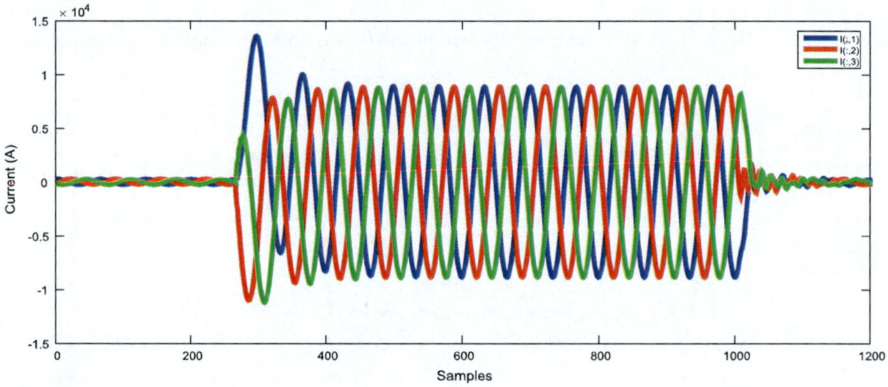

Fig. 7 ABC-G fault at 50% line length at 0.08 s among V_S = 385-kV and F_S = 49.90-Hz

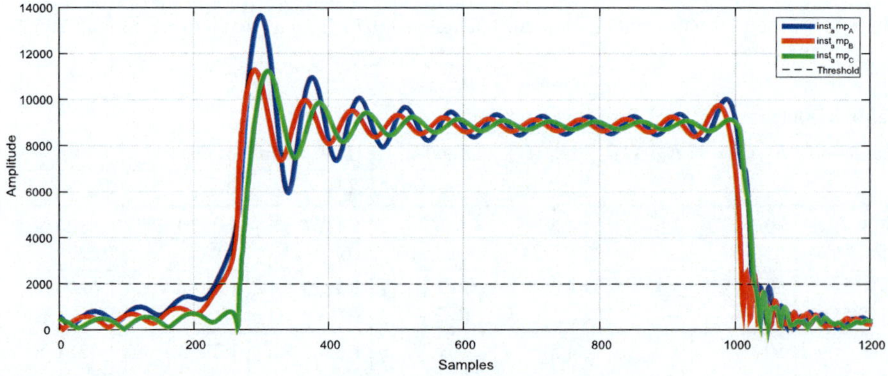

Fig. 8 HHT outputs for ABC-G fault at 50% line length at 0.08 s among V_S = 385-kV and F_S = 49.90-Hz

Table 3 Outcomes of HHT for different V_S and F_S

Fault type	V_S (kV)	F_S (Hz)	HHT outputs of phases		
			A	B	C
ABCG	385	49.90	1.3627×10^4	1.1250×10^4	1.1217×10^4
AG	395	50.10	6.0731×10^3	589.5165	672.6820
BG	410	49.85	749.0318	5.4989×10^3	678.2483
CG	390	50.45	585.9067	679.2656	5.4353×10^3
ABG	405	50.05	1.2514×10^4	1.1240×10^4	973.0243

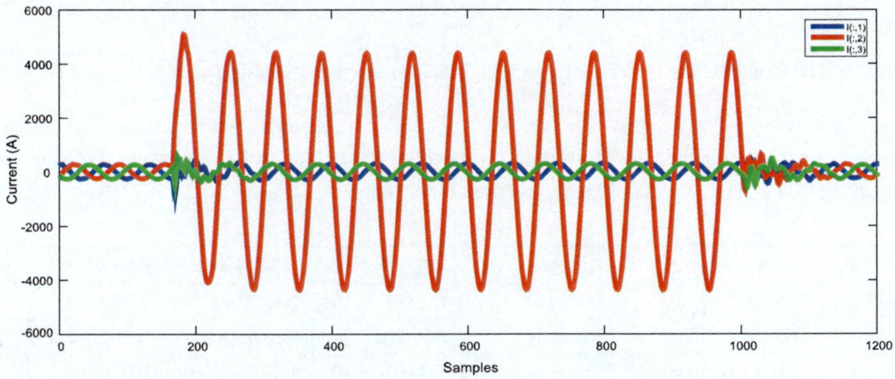

Fig. 9 B-G fault at 50% before capacitor bank among $F_R = 5.5\ \Omega$ and $G_R = 2.5\ \Omega$

4.4 The Effectiveness of HHT for Faults Around Capacitor Bank

The HHT has been inspected for varying fault position around the series capacitor bank. Figure 9 exemplifies the B-G fault ($F_R = 5.5\ \Omega$ and $G_R = 2.5\ \Omega$) before the capacitor bank at 50% at 0.05 s. Figure 10 shows the HHT outputs for the B-G fault. It is explored from Table 4 that the HHT-technique well recognizes all the faults.

4.5 The Effectiveness of HHT for Different Faults

The HHT-technique has also been investigated for different faults. Figure 11 exemplifies the AB-G fault ($F_R = 4.5\ \Omega$ and $G_R = 3.5\ \Omega$) created at 50% TL length at 0.06 s. Figure 12 depicts the HHT outputs for the AB-G at 50%. The HHT outcomes are reported in Table 5. It is explored from Table 5 that the HHT efficiently recognizes all the faults.

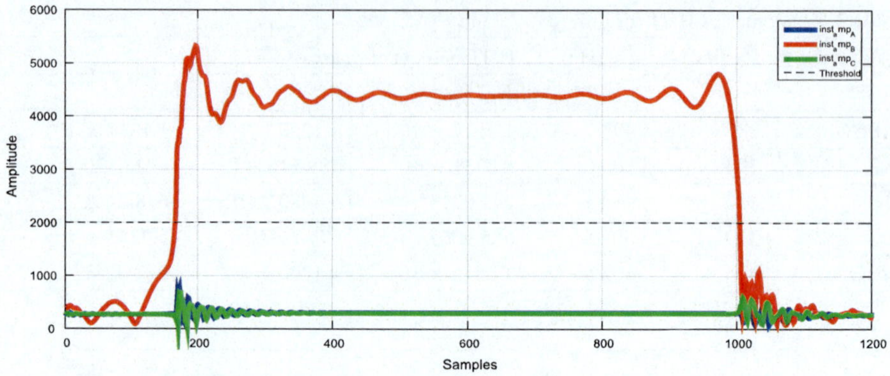

Fig. 10 HHT outputs for B-G fault at 50% before capacitor bank among $F_R = 5.5\ \Omega$ and $G_R = 2.5\ \Omega$

Table 4 Outcomes of HHT for faults around capacitor bank in forward and in reverse direction

Fault type	F_R (Ω)	G_R (Ω)	Fault direction	HHT outputs of phases		
				A	B	C
BG	5.5	2.5	Reverse	762.1567	5.3230×10^3	676.9088
AG	5.5	2.5	Forward	6.1373×10^3	637.2197	737.4434
CG	5.5	2.5	Reverse	616.8017	762.9509	5.5020×10^3
ABG	5.5	2.5	Forward	1.2391×10^4	1.1198×10^4	930.6960
ABCG	5.5	2.5	Reverse	1.4143×10^4	1.1429×10^4	5.6249×10^4

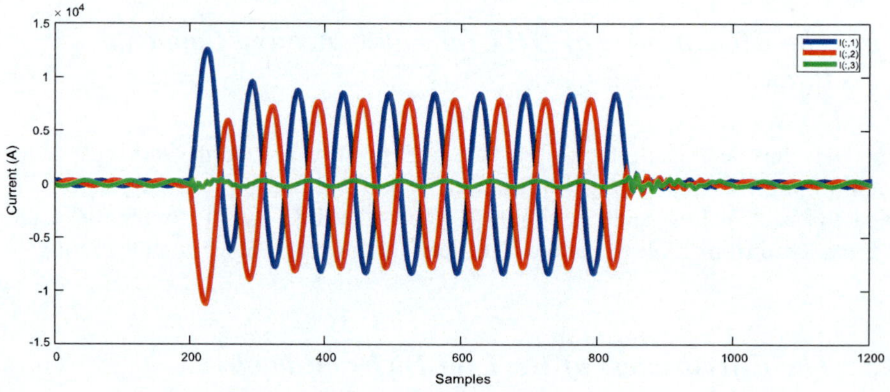

Fig. 11 AB-G fault at 50% at 0.06 s among $F_R = 4.5\ \Omega$ and $G_R = 3.5\ \Omega$

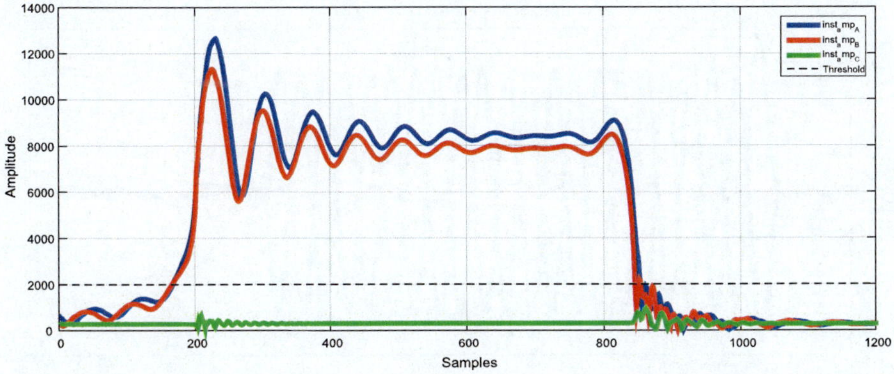

Fig. 12 HHT outputs for AB-G fault at 50% at 0.06 s among $F_R = 4.5\ \Omega$ and $G_R = 3.5\ \Omega$

Table 5 Outcomes of HHT for different faults

Fault	F_L (%)	FST (S)	F_R (Ω)	G_R (Ω)	HHT outputs of phases		
					A	B	C
ABG	50	0.06	4.5	3.5	1.2643×10^4	1.1301×10^4	948.7518
ACG	50	0.06	4.5	3.5	1.2214×10^4	949.9702	1.2344×10^4
BCG	50	0.06	4.5	3.5	1.1802×10^3	9.2824×10^3	8.9673×10^3
CG	50	0.06	4.5	3.5	595.9531	747.5821	5.5686×10^3
AG	50	0.06	4.5	3.5	6.3153×10^3	585.5229	668.3924

4.6 The Effectiveness of HHT for $\mathbf{F}_R$ (Fault Resistance)-Variation

The HHT has been inspected for varying F_R. Figure 13 exemplifies the AC-G fault ($F_R = 15\ \Omega$ and $G_R = 4\ \Omega$) at 50% TL length at 0.0315 s. Figure 14 shows the HHT outputs for the AC-G fault. The HHT outcomes for different F_R's are reported in Table 6. It is explored from Table 6 that the HHT successfully recognizes all the faults.

4.7 The Effectiveness of HHT for Variation in Wind-Turbine Units

The HHT-technique is tested for modification in the wind-turbine units of the wind farm. Figure 15 exemplifies the BC-G fault ($F_R = 7\ \Omega$ and $G_R = 3\ \Omega$) switched at 50% TL length at 0.09 s among three wind-turbine units. Figure 16 shows the HHT outputs for the BC-G fault. It is examined from Table 7 that modifications of the wind-turbine units with different faults does not affect the response of HHT-technique.

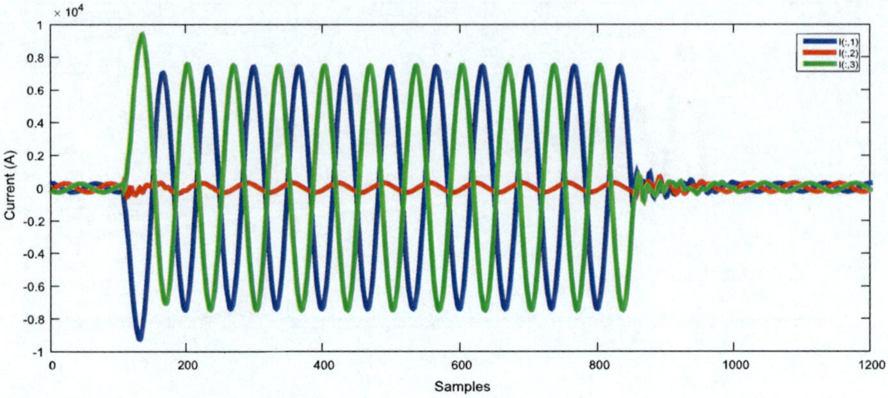

Fig. 13 AC-G fault at 50% length at 0.0315 s among $F_{\mathrm{R}} = 15\ \Omega$ and $G_{\mathrm{R}} = 4\ \Omega$

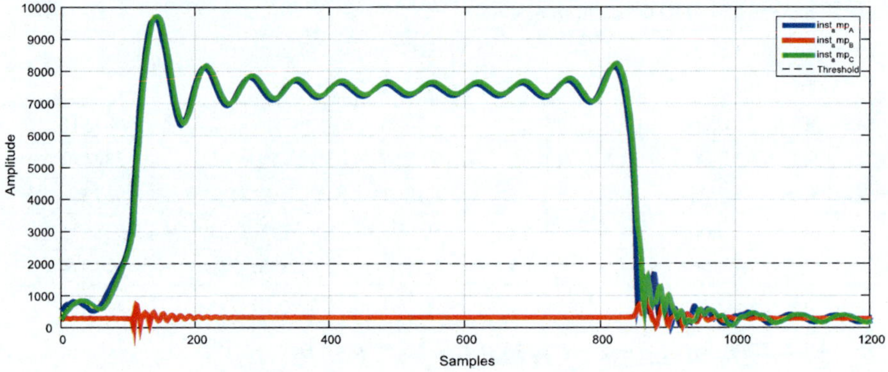

Fig. 14 HHT outputs for AC-G fault at 50% length at 0.0315 s among $F_{\mathrm{R}} = 15\ \Omega$ and $G_{\mathrm{R}} = 4\ \Omega$

Table 6 Outcomes of HHT for faults among different F_{R}

Fault	F_{L} (%)	FST (S)	F_{R} (Ω)	G_{R} (Ω)	HHT outputs of phases		
					A	B	C
ACG	50	0.0315	15	4	9.6562×10^3	839.5712	9.7095×10^3
AG	50	0.0315	35	4	4.3713×10^3	528.1566	618.2553
ABG	50	0.0315	55	4	4.5397×10^3	4.7716×10^3	645.6247
BG	50	0.0315	75	4	722.3199	3.0190×10^3	629.8424
ACG	50	0.0315	105	4	3.2232×10^3	615.5234	2.7448×10^3

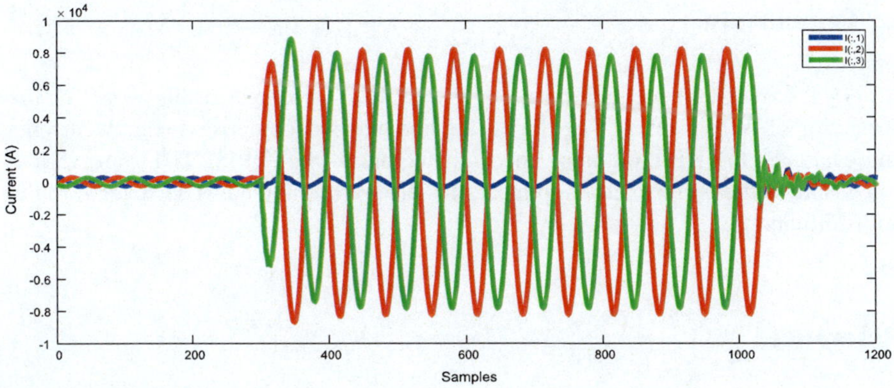

Fig. 15 BC-G fault at 50% length at 0.09 s among three wind-turbine units

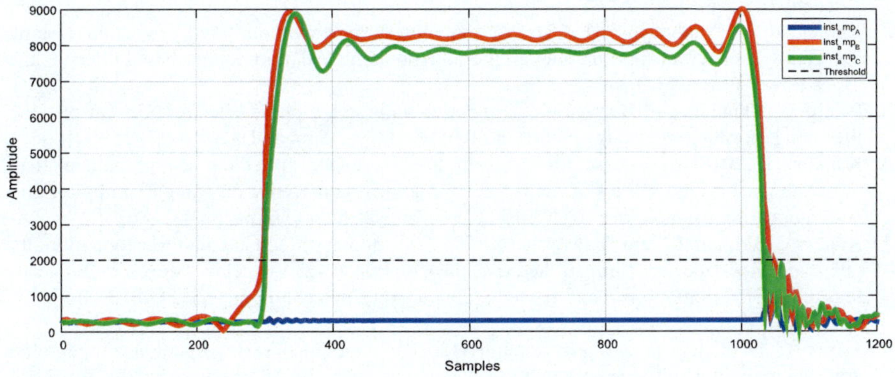

Fig. 16 HHT outputs for BC-G fault at 50% length at 0.09 s among three wind-turbine units

Table 7 Outcomes of HHT for variation in the wind-turbine units

Fault	WT units	F_L (%)	HHT outputs of phases		
			A	B	C
BCG	3	50	963.5365	8.9985×10^3	8.8759×10^3
ACG	4	50	1.1465×10^4	920.9843	1.1413×10^4
AG	1	50	6.1504×10^3	597.1395	681.6256
BG	5	50	777.2727	5.1156×10^3	687.9359
CG	2	50	595.7065	743.9358	5.4592×10^3

5 Conclusion

An HHT based fault recognition method has been presented in this work for the protection of WPISCTPTL. The HHT is applied which decomposes the fault currents and evaluates the HHT outputs. The fault factors of the WPISCTPTL are varied. According to the performance appraisal, it is discovered that the HHT detects all the faults efficiently.

References

1. Kapoor G (2019) Wavelet transform based detection and classification of multilocation double line to ground faults in twelve phase series capacitor compensated transmission line. In: Proceedings of the 5th international conference for convergence in technology (I2CT). IEEE, Bombay, India, pp 1–7
2. Khodadadi M, Shahrtash SM (2013) A new non communication-based protection scheme for three-terminal transmission lines employing mathematical morphology-based filters. IEEE Trans Power Deliv 28(1):347–356
3. Salehi M, Namdari F (2018) Fault classification and faulted phase selection for transmission line using morphological edge detection filter. IET Gener Transm Distrib 12(7):1595–1605
4. Kapoor G (2018) Six phase transmission line boundary protection using mathematical morphology. In: Proceedings of the IEEE international conference on computing, power and communication technologies (GUCON). IEEE, Greater Noida, India, pp 857–861
5. Kapoor G, Tripathi S, Jain G, jayaswal K (2019) Detection of fault and identification of faulty phase in series capacitor compensated transmission line using wavelet transform. In: Proceedings of the 5th international conference for convergence in technology (I2CT). IEEE, Bombay, India, pp 1–8
6. Kapoor G, Gautam N, jayaswal K, Tripathi S (2019) Protection of series capacitor compensated double circuit transmission line using wavelet transform. In: Proceedings of the IEEE 5th international conference for convergence in technology (I2CT). IEEE. Bombay, India, pp 1–8
7. Kapoor G, Merotha M, Shrivastava S, Mukherjee D (2019) Hilbert Huang transform based protection technique for DSTATCOM compensated transmission line. In: Proceedings of the 5th international conference for convergence in technology (I2CT). IEEE, Bombay, India, pp 1–7
8. Ashok V, Yadav A (2019) A real-time fault detection and classification algorithm for transmission line faults based on MODWT during power swing. Int Trans Electr Energy Syst 30(1):1–27
9. Manohar M, Koley E, Ghosh S (2018) Microgrid protection under wind speed intermittency using extreme learning machine. Comput Electr Eng 72:369–382
10. Kumar D, Bhowmik PS (2018) Artificial neural network and phasor data-based islanding detection in smart grid. IET Gener, Transm Distrib 12(21):5843–5850

Simulation of Five-Channel De-multiplexer Using Double-Ring Resonator Photonic Crystal-Based ADF

Neha Singh and Krishna Chandra Roy

Abstract Photonic crystal has equally influenced the light propagation as the atom of crystal has on electrons. The optical filters are one of the remarkable applications of the photonic crystal. In the proposed paper an add/drop filter has been designed using two ring resonators. The shape of the resonators is a curved edge shape. The design layout has been performed using Opti-FDTD and the simulation results show 100% of add efficiency and 90% of drop efficiency. In addition to this MATLAB software has been also used and with further enhancement, a five-channel de-multiplexer has been proposed with a transmitted efficiency of more than 97% and the average quality factor of 781 has been determined.

Keywords Photonic crystal · Ring resonator · Add/drop filter · Finite difference time-domain method · Dimensional

1 Introduction

The scientific knowledge of semiconductor has been an influencing material from the last 50 years and has played a major role in most of the application which is related to day to day life. In the entire world of technology, the compact size and speedy performance of electronic circuits based on the integration mechanism has remarkably accelerated the research achievement. However, the resistance of a circuit increases along with a deficiency in power whenever the compact size of the circuits is considered. In addition to this when immense speed is considered then it will result in more sensitivity to the synchronization of the signal [1].

N. Singh (✉)
Electronics and Communication Jaipur Engineering College & Research Centre, Jaipur, India
e-mail: nehasingh.ece@jecrc.ac.in

K. C. Roy
Electrical and Electronics Engineering, Kautilya Institute of Technology and Engineering, Jaipur, India
e-mail: dr.kcroy@gmail.com

K. Ray et al. (eds.), *Proceedings of International Conference on Data Science and Applications*, Lecture Notes in Networks and Systems 148,
https://doi.org/10.1007/978-981-15-7561-7_13

In recent times to transmit the information, the researchers have inclined toward the light rather than an electron in order to enhance great density in the integration and performance of the system. The reason behind it is that there are various advantages of light as compared to electron such as, it can transit in a dielectric material at a more high speed than electron which travels in a metallic wire, light also has the capability to carry huge amount of data per second and the bandwidth of dielectric material is more in contrast to metals. Besides this, the particles of light commonly known as photons are not able to interact in a strong manner as compare to electrons and this, in turn, decrease the energy losses.

The devices constructed with photonics have the capability to minimize the power loss by appropriate choosing the medium and thus the light coherence is maintained through traveling [2]. The material is known as 'crystal' as it created by a repeated adjustment of fundamental building block as shown in Fig. 1. The word photonic has been used along with crystal as the photonic crystal is created to influence the propagation characteristic of photons.

The photonic crystal is a usual array of materials that consists of a variant refractive index. The photonic crystal is categorized according to the dimensions as one-dimensional, two-dimensional and three-dimensional photonic crystals.

The one-dimensional crystal is the uncomplicated and the most convenient photonic crystal that can be constructed. This type of crystal is composed of two layers of dielectric material which are variant nature and the perfect mirror as the best-suited example of this dimensional crystal.

The two-dimensional crystal is composed of a many square lattice array of parallel rods which are dielectric in nature. In this type of dimension, the periodicity lies along two axes and homogeneity lies along the third side. Super-prism is the best example of this category.

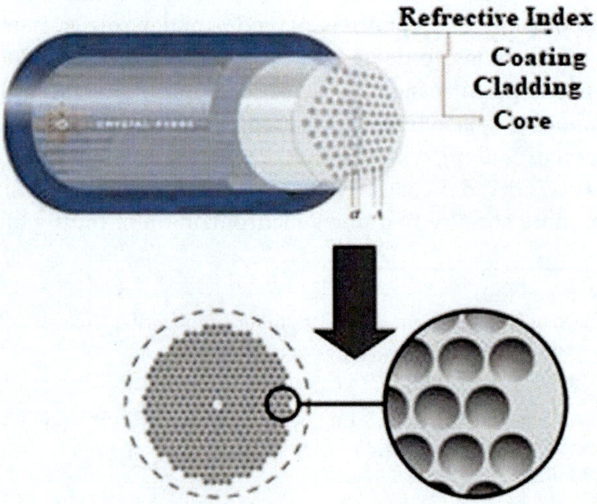

Fig. 1 Solid core fiber based on photonic crystal [2]

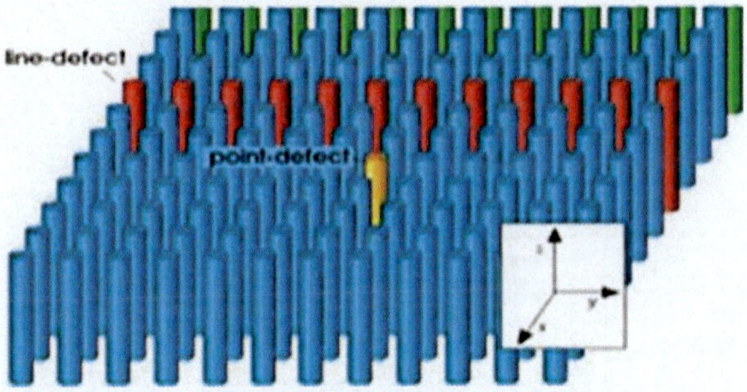

Fig. 2 Defects in photonic crystal

The three-dimension photonic crystal is constructed in order to create structures that have the ability to emit in spontaneous manner. The best example of this type is the inverse opal which displays a range of frequency above which the linear traveling is restricted [3].

The photonic devices have shown remarkable development as they have the advantage to confine, control and direct the light on a nanometer scale and therefore in recent time circuit based on photonics instead of modern circuits based on electronics are more in demand. In addition to these numerous varieties of devices can be constructed by using the photonic crystal as the prime material.

The photonic crystal makes use of appropriate material, like silicon in order to enhance the properties of the device on which it has to be embedded. In order to create a discontinuity for the propagation and controlling of light within the band gap, the line and point defects can be introduced inside the structure of photonic crystal as shown in Fig. 2.

In today's era, optical filters are used often for communication purposes, alike conventional filter, as the former one has the ability to send the certain light in a certain spectrum while blocking the remaining. There are two types of optical filters which are commonly used, namely absorptive and dichroic filters. The former blocks the light depending upon the glass substrate's absorption properties. While the latter filter transfers the particular wavelength while rejecting the undesired light by using the principle of interference [4].

A Channel drop filter is one of the significant applications in the field of optical filters. Besides this, the frequency response, which describes the manipulation of a coming signal related to magnitude and phase is important parameter to be considered as the characteristics of optics are depending upon this feature. In order to build an add/drop filter, the most essential requirement is to have the ring resonator and any shape of ring can be used to construct add/drop filter by using different types of ring resonators like the ring resonator square ring, hexagonal ring, quasi-ring, diamond-shaped ring [5–7].

Considering various types of ring resonators, in the proposed paper a curved edge type ring resonator, has been created, and with this ring resonator a channel drop filter has been designed in which dual-ring resonator has been used. Figure 3 shows the

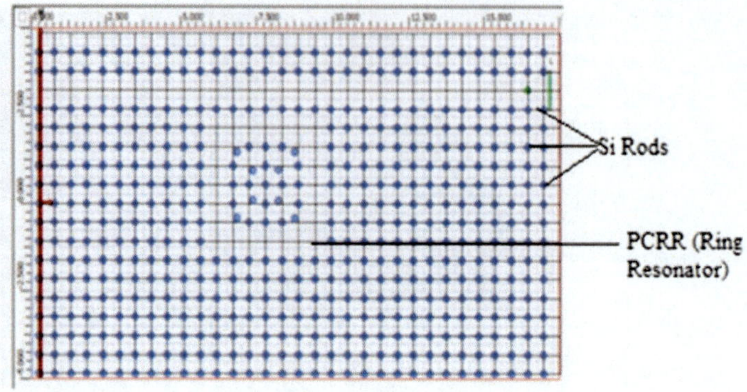

Fig. 3 Underlying structure of ring resonator

design layout of a basic ring resonator with the help of which the dual-ring resonator has been created [8].

The arrangement of paper is as: In Sect. 2, the literature review has been described, Sect. 3 consists of layout design of double-ring resonator, and the simulation results have been shown in Sect. 4.

2 Literature Review

Banaei et al. [9] designed a drop filter in the shape of '*T*' and also the arrangement of the resonator in the shape of ring has also been done simply by situating the crystal as 12-fold quasi in the center of a cavity which is in the form of square having size 7 * 7. The proposed paper has obtained approximate dropping efficiency of 90% at a central wavelength of 1551 nm and also the QF of the designed structure is 387.

Birjandi et al. [10] surveyed many papers in which the contribution of photonic crystal has demonstrated remarkable achievement in the commercial market where excessive immediate acknowledgment, consuming less energy in terms of optics and capability to integrate communication signals are crucial gateways for switching and other types of equipments related to this technology. Therefore in order to enhance more in photonics, the authors of the proposed paper have constructed add/drop filter which basically is a tunable filter based on optical technology.

Rashki et al. [11] demonstrated a channel drop filter based on ring-shaped resonator in photonics which is composed with 2D behavior. The filter consisting of optical technology is the most crucial device in the field of the communication system. Besides this, the filters known as channel drop filters are the most required filters in order to design integrated circuits and wavelength division multiplexers as well in a communication system for optical technology. The filter basically is a device which possesses characteristic to add or drop a specific wavelength of a channel from various wavelengths transmitted in a signal.

Robinson et al. [12] Designed add and drop filter which is based on a photonic crystal ring resonator having the 2D characteristics. The circular-shaped rods have been used by the authors for the designing of structure. The proposed designed has been made for the ITU-T G.694.2 consisting of eight number of channels in a CWDM network. The structure has been made to add or drop a channel at a central wavelength of 1491 nm. The observation of 99% of drop efficiency has been seen at the refereed wavelength. And the quality factor of the structure came out to be 114.69.

Venkatachalam et al. [13] proposed a paper on wavelength division demultiplexer consists of four channels consisting of wavelength 1477, 1482, 1487, and 1492 nm and the designing is based on photonic crystal having two-dimensional nature. In the proposed paper the dropping of every channel is done by changing the radii of the silicon rods and also the size of every cavity corresponding to the ring shape resonator. The two important techniques such as PWE and FDTD have been used in order to evaluate the value of the band gap of photonic crystal along with the spectral output for the suggested structure.

On the substructure of the above review of literature, our proposed paper demonstrated the structure of dual-ring resonator ADF, which has been designed on the silicon platform.

3 Design Layout of ADF Using Double-Ring Resonator

In the proposed paper design of ADF has been shown using two ring resonators and for the design, the silicon has been taken as prime material. The no of silicon rods that are suspended in air is approximate 20 * 34. The r is denoted as the radius for rod and it has been taken 0.110 μm and the value for lattice constant, $a = 540$ nm and is 3.59 is then taken as the dielectric constant of the proposed structure. The input optical wavelength, $\lambda = 1.562$ μm. The width is 16.5*10 μm^2 and 1.82 μm is the thickness of the proposed design. The use of defects like line and point has been introduced within the structure for the traveling of light inside the structure of the designed photonic crystal.

The first proposed design exhibit a basic structure of ring resonator, as shown in Fig. 3, and this basic triple-layer ring resonator is coupled with linear bus waveguide. The resonator in a photonic crystal can be constructed by introducing irregularity either by modifying the size of every rod or by manipulating their dielectric constant and these irregularities is used to propagate the light which has the frequency equivalent to the frequency of the defected region.

The ADF based on dual-ring concept has been designed in order to arrange double filtering of the propagating wave, and in addition to this, the filter is capable to drop two wavelengths. Figure 4 shows the ADF comprising of four ports: Port A is named as an input port, port B is termed as transmitting port, and port C and port D are named as forward and dropping ports, respectively. The figure shows two types of waveguide known as bus and drop waveguide. These waveguides are coupled with the help of two rings. To prevent the scattering of the wave the eight scattering rods

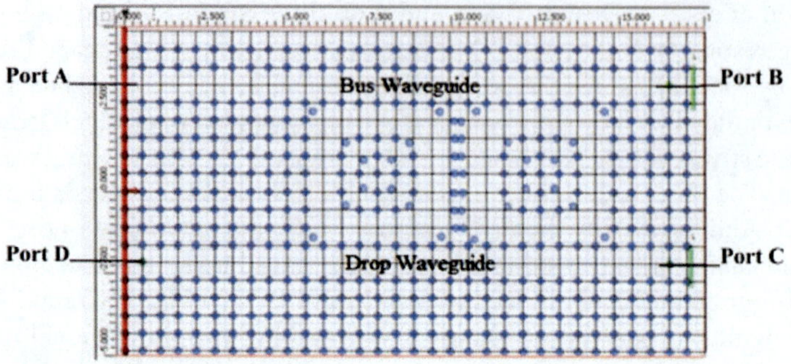

Fig. 4 Layout of double-ring resonator ADF

have been provided at each corner of the ring resonators and also for the coupling 14 silicon rods have been used between the two resonators. In a designed structure, through port A wave moves inside the structure and departs through port B. Although at a certain wavelength the input wave descends at drop waveguide and moves toward C port. As mentioned earlier the input source adjusted at 1562 nm in order to maintain the propagation and coupling of wave with the ring resonator.

4 Simulation of Dual-Ring Resonator ADF Using Opti-FDTD and MATLAB

After designing the structures, the simulation of the respective designs has been performed by using two methods, i.e., plane-wave expansion method for band gap calculation and for field distribution, finite difference time-domain method. The Field Distribution of the entire design layout has been carried out in the form of OFF resonant and ON resonant wavelength. The former one is also known as add wavelength and is denoted by λ_{OFF}, while the latter is known as drop wavelength and is denoted by λ_{ON}. The add wavelength basically represents the wavelength that is propagated inside the band gap of the resonator from the input side as shown in Fig. 5 and the drop wavelength is that wavelength which is obtained at the output of the resonator and through which various devices can be operated as shown in Fig. 6.

The above all simulation results have been obtained using Opti-FDTD software and the results show that the double-ring resonator is far better. Therefore in this part, further enhancement has been done and for this, the double resonator has been simulated on three-dimension photonic crystal using MATLAB as different software. Besides this, five ADF with dual-ring resonators has been placed in a photonic crystal which has been suggested as a five-channel de-multiplexer and the CWDM technique has been used in order to design the proposed structure providing five different output wavelengths with a spacing of approximate 20 nm.

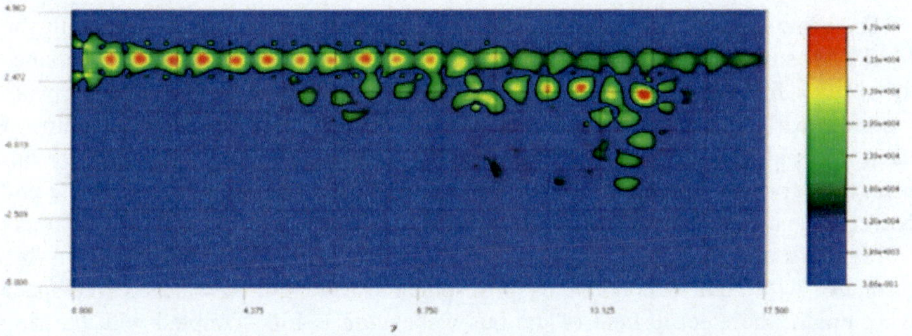

Fig. 5 Field distribution of DRR ADF at 1554 nm

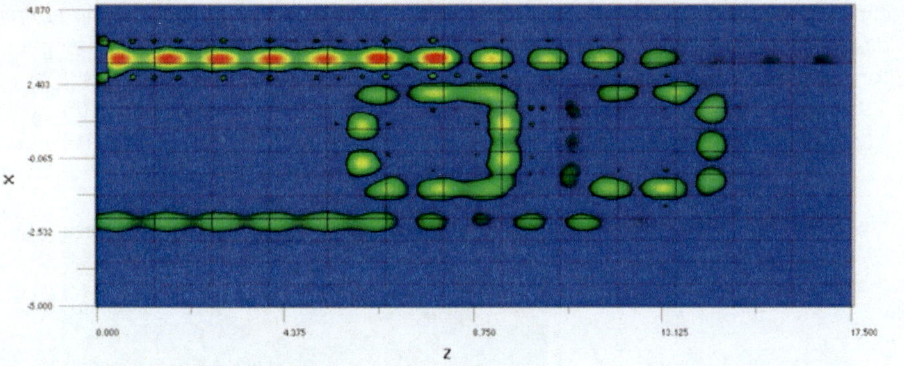

Fig. 6 Field distribution of DRR ADF at 1585 nm

In the diagram, the cascade method is used to arrange dual rings. The wavelength of a particular channel is determined by modifying the inner rod radius of every dual ring resonator. The size and thickness of the device have been taken $22.5 \times 10\ \mu m^2$ and $2\ \mu m$ respectively. The inner rod radius of the first channel (λ_1) is taken as 96 nm and consecutively for the other four channels the radius of the inner rod has been taken with 2 nm step size such as

$$(r, r \pm \Delta r, r + 2\Delta r, r + 3\Delta r, r + 4\Delta r)^{+}5\%\ \text{tolerance}$$

Therefore, second channel (λ_2), third (λ_3), fourth (λ_4), and fifth (λ_5) have 98 nm, 100 nm, 102 nm and 104 nm radius for inner rod respectively. In the same manner, the radius of the scattering rod for the first dual ring resonator has been taken 146 nm and for all other, the radius differs by 2 nm. Thus from second to the fifth channel, the scattering rod radius is 148–154 nm and also the size of the gap in every ring resonator is varied from 1.032 μm for the first channel to 1.024 μm for the fifth channel.

The simulation of the proposed de-multiplexer has been carried out with the MATLAB software and the 3-D photonic crystal has been used in order to enhance the results. In contrast to MATLAB for 3-D simulation, the Opti-FDTD software requires extra time for simulation as compared to 2-D and this huge amount of memory is also needed for high preciseness. Thence the MATLAB software has been used for the simulation of 3-D photonic crystal as it is one of the flexible and power tools used in various applications of science, engineering, and mathematics.

Figures 7 and 8, shows the electric field pattern of pass region and stop Ez in ring resonator with ADF ML boundary at resonant wavelength, $\lambda = 1.583$ free-space wavelength, the electric field of the bus waveguide is fully coupled with the ring and reached to one of its output port or in other words the signal directly reached to transmission terminal. In addition, the resonant wavelength can be shifted by varying the dielectric constant of the structure and also the increased resonant wavelength can be shifted linearly into appropriate values, which can be added or drop at the desired WDM channel for inner core radius ra = 96 nm. The same simulation goes for the other four channels.

Figure 9 represents the transmission spectrum of five-channel de-multiplexer. The graph has been plotted between the inner rod radius of the entire five dual

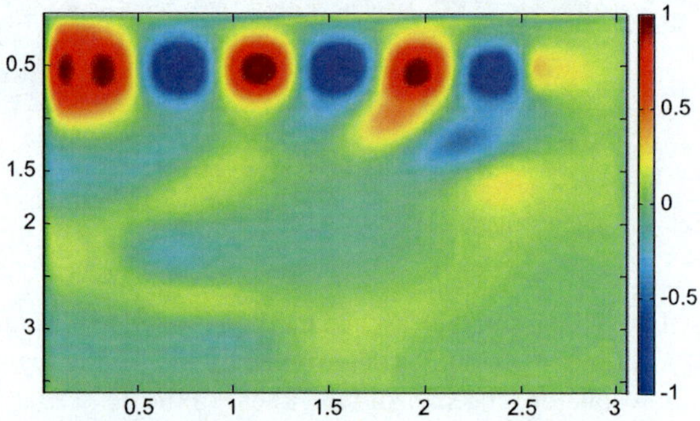

Fig. 7 Initial color scaled image plot for Ez in ring resonator

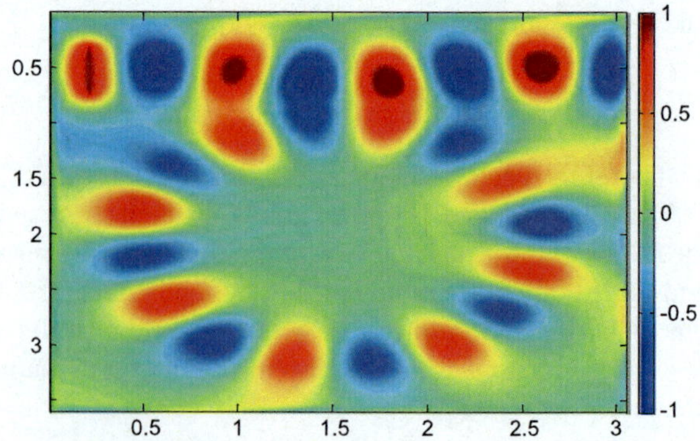

Fig. 8 After color scaled image plot for Ez in ring resonator

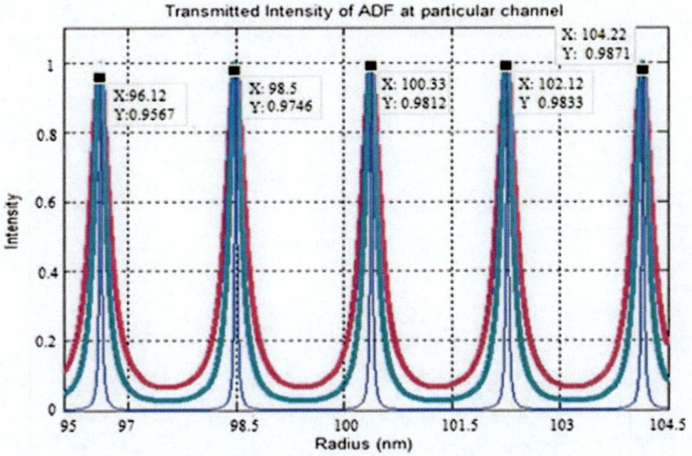

Fig. 9 Transmitted intensity of proposed ADF

ring resonators and the intensity to transmit the wavelength at the output port. The different intensity corresponding to different radius has been marked in the graph. The average transmission efficiency of the proposed de-multiplexer is nearly about 98% and the average quality factor of the system is 781.

The MATLAB tool calculated the desired output wavelength for each channels with help of equation written above, like for channel 1 the obtained wavelength is 1523.2 nm and for channel 2, 3, 4 and 5 the calculated wavelength is 1542.6 nm, 1563.1 nm, 1582.9 nm, and 1602.6 nm respectively. Figure 10 shows the graph which has been frame between the desired output wavelength in microns or micrometer and inner rod radius in nm. The graph depicts that with the change in inner rod radius there is a corresponding change in the output wavelength and Table 1 represents the resonant wavelength, transmission efficiency and quality factor for every channel.

5 Conclusion

The present work is imperative as the Add/Drop filters designed from photonic crystal is an influencing applicant in the area of filtering and also it has the capability to provide good response as it is composed of resonating modes of various shape and sizes. In addition to this, it is more advantageous to as compare another type of optical filter because their size is not applicable to photonic integrated circuits.

The design layout of dual ring resonator type ADF has been constructed with the help of a basic optical ring resonator. The shape of the ring resonator has been taken in the form of a curved edging structure with triple layers in a square lattice. The PWE band gap solver has been used to calculate the band gap region and the transverse band gap lies from 1255 to 1656 nm.

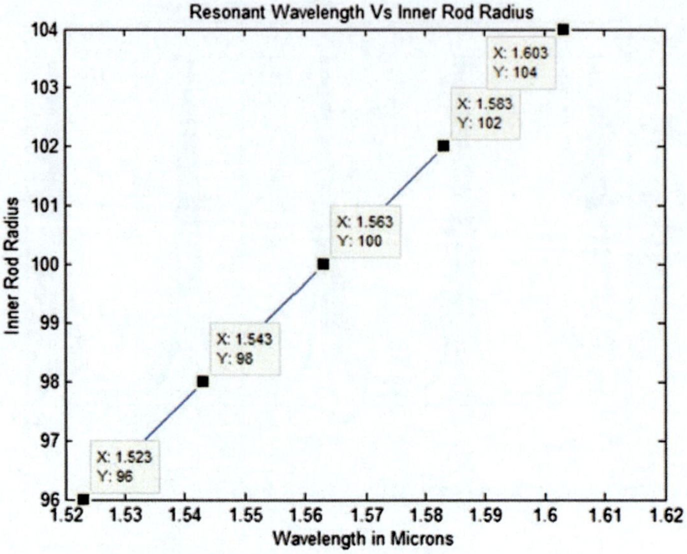

Fig. 10 Resonant wavelength for particular inner rod radius

Table 1 Resonant wavelength, transmission efficiency and quality factor for eve channel

Number of channels	Resonant wavelength (λ_0) in nm	Transmission efficiency (%)	Quality factor (QF)
Channel 1	1523.2	95.67	761
Channel 2	1542.6	97.46	771
Channel 3	1563.1	98.12	781
Channel 4	1582.9	98.33	791
Channel 5	1602.6	98.71	801

The further enhancement of ADF with dual ring resonator have been done based on 3D photonic crystal and simulated using MATLAB as different software. The five ADFs with dual ring resonators have been placed in a photonic crystal which has been suggested as a five-channel de-multiplexer and the CWDM technique has been used in order to design the proposed structure providing five different output wavelengths with a spacing of approximate 20 nm. In addition to this, the silicon nitride rods are used with air rather than Silicon rods. The transmission spectrum of five-channel de-multiplexer has been obtained and analyzed between the inner rod radius of the entire five dual ring resonators and the intensity to transmit the wavelength at the output port. The average transmission efficiency of the proposed de-multiplexer came out to be nearly about 98%. The calculated desired output wavelength for each channel with help MATLAB for channel 1 is 1523.2 nm and for channel 2, 3, 4 and 5 the calculated wavelength is 1542.6 nm, 1563.1 nm, 1582.9 nm, and 1602.4 nm respectively and the average quality factor of the system is 781.

References

1. Joannopoulos JD, Villeneuve PR, Fan S (1997) Photonic crystal: putting a new twist in light. Nature 386:143–146
2. Bowden CM, Dowling JP, Everitt HO (1993) Special issue on development and application of materials exhibiting photonic band gaps. J Opt Soc Am B 1:280–408
3. Merkis A, Chen JC, Kurland I, Fan S, Villeneuve PR, Joannopoulos JD (1996) High transmission through sharp bends in photonic crystal waveguides. Phys Rev Lett 77:3787–3790
4. Shih TT, Wu YD, Lee JJ (2009) Proposal for compact optical triplexer filter using 2-D photonic crystal. IEEE Photon Technol Lett 2:18–21
5. Mai TT, Hsiao FL, Lee C, Xiang W, Chen CC, Choi WK (2011) Optimization and comparison of photonic crystal resonator for silicon micro cantilever sensors. Sens Actuators A: Phys 16:16–25
6. Li L, Liu GQ, Lua HD, Branch De AY(2013) Multiplexer based on kagome lattice photonic crystals. Opt- Int J Light Electron Opt 124(17):2913–2915
7. Saranya T, Robinson S, Vijaya Shanthi K (2014) Design and simulation of two dimensional photonic crystal ring resonator based four port wavelength de-multiplexer. Int J Innov Sci Eng Technol 1(2):255–261
8. Singh N, Roy KC (2016) Designing of photonic crystal ring resonator based ADF filter for ITUT- G.694.2 CWDM system. In: SMARTCOM CCIS, vol 628. Springer Nature Singapore, pp 39–46
9. Banaei HA, Jahanara M, Mehdizadeh F (2014) T-shaped channel drop filter based on photonic crystal ring resonator. Optik Int J Light Electron Opt 125(18):5348–5351
10. Birjandi MAM, Tavousi A, Ghadrdan M (2016) Full-optical tunable add/drop filter based on nonlinear photonic crystal ring resonators. Photonic Nanostruct Fundam Appl 21:44–51
11. Rashki Z, Javed S, Chabok SM (2016) Novel design of optical channel drop filters based on two dimensional photonic crystal ring resonator. Opt Commun 395:231–235
12. Robinson S, Nakkeeran R (2013) Two dimensional photonic crystal ring resonator based add drop filter for CWDM systems. Optik 124:3430–3435
13. Venkatachalam K, Robinson S, Kumar DS (2017) Design and analysis of dual ring resonator based 2D-photonic crystal WDDM. AIP 1849:020016(1)–020016(6)

Optimization of Surface Roughness and Material Removal Rate in Turning of AISI D2 Steel with Coated Carbide Inserts

Anil Kumar Yadav and Bhasker Shrivastava

Abstract In metal cutting industry, the main challenges during turning processes is to obtain high amount of productivity & achieve good qualities of machined part nowadays. In the turning process, optimization is considered to be an important function for the continuous development of the final output qualities in methods or products. This includes the determination for input & output modeling, create relation between process functions & obtain optimal condition for cutting. Attempts have been made to find importance of the insert in turning process which is coated carbide, basically used for machining hard material AISI D2 steel. In this study, we use ANOVA analysis, and this analysis mainly examines influence of machining parameter that are (feed, depth of cut & cutting speed) on roughness parameters (Ra and Rz) & MRR. The experiment is performed on Taguchi L9 OA design. Experimental values are examined to find the range for the surface roughness parameters & MRR.

Keywords AISI D2 steel · ANOVA analysis · Taguchi · Surface roughness · MRR · Coated carbide inserts

1 Introduction

As we know, turning operation is one of the machining process that is widely used in manufacturing companies which basically deals in the process of metal cutting. Main obstacle in the metal cutting and machining industry is to obtain better quality in terms of good surface finish, better accuracy of work piece, high productivity, less wear and tear of cutting tool & achieve better product function, but have fewer effects on environment [1, 2]. Traditionally, the grinding process is used due to hard

A. K. Yadav · B. Shrivastava (✉)
Department of Mechanical Engineering, Yagyavalkya Institute of Technology, Jaipur, India
e-mail: mechbhasker@gmail.com

A. K. Yadav
e-mail: anilkumarout@gmail.com

K. Ray et al. (eds.), *Proceedings of International Conference on Data Science and Applications*, Lecture Notes in Networks and Systems 148,
https://doi.org/10.1007/978-981-15-7561-7_14

metals for machining because it has good mechanical properties, but critical time is needed for the operation of grinding. During the turning, we use cutting tool for removing excess material from workpiece which does not affect surface quality [3, 4]. In this process, selection of effective tool determines the optimum parameter and has good tool geometry for obtaining desired surface finish as compared with the surface obtained in grinding process [5]. The mathematical method and design of experiment (DOE) are used for determine surface quality. For achieving effective and targeted conclusion, the DOE method is planned with suitable data's. After that the Taguchi design, ANOVA method, response surface methodology (RSM), is performed which requires lot of time and is very expensive [6].

Objectives and Approaches for Research

- During turning operation, determine the effect of cutting parameter on surface roughness & MRR by coated carbide inserts on AISI D2 steel by executing the experiment.
- To determine the optimum value for surface roughness parameter & MRR when the turning process is performed on AISI D2 steel at the 95% of CI level.

To achieve the objectives, the work is done through following approaches

- To create a experimental setup with the various input parameters in turning and then examine their responses.
- Taguchi's OA is used to examine impact of machining factors. ANOVA method is performed to analyze the % contribution and importance of all the processing parameters.
- To know the impact of machining factors on surface roughness & MRR through main effects plot, normal probability graph, and contour plot.

2 Experimentation Details

Work Materials: The workpiece material we use is D2 steel which is in cylindrical shape and have the dimensions of 80 mm length & 30 mm diameter. Since this D2 steel material has wide applications in manufacturing industries.

Typical applications for D2 steel: Burnishing roll, gauges, shear blades, drawing dies, thread rolling, forging, molding, and blanking.

The chemical composition for AISI D2 steel is given in Table 1.

Experimental procedure

Experiments are performed to determine the effect of machining parameters for the roughness parameters (Ra, Rz) & MRR. Experiments are carried out with the help of three parameters which are mainly divided into three levels, which are given in Table 2. Experiments are basically planned to execute with Taguchi L_9 (3^3) orthogonal

Table 1 Chemical composition for AISI D2 steel (%)

S. no.	Metal	Contribution (%)
1	Iron (Fe)	71.58
2	Carbon (C)	19.57
3	Chromium (Cr)	8.20
4	Silicon (Si)	0.66

Table 2 Cutting parameters and levels

Parameters	Unit	Different levels		
		1	2	3
Cutting speed (*v*)	m/min	180	210,240	
Feed (*f*)	mm/rev	0.06	0.09	0.12
Depth of cut (*d*)	mm	0.30	0.45	0.60

array (OA) with 8 DOF. In CNC lathe machine, we performed the turning process on workpiece in dry conditions which have the 350 rpm optimum spindle speed & 16 kW maximum power. Before processing, we remove the layer of rust from all the workpieces by a cutting insert for reducing any effect on experimental results. Roughness of the work piece after turning operation was measured with the help of roughness tester (Table 3).

The developed experimental design is basically used to distinguish the effects of depth of cut (d), feed (f) & cutting speed (v) on basic parameters of roughness values like (Ra &Rz) & MRR. After the experimental tests, we get the MRR & roughness result which has been represented in tabulated form which is given in Table 4. At $d = 0.30$ mm, $f = 0.09$ mm/rev & $v = 240$ m/min, and hence, the lowest value of roughness Ra &Rz is finally achieved at test no. 8; and at $d = 0.60$ mm, $f = 0.12$ mm/rev & $v = 180$ m/min, we obtain maximum value of roughness Ra &Rz,

Table 3 Orthogonal array L_9 of Taguchi experiment design and experimental results

Test no.	*v*	*f*	*d*	Ra	Rz	MRR
1	180	0.06	0.30	1.62	11.48	0.09737
2	180	0.09	0.45	1.70	11.55	0.16501
3	180	0.12	0.60	1.81	11.82	0.22786
4	210	0.06	0.45	1.75	10.81	0.1269
5	210	0.09	0.60	1.68	10.52	0.2166
6	210	0.12	0.30	1.50	9.63	0.19206
7	240	0.06	0.60	0.56	4.12	0.15473
8	240	0.09	0.30	0.50	3.80	0.22273
9	240	0.12	0.45	0.52	3.98	0.28985

Table 4 ANOVA of surface roughness (Ra)

Source	DOF	Adj. SS	Adj. MS	*F*-value	*P*-value	Cont. (%)
v	2	2.6516	1.3258	157.8	0.006	98.03
f	2	0.0016	0.0008	0.10	0.910	0.06
d	2	0.0348	0.0174	2.08	0.325	1.29
Error	2	0.0168	0.0084			0.62
Total	8	2.7050				100
$S = 0.0916515$		R-sq = 99.38%		R-sq (adj) = 97.52%		

and hence, maximum MRR is achieved at $d = 0.45$ mm, $f = 0.12$ mm/rev & $v =$ 240 m/min.

3 Results and Discussion

Results of ANOVA: The results are finally experimented and evaluated by ANOVA method. Surface roughness & MRR result for ANOVA are represented below (see Tables 4, 5 and 6). The following analysis has performed for the 95% of confidence interval (CI) level at the acceptance level of $a = 0.050$. Below table shows percent

Table 5 ANOVA of surface roughness (Rz)

Source	DOF	Adj. SS	Adj. MS	*F*-value	*P*-value	Cont. (%)
v	2	100.569	50.2843	466.1	0.002	99.14
f	2	0.161	0.0803	0.74	0.573	0.16
d	2	0.496	0.2479	2.30	0.303	0.49
Error	2	0.216	0.1079			0.21
Total	8	101.441				100
$S = 0.328448$		R-sq = 99.79%		R-sq (adj) = 99.15%		

Table 6 ANOVA of material removal rate (MRR)

Source	DOF	Adj. SS	Adj. MS	*F*-value	*P*-value	Cont. (%)
v	2	0.00564	0.00282	3.77	0.210	20.45
f	2	0.01903	0.00951	12.71	0.073	69.00
d	2	0.00141	0.00070	0.94	0.514	5.12
Error	2	0.00149	0.00074			5.43
Total	8	0.02758				100
$S = 0.0273619$		R-sq = 94.57%		R-sq (adj) = 78.29%		

SS—sum of square, DOF—degree of freedom, MS—mean square

pittance or contribution of the important element and that signify the value of impact for results obtain.

After the application of ANOVA software, experimentation has been tabulated in Tables 4, 5 and 6 which represents surface roughness (Ra & Rz) and MRR. It is seen that with reference to the surface roughness (Ra), the first important criteria is cutting speed which is having the contribution of 98.03% succeeded by depth of cut with the contribution of 1.29% and at last 0.06% contribution of feed.

For Rz, we observed that most affecting criteria is cutting speed with contribution of 99.14%, second criteria is depth of cut which affects the Rz with contribution of 0.49%. At lastm feed has the minimal contribution of 0.16% on the surface roughness (Rz).

For MRR, the first important criteria is feed which is having the contribution of 69.00% succeeded by cutting speed with the contribution of 20.45% and at last 5.12% contribution of depth of cut.

Contribution of error for roughness Ra, Rz & MRR is 0.62%, 0.21%, and 5.43%, respectively. Since the negligible contribution of error, it is suggested that the important factor is not omitted, and measurement error has no involvement.

Plots Interpretation
It is used for determining main effect of machining criteria, and data are calculated further and we use MINITAB17 software to show the main effect plot, contour plot, and probability graph for Ra, Rz & MRR in Fig. 1. From these graphs, it represents the deviation of response among all three machining criteria that are depth of cut, feed & cutting speed.

Optimal design
It represent plot of significant processing criteria and is employed to evaluate the mean values for surface roughness & predicts conditions for optimal design. From these figures, it is suggested that the cutting speed & depth of cut are the two main affecting criteria of surface roughness, and for MRR, feed & cutting speed are the two main affecting parameters. When we set cutting speed at 240 m/min (level 3) & depth of cut at 0.30 mm (level 1) gives smaller values for surface roughness & when we set the feed at 0.12 mm/rev (level 3) & cutting speed at 240 m/min (level 3) gives higher value of MRR.

For the evaluation of the mean values, it basically depends on the factorial effect of additivity. When we add the 1 factorial effects to the other mainly for the forecasting of outcome, then we have good additivity.

From Tables 7, 8 and 9, the estimated mean value of roughness (Ra) are calculated as:

$$\begin{aligned}\mu_{\mathrm{Ra}} &= \bar{v}_3 + \bar{d}_1 - \overline{T}_{\mathrm{Ra}} \quad (\text{from Table 3, } \overline{T}_{\mathrm{Ra}} = 1.2933)\\ &= (0.5267 + 1.2067) - 1.2933\\ &= 0.4401\end{aligned}$$

95% C.I. level of surface roughness is represented by:

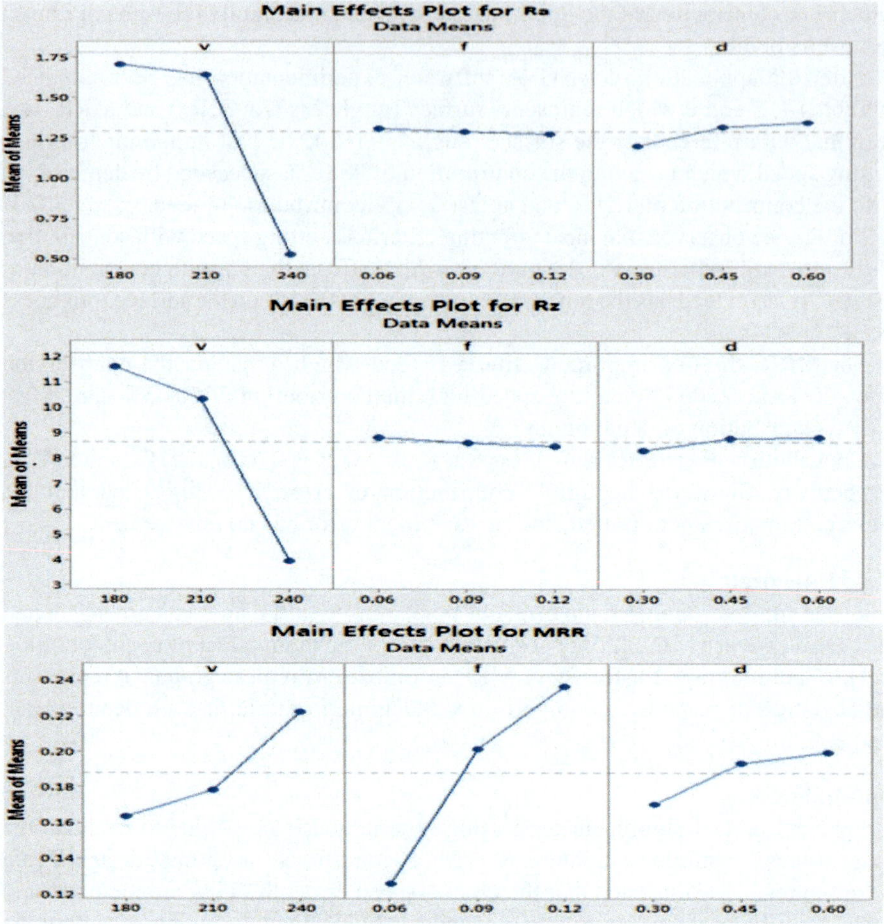

Fig. 1 Main effect graph for Ra, Rz & MRR

Table 7 Mean value for surface roughness (Ra) at different level

level	$\bar{v}$	$\bar{f}$	$\bar{d}$
1	1.7100	1.3100	**1.2067**
2	1.6433	1.2933	1.3233
3	**0.5267**	1.2767	1.3500
Delta	1.1833	0.0333	0.1433
Rank	1	3	2

Table 8 Mean value for surface roughness (Rz) at different level

Level	$\bar{v}$	$\bar{f}$	$\bar{d}$
1	11.617	8.803	**8.303**
2	10.320	8.623	8.780
3	**3.967**	8.477	8.820
Delta	7.650	0.327	0.517
Rank	1	3	2

Table 9 Mean values for material removal rate (MRR) at different level

Level	$\bar{v}$	$\bar{f}$	$\bar{d}$
1	0.1634	0.1263	0.1707
2	0.1785	0.2014	0.1939
3	**0.2240**	**0.2366**	0.1997
Delta	0.0590	0.1103	0.0290
Rank	2	1	3

Highlighted value demonstrates level for important parameter for getting suitable result & then calculation for optimum design is done

$$\text{C.I.} = \sqrt{\frac{F_{95\%;(1,\text{DOF}_{\text{error}})} \times V_{\text{error}}}{\eta_{\text{eff}}}}$$

where

$$\eta_{\text{eff}} = \frac{\text{N}}{1 + \text{DOF associated to that level}} = \frac{9}{1+2+2} = 1.8$$

$F_{95\%(1,2)} = 18.51$ & $V_{\text{error}} = 0.00840$ (from Table 4).
Hence,

$$\text{CI}_{\text{Ra}} = \sqrt{\frac{18.51 \times 0.00840}{1.8}}$$

$$\text{CI}_{\text{Ra}} = 0.2939$$

At 95% C.I. estimated limit of Ra is $[\mu_{\text{Ra}} - \text{CI}_{\text{Ra}}] \leq \mu_{\text{Ra}} \leq [\mu_{\text{Ra}} + \text{CI}_{\text{Ra}}]$
i.e., $(0.4401 - 0.2939) \leq \mu_{\text{Ra}} \leq (0.4401 + 0.2939)$
$0.1462 \leq \mu_{\text{Ra}} \leq 0.7340\ \mu\text{m}$.
Similarly, the estimated mean value for the surface roughness (Rz) is calculated as:

$$\begin{aligned}\mu_{\text{Rz}} &= \bar{v}_3 + \bar{d}_1 - \overline{T}_{\text{Rz}} (\text{from Table 3, } \overline{T}_{\text{Rz}} = 8.6344) \\ &= (3.967 + 8.303) - 8.6344 \\ &= 3.6356\end{aligned}$$

where

$$\eta_{\text{eff}} = \frac{9}{1+2+2} = 1.8$$

$F_{95\%(1,2)} = 18.51$ & $V_{\text{error}} = 0.1079$ (from Table 5).
Hence,

$$\text{CI}_{\text{Rz}} = \sqrt{\frac{18.51X0.179}{1.8}}$$
$$\text{CI}_{\text{Rz}} = 1.0533$$

Finally, at 95% CI the estimated range of Rz is $[\mu_{\text{Rz}} - \text{CI}_{\text{Rz}}] \leq \mu_{\text{Rz}} \leq [\mu_{\text{Rz}} + \text{CI}_{\text{Rz}}]$
i.e., $(3.6356 - 1.0533) \leq \mu_{\text{Rz}} \leq (3.6356 + 1.0533)$
$2.5823 \leq \mu_{\text{Rz}} \leq 4.6889$ µm.
Similarly, the estimated mean value for the material removal rate is determine as:

$$\begin{aligned} \mu_{\text{MRR}} &= \bar{\text{f}}_3 + \bar{\text{v}}_3 - \overline{T}_{\text{MRR}} (\text{from Table 3}, \overline{T}_{\text{MRR}} = 0.1881) \\ &= (0.2366 + 0.2240) - 0.1881 \\ &= 0.2725 \end{aligned}$$

where

$$\eta_{\text{eff}} = \frac{9}{1+2+2} = 1.8$$

$F_{95\%(1,2)} = 18.51$ & $V_{\text{error}} = 0.000749$ (from Table 6).
Hence,

$$\text{CI}_{\text{MRR}} = \sqrt{\frac{18.51 \times 0.000749}{1.8}}$$
$$\text{CI}_{\text{MRR}} = 0.08774$$

Finally, at 95% CI the estimated range of MRR is $[\mu_{\text{MRR}} - \text{CI}_{\text{MRR}}] \leq \mu_{\text{MRR}} \leq [\mu_{\text{MRR}} + \text{CI}_{\text{MRR}}]$
i.e., $(0.2725 - 0.08774) \leq \mu_{\text{MRR}} \leq (0.2725 + 0.08774)$
$0.18476 \leq \mu_{\text{MRR}} \leq 0.36024$ gm/s (Fig. 2).

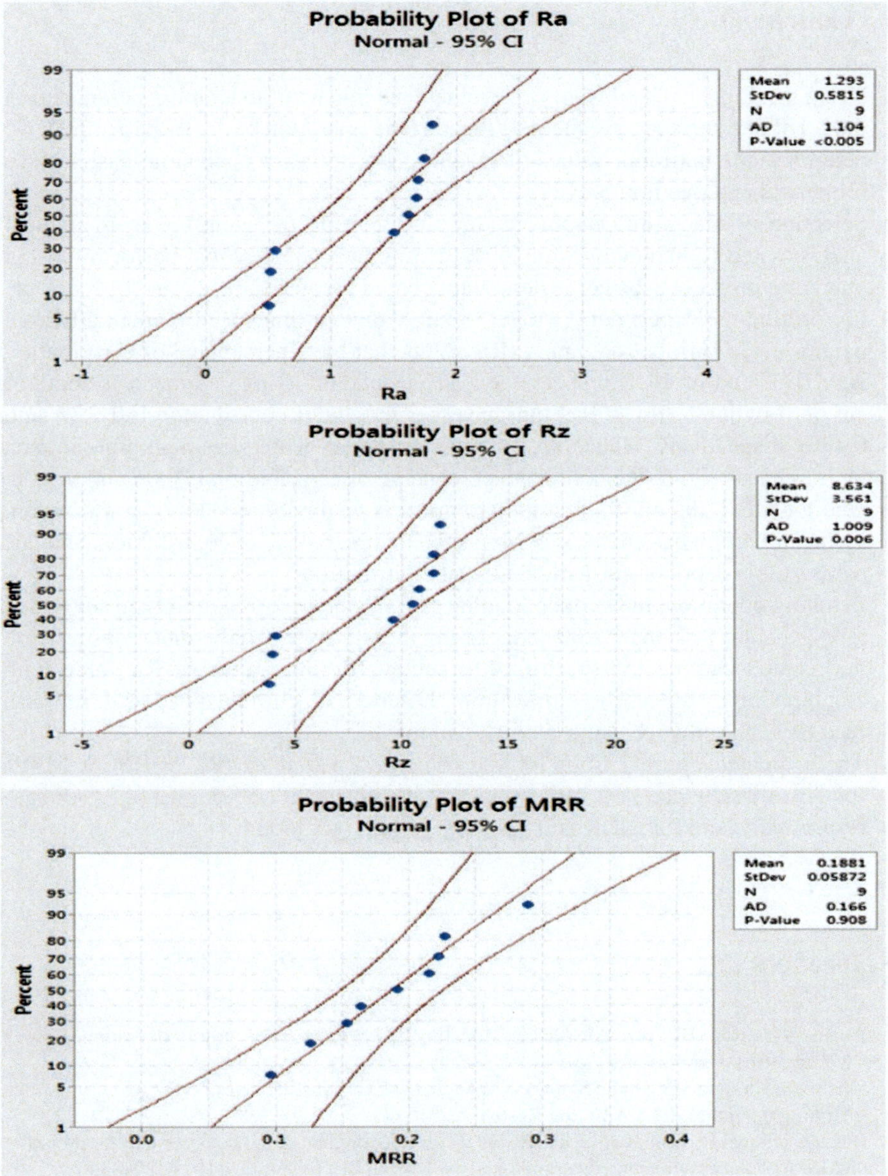

Fig. 2 Probability graph for Ra, Rz & MRR

4 Conclusions

- Using the Taguchi OA design, we evaluate the impact of machining parameter on the MRR and surface roughness. When turning operation basically performed, we determine the optimum values of machining criteria for achieving least surface roughness and maximum MRR.
- Selection of machining parameter may be feasible by using the mathematical analysis, and to achieving high surface roughness quality, we perform turning operation on AISI D2 steel is done with coated carbide tool.
- Performing the experiment, we get to know that cutting speed is main affecting parameter on surface roughness (Ra & Rz) that have percentage of contribution is 98.03% and 99.14%, respectively. The second affecting parameter is depth of cut on roughness (Ra & Rz) that have percentage of contribution is 1.29% and 0.49%, respectively. Hence, last parameter is feed which has negligible impact on roughness (Ra & Rz) that have percentage of contribution is 0.06 and 0.16%. And for MRR, the most affecting parameter is feed with 69.00% of contribution, second affecting parameter is cutting speed with 20.45% of contribution, and last parameter is depth of cut with 5.12% of contribution.
- In tuning operation, the surface roughness Ra value occurred mainly in suggested range that is less than 1.60 μm. From the study, we have clear observation that coated carbide tool performs to achieve desired range of Ra value with combination of the all three machining criteria, i.e., cutting speed (240 m/min), feed (0.12 mm/rev) & depth of cut (0.30 mm).
- The evaluated optimal limit for the roughness variables (Ra & Rz) & MMR for given workpiece is $0.1462 \leq \mu_{\mathbf{RA}} \leq 0.7340$ and $2.5823 \leq \mu_{\mathbf{RZ}} \leq 4.6889$, respectively, and for MRR is $0.18476 \leq \mu_{\mathbf{MRR}} \leq 0.36024$.

References

1. Swamy MK, Raju BP, Teza BR (2012) Modeling and simulation of turning operation. J Mech Civ Eng 3(6):19–26
2. Singhvi S, Khidiya MS, Jindal S, Saloda MA (2016) Experimental investigation of cutting force in turning operation. Int J Adv Eng Res Dev 3(3)
3. Das SR, Dhupal D, Kumar A (2015) Study of surface roughness and flank wear in hard turning of AISI 4140 steel with coated ceramic inserts. J Mech Sci Technol 29(10):4329–4340
4. Lalwani DI, Mehta NK, Jain PK (2008) Experimental investigations of cutting parameter influence on cutting forces and surface roughness in finish hard turning of MDN250 steel. J Mater Process Technol 206:167–179
5. Goyal RK (2006) Production engineering—II. Ashirwad Publication
6. Kaladhar M, Subbaiah KV, Rao CS, Rao KN (2010) Optimization of process parameters in turning of AISI 202 austenitic stainless steel. ARPN J Eng Appl Sci 5(9)

An Approach to Improved MapReduce and Aggregation Pipeline Utilizing NoSQL Technologies

Monika and Vishal Shrivastava

Abstract Technology has been changed in the most recent decade. Along these lines, we wanted proficiently systems that help ongoing examination over tremendous information. NoSQL got popularity in the mid-twenty-first century because of the improvement of disseminated computing, issues of organizations that requires the usage of Web and data genuine organizations. NoSQL (Not Only SQL) development consolidates a wide scope of database propels that were made in light of the limit of a significant volume of customer data. We start up this structure MongoDB which is appropriate for constant information because of aggregation systems, question execution, and circulated engineering, which Resulting in an assortment of non-social databases, for example, JSON-report databases. In versatile applications, database highlights can be particularly broke down for the customer and the server layers. The real-time huge-scale information investigation procedure is significant for continuous applications, dealing with high access repeat, execution of system and treatment of the data. RDBMS has a sorted out kind of data that are used in various applications for a long time, the data is taken care of in Relations and it is taken care of in a huge way, anyway now there is need to store and manage a great deal of data which cannot be taken care of by standard RDBMS. NoSQL development is a response for crushed this segment of the RDBMS by giving a capable strategy for taking care of and regulating various types of data with an immense proportion of datasets. In this paper, an execution examination is done on chronicle orchestrated databases: MongoDB. The record arranged database is a class of the NoSQL databases where the data is taken care of in JSON like reports.

Keywords NoSQL Databases · MongoDB · MapReduce · Aggregation pipeline big data

---

Monika · V. Shrivastava (✉)
Department of Computer Science, Arya College of Engineering and It, RTU, Jaipur, Rajasthan, India
e-mail: Vishal500371@yahoo.co.in

Monika
e-mail: Monikabishnoi3@gmail.com

K. Ray et al. (eds.), *Proceedings of International Conference on Data Science and Applications*, Lecture Notes in Networks and Systems 148,
https://doi.org/10.1007/978-981-15-7561-7_15

1 Introduction

The NoSQL databases give a medium to limit and recuperation of immense information that is shown in the non-forbidden structure instead of inconceivable relations used in social databases. The information structures used in NoSQL databases are of various types rather than unthinkable structures used by social database the official's systems, e.g., record, graph, key-esteem, etc. These databases are pattern less, which makes them radiantly mind-blowing in execution.

NoSQL databases are better than social databases in light of the fact that these are dynamically versatile and give extraordinary execution [1]. There are principally three portrayals of information sorted out, semi-composed and unstructured information. Social databases can simply manage composed information; they cannot manage the other two sorts of information. Social databases need information of the information before truly taking care of it in kind of examples, which fits deficiently with agile headway approaches in light of the way that there will be a need to change the outline each time at whatever point there is need of new features which log sticks the methodology if the database is generous. This report makes an undertaking to separate the execution time of inquiries (of expelling or embedding's information) into the two record document-oriented databases MongoDB.

For associations like Google, Facebook, Amazon, LinkedIn which gathers information in Terabytes (1e+12) every single day. It is insufficient to manage this gigantic information with customary databases (RDBMS). In most of cases, the information got by these associations is not sorted out which again confines the use of social databases.

This issue is clarified by NoSQL databases by giving level scaling which infers a growing number of servers (center points) in the gathering, which diminishes the commitments of each server by giving additional endpoints to the client affiliations. The idea behind level scaling is to extend the sharing of weight among the centers brings about a development in dick limit [1].

1.1 Document Database

Document databases are a champion among the most used and common NoSQL structures, where each record is thought of as a "document". The document is used to store a social event of information that changes over a sort of customer understandable information to the standard associations, for instance, JSON, XML, BSON, etc. These databases are a subclass of key-esteem databases [2]. To keep up a locale of the information report is the best alternative in light of the way that these are self-governing units which improve execution in light of the fact that the related information is scrutinized adjoining off a circle. It also makes the course of information over different servers straightforward. The limit of unstructured information is straightforward in light of the way that the record contains only those keys and characteristics which the application method of reasoning requires. It in like manner gives unbelievable flexibility by not realizing the information design early. [2] We are working at document databases MongoDB on an alternate strategy.

1.1.1 MongoDB

MongoDB is cross-arrange NoSQL document database. MongoDB is written in C, C++, JavaScript. It was first made by the item association 10gen (by and by MongoDBInc) and moved to open source arrange in 2009. MongoDB is commonly used by various associations, for instance, Forbes, Bosch, MetLife, etc. The fig shows the MongoDB information model which delineates how the semi-composed information is taken care of in records as various fields. Accumulation is a social occasion of documents however the database is a get-together of gathering. This improves the perception of the databases [3].

MongoDB stores the information in chronicles like JSON. These fields can join clusters, double information, and sub-documents. The selection of fields in an application can move according to the essential. This component empowers architects to change the information model a great part of the time as their application requires.

Documents can be traversed rich drivers available in essentially all renowned programming tongues, for instance, Java, PHP, etc. MongoDB clears the requirement for a free ORM layer which infers architects need not manage the mapping of articles from a database to an application that makes them continuously productive.

MongoDB grants auto-sharding for scaling of the database. To give high availability across over server ranches it gives replication. Replication suggests taking care of a copy of the information (second set) in servers. If at whatever point the basic game plan of information goes down, the assistant set thus takes over as the basic game plan of information. MongoDB also gives an in-memory instrument to quicken the assignment by wide use of RAM (Random Access Memory). The working of MongoDB further can be explained by going with Fig. 1.

The fig showed up above exhibits the normal auto-sharding used by MongoDB to ensure the high openness of information. It in like manner exhibits a replication system that is used for taking care of similar information copied in at any rate one server.

1.2 *JSON*

JavaScript Object Notations (JSON) is an open standard association that gives a way to deal with transmitting information items involving property estimation consolidates in the kind of clear substance. It is used as a choice as opposed to Extensible Markup Language (XML). It is a strategy for sending and tolerating information between Web applications and servers. JSON was first dictated by Douglas Crockford. The information is taken care of in records with .json enlargement [4]. These

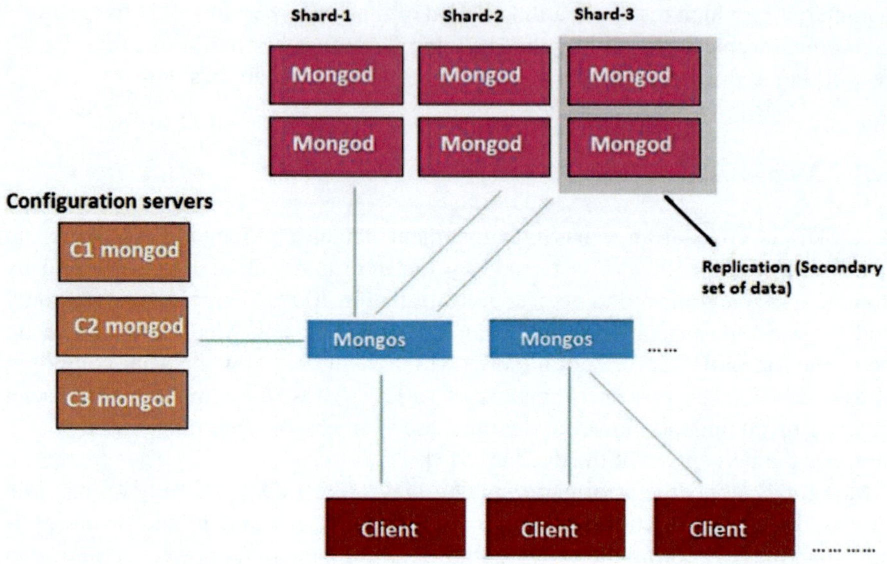

Fig. 1 Working of Mongo DB

goals are constrained by AJAX. These areas work without conceding page rendering process. This licenses trading up the substance of explicit segments back to front plans without the need for stimulating the page.

Starting late the extension in the predominance of web-based life, various locales rely upon the substance given by Facebook, Twitter, Flickr, and others. These destinations give RSS channel, which is nothing however hard to work with AJAX. JSON handles the cross-space issue. In JSON configuration secluded in segment bunch by using wavy props. It is definitely not hard to scrutinize and get it.

1.3 MapReduce

MapReduce is a creating example of uses that handle enormous information. Huge information separating is a troublesome issue today. Enormous information applications, the MapReduce framework has starting late pulled in a lot of thought. Google's guide diminishes or its open-source partners Hadoop is a helpful resource for a structure such applications. In this instructional exercise, we will display the guide's lesson structure. Guide diminish is a framework that uses the record requesting with mapping, organizing, revising, decreasing by two limits that are Map work and lessen work. The MapReduce work is done in two basic exercises: Map and Reduce. The Map work read sets of information and performs computations on it. By then, the came-about widely appealing (key, esteem) sets are furthermore passed to Reduce

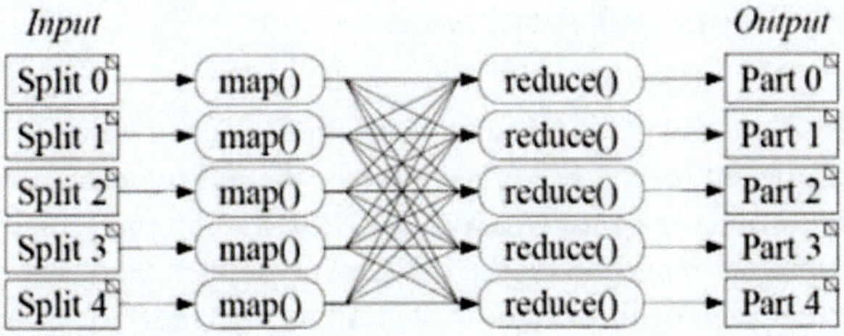

Fig. 2 MapReduce architecture

work. Diminish work bundles all of the characteristics for each remarkable key made by Map work.

To grasp the working, the model MapReduce work has shown up. In this point of reference, the Map work scrutinizes the information of the record and parses out the words. In guide adventure for each word, (key, esteem) pair is delivered for instance (word, 1). Here the word is a key and worth 1 shows different occasions of a word is one in the file [5] (Fig. 2).

2 Literature Review

Qi Lv and Wei Xie had a novel execution approach for an event-driven document log analyzer are presented, and execution assessment of an inquiry, clear and gathering assignments over MongoDB, HBase, and My SQl is bankrupt down. In their test outcomes exhibit that HBase performs best balanced in all assignments, while MongoDB gives under 10 ms inquiry speed in specific exercises which are most sensible for continuous applications.

HamoudAlshammari and Hassan Bajwa form a paper and in this paper, they present a survey of a made MapReduce figuring and its reenactment using HBase as a NoSQL database. They depicted and complete the proposed Enhanced Hadoop Architecture with specific movements on the primary intent to meet different cases.

ChiragPatal and MosinHasan create a paper about the Benchmark of Map-Reduce Framework in MongoDB. The basic edges behind this paper are to show the Design of "Benchmark of Map-Reduce Framework in MongoDB" for instance "MongoMR-Bench" and "MongoGen" is a structure that produces semantic gigantic information that is supported by MongoDB which is Document based database.

Jeffrey Dean and Sanjay Ghemawat showing about MapReduce used on Large groups. Their execution of MapReduce continues running on a gigantic bunch of creation machines and is exceptionally adaptable: a regular MapReduce figuring structures various terabytes of information on an enormous number of machines. Engineers and the structure easy to use: numerous MapReduce tasks have been realized and as much as one thousand MapReduce vocations are executed on Google's bunches every day.

3 Proposed Approach

MongoDB is NoSQL databases that are used when information is enormous. In our proposed research work we study the aggregation activity at that point look at the hour of activity for two MongoDB works, in particular, MapReduce and aggregation pipeline in python on the equivalent dataset. Here JSON document is used to store a great deal of information. We Analysis MapReduce and Match Aggregate Pipeline Performance for Document database. We will use differing collections (i.e., 5, 10, 50, and 100 k records) in it. At that point, we include parallelizing capacity with the aggregation work over subsets of the accumulation, utilizing various strings with total pipeline and called as altered Aggregate Pipeline. A capacity is characterized with two parameters x and y, x is utilized for the cutoff inquiry and y is utilized for skip question.

The consequence of this aggregation activity was documented which contained two fields _id which is Customer Id and whole which is the total of a repeat of the given Customer Id where the condition is that the Customer Id ought to be from the United Kingdom. The framework utilized had eight strings so the all-out documents per gathering were isolated into these eight strings and after that joined. Each string does the activity in this way utilizing the greatest capability of the framework and doing it productively.

I exhibited this alteration with the total pipeline to give a superior outcome in the MongoDB Database in an unstructured organization. This procedure plays out the way toward gathering metadata parameters present in the document, gathering and every one of the records into the parameters that have a place with each record. The consequence of the coordinating procedure is ID, record data, status record, yield apparatus, statistic, and metadata bunch in a class document.

We will use the Windows 8 phase to play out the assignments on MongoDB. These databases should give quick and high throughput when appeared differently in relation to social databases.

4 Results and Discussions

Assessments were led on the "Online business Data" dataset that concentrated on the document database. The informational collection contains Invoice No, Stock Code, Description, Quantity, Invoice Date, Unit Price, Customer ID, and Country. Then the procedure for setting up MongoDB was done (Windows 10 framework). MongoDB was downloaded from its official site. At that point, a database "mydb" was made and accumulations were made in particular "fivek", "tenk", "twentyk" and "fiftyk. First, the CSV record dataset was opened then it was opened in a document iterator.

At that point in a rundown, the distinctive section names of the table were put away. Utilizing the iterator each line of the table was spared as a document. The initial 5000 lines were spared in the "fivek" gathering, initial 10,000 columns were

Table 1 Comparison between MapReduce and aggregation pipeline

Data records (k)	MapReduce (s)	Aggregation pipeline (s)	Aggregation pipeline with parallelization (s)
5	1.416	0.223	0.396
10	2.553	0.412	0.387
20	5.240	0.842	0.384
50	12.102	2.013	0.401

spared in the "tenk" accumulation, first 20,000 lines were spared in the "twentyk" gathering and initial 50,000 lines were spared in the "fiftyk" gathering (Table 1).

First, we perform the aggregation function on the dataset. This aggregation operation took 0.223 s to happen for "fivek" collection. This aggregation operation took 0.412 s to happen for "tenk" collection. This aggregation operation took 0.842 s to happen for "twentyk" collection. This aggregation operation took 2.013 s to happen for "fiftyk" collection.

Next we perform the MapReduce operation on the documents. The MapReduce operation took 1.416 s to happen for "fivek" collection. The MapReduce operation took 2.553 s to happen for "tenk" collection. The MapReduce operation took 5.240 s to happen for "twentyk" collection. The MapReduce operation took 12.102 s to happen for "fiftyk" collection. Thus we can say that the aggregation pipeline function is much faster as compared to the MapReduce function and produces equally accurate results.

Then we try to optimize or improve the performance of aggregation pipeline operation by using parallelization by using a multithreading approach. The Aggregation pipeline with parallelization took 0.396 s to happen for "fivek" collection. The Aggregation pipeline with parallelization took 0.387 s to happen for "tenk" collection. The Aggregation pipeline with parallelization took 0.384 s to happen for "twentyk" collection. The Aggregation pipeline with parallelization took 0.401 s to happen for "fiftyk" collection. Thus the optimized aggregation performs the best specially for large datasets.

5 Conclusion

In This paper we depict an assignment where we directed an investigation to look at the hour of activity for two Mongo Db works to be specific MapReduce and aggregation pipeline in python on a similar E-business Data dataset. that dataset is download from Kaggle which was taken from UCI AI storehouse this information is a lot of genuine exchanges between the years 2010 and 2011. The informational index contains Invoice No, Stock Code, Description, Quantity, Invoice Date, Unit Price, Customer ID, and Country. In our assessment procedure, we will perform curd activity on the dataset by applying MapReduce, Aggregate Pipeline and adjusted Aggregate Pipeline strategy.

The altered Aggregate Pipeline strategy has better execution in handling information in the bigger square size contrasted with both Match Aggregate Pipeline and MapReduce. Because of the execution time can be diminished considerably further, by parallelizing the aggregation work over subsets of the accumulation, utilizing numerous strings. I showed this for a solitary non-sharded database.

6 Future Scope

The present and just as future extent of NoSQL innovations are splendid. As there are numerous chances and incredible difficulties which should be defeated in Big Data. Later on, we can play out a correlation of these two document-based databases with the diverse sorts of records, for example, CSV, XML, and JSON. We can likewise contemplate other report situated databases, for example, Redis, HBase, and Postgresql and so forth. NoSQL advancements are likewise brilliant if there should arise an occurrence of Pattern Matching Algorithms on account of Big Data.

References

1. Kumar N, Saxena S (2015) A preference-based resources allocation in cloud computing systems. In: 3rd international conference on recent trends in computing 2015. Procedia computer science, vol 57, pp 104–111
2. Xu Q, Arumugam RV, Yong KL, Wen Y, Ong YS, Xi W (2015) Adaptive and scalable load balancing for metadata server cluster in cloud-scale file system. Front Comput Sci 9(6):904–918
3. Teng F (2012) Management Des Donnees Et Ordinnnancement Des Taches Sur Architectures Distributes, Desertation, Ecole Cenrale Paris Et Manufactures, Centrale Paris
4. Yu Z (2012) Research of conversion method of entity object and JSON data. In: The 2nd international conference on computer application and system modeling, Guizhou University for nationalities, Guiyang, China
5. Dean J, Ghemawat S (2008) MapReduce: simplified data processing on large clusters. Commune. ACM, pp 107–113
6. Pandey S (2010) Scheduling and management of data intenesive application workflows in grid and cloud computing environments. Dissertation, Department of Computer Science and Software Engineering, The University of Melbourne, Australia
7. Sedaghat M, Rodriguez FH, Elmroth E (2013) A virtual machine Re-packaging approach to the horizontal versus vertical elasticity trade-off for cloud autoscaling. In: The 2013 ACM cloud and autonomic computing conference
8. Manyinka J, Chui M, Brown B, Bughin J, Dobbs R, Roxburgh C, Byers AH (2011) Big Data: the next frontier for innovation, competition, and productivity, McKinsey Global Institute 2011 Report, [Online]
9. Schmitt O, Majchrzak TA Using document-based databases for medical information systems in unreliable environments. University of Münster, Germany
10. Murtaza S Implementation and evaluation of a JSON binding for mobile web services with IMS integration support. KTH School of Electrical Engineering. Available http://www.diva-portal.org/smash/get/diva2:541461/FULLTEXT01.pdf

11. Kadebu P Innocent Mapanga, A security requirements perspective towards a secured NOSQL database environment. In: International conference of advance research and innovation (ICARI), 2014
12. https://docs.mongodb.com/manual/introduction/
13. Edlich S, Friedland A, Hampe J, Brauer B (2010) NoSQL: Einstieg in dielt Nichhtrelationaler Web 2.0 Datenbanken, 2nd edn. Hanser Fachbuchverlag
14. Padhy RP, Patra MR, Satapathy SC (2011) RDBMS to NOSQL: reviewing some next generation non realtional databases. Int J Adv Eng Sci Technol 11(1)

Evaluation of Bio-movements Using Nonlinear Dynamics

Sergio Mejia-Romero, J. Eduardo Lugo, Delphine Bernardin, and Jocelyn Faubert

Abstract Biological movement analysis describes specific characteristics of the person's health status, which motivates the analysis of the dynamic aspects related to the rhythms and movements of our body. Natural, free movements can be periodic or irregular in time and space, and each type of dynamic behavior can be related to efficient or altered movements. This research describes an overview of nonlinear dynamics and concepts of chaos applied to bio-motion paths as a way to describe and analyze a bio-movement, for example, the eye movement presents different rhythms depending on the demand for exploration or the state of health of the person. For this research, 20 subjects with normal vision were involved. Their eye movements and head movements were registered using an eye tracker and head-mounted tracker, measuring the position and rotation of the eye and head. The results demonstrate that nonlinear analysis can be applied to evaluate alteration in the biological system showing more sensitivity than traditional spectral analysis. This type of evaluation can provide a tool to find the relationship between movement dynamics and physiological phenomena, which might be useful to describe any alteration in the biological system.

Keywords Bio-movements · Eye movement · Head movement · Nonlinear dynamics · Entropy · Chaos

---

S. Mejia-Romero · J. Eduardo Lugo (✉) · D. Bernardin · J. Faubert
FaubertLab, School of Optometry, Université de Montréal, Montréal, QC, Canadá
e-mail: eduardo.lugo@gmail.com

S. Mejia-Romero
e-mail: sergio.mejia.romero@umontreal.ca

D. Bernardin
Essilor Canada Ltd., Montréal, QC, Canadá

K. Ray et al. (eds.), *Proceedings of International Conference on Data Science and Applications*, Lecture Notes in Networks and Systems 148,
https://doi.org/10.1007/978-981-15-7561-7_16

1 Introduction

Biological movement signals are generated by complex self-regulating systems that process inputs with a broad range of characteristics. Many physiologic movements time series are extraordinarily inhomogeneous and non-stationary, fluctuating in an irregular and complex manner [1].

The goal of several studies is to construct a realistic model that can be used to obtain useful information associated with the signal. On this basis, it should be pointed out the importance of the analysis of high-dimensional dynamical systems. Recently, the spatiotemporal chaos has attracted so much attention due to its theoretical and practical applications [2–5]. In bio-movement signals, spatiotemporal chaos has been analyzed to investigate the interaction between dynamic movements and diseases.

Time series analysis considers just a scalar time series, usually associated with an experimental acquisition, to understand the dynamical system behavior. The essential point of this analysis is that a time series contains information about unobserved variables of the biological system, which allows the system analysis performing state space reconstruction [6, 7].

The complexity of a signal [8] can be described by different measures [9]: The Lempel–Ziv complexity [10], entropy [11], and chaos. The Lempel–Ziv complexity indicates the degree of regularity of time series; as an example, the relationship between brain activity patterns in patients [12].

The relative complexity values of a signal may be calculated to describe whether a signal of lower complexity has an excellent efficiency or not; another indicator of an efficient strategy is the entropy time evolution. An entropy increases in the signal means that the movement pattern is not constant, and consequently, the wearer or subject doing the movement gets more tired or uncomfortable. The evolution of entropy can also reveal the fragility of a strategy when the task becomes too complex or uncertain [13].

Moreover, the system presents a sensitive dependence on initial conditions [14]. This property characterizes the chaotic behavior of a dynamical system. This sensitive dependence represents the butterfly effect described in Lorenz's work [15].

2 Materials and Methods

2.1 Dataset Description

The research goal was realized with the use of eye and head movement data collected from 20 healthy participants, with normal vision, during two sessions, separated by a two-week break. The participants were expected to drive in the Virage simulator driving test [16] follow the same methodology used by Michael's study [17], the length of the series is almost 6 min each session.

Eye movements were registered with a 120 Hz sampling rate SMI eye tracker [18]. It uses direct infrared oculography sensors measuring the resultant position of the left and right eye. The head movement was registered with a 120 Hz sampling rate OptiTrack system [19], and the signal was processed by Motive software resulting in the position and rotation of the head.

The calibration and validation were carried out in a 52-in. section of the screen with a resolution of 1200 × 720 (0.438 m × 0.263 m) at a distance of 1.15 m from the participants, and considering the head position as the reference center, the center of the reference pattern is at mean vertical angle of ~4.2° and horizontal mean angle of ~2.8°. The maximum horizontal eye movement angle was within a range from ~−18.3° to ~18.3°, and the vertical eye movement angle ranges from ~−6° to ~10°. Since the measurement offset error and the magnitude of the measurement noise is well above the length of a pixel, the quantization error effect is negligible.

All recordings collected during this experiment were identified by the user; labels were placed in the first and the second session. Considering the number of participants and the two sessions, we have forty series, but three series were excluded, for two participants, from the second session of the experiment, since an abnormal behavior of head and eye movement was observed. It was probably introduced by the eye tracker, as unexpected data was discovered for these participants. Thus, the 36-time series were finally analyzed. Figure 1 shows examples of two-time series recorded for the same participants, but during different sessions.

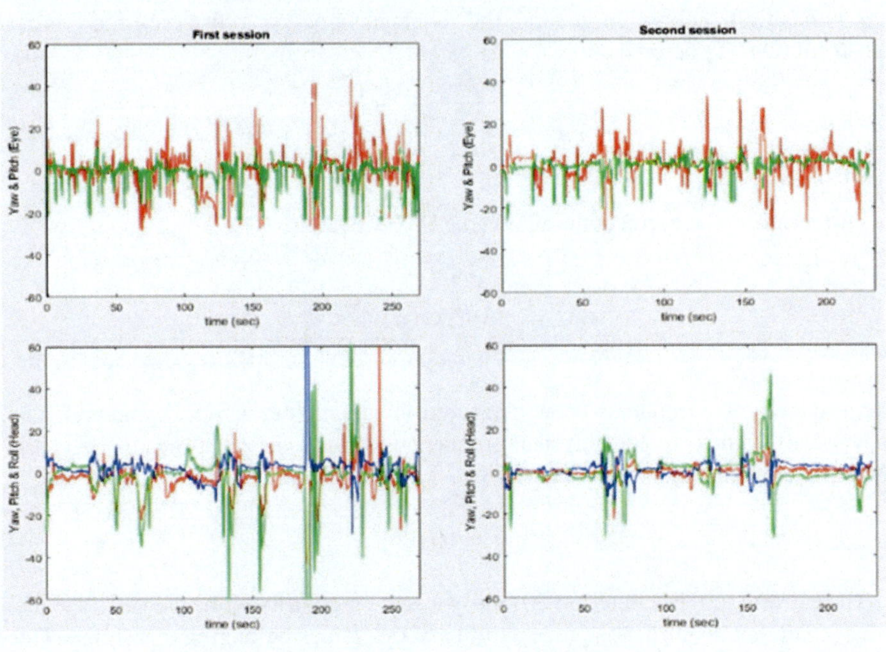

Fig. 1 Signal of the movement of eyes and head for the pilot test during the corresponding two sessions

2.2 Methods I

The real signals can be characterized as being random (from the observer's viewpoint). This means that the variation of such a signal outside the observed interval cannot be determined precisely but only specified in statistical terms of averages. We will be interested in estimating the spectral characteristics of random signals. As we will show, it is advantageous considering the spectral analysis of signals.

A sinusoidal signal $s(t) = \alpha \cos(\omega t + \phi)$, can be rewritten as a linear combination of two complex-valued sinusoidal signals, $s(t) = \alpha_1 e^{i(\omega_1 t + \phi_1)} + \alpha_2 e^{i(\omega_2 t + \phi_2)}$, the fact that we need to consider two constrained complex sine waves to treat the case of one unconstrained real sine wave which shows that the real-valued case of sinusoidal signals can actually be considered to be more complicated than the complex-valued case.

If we consider discrete signals (eye and head signal detected), such signals are most commonly obtained by the temporal or spatial sampling of a continuous (in time or space) signal.

$$\tilde{g}(\rho, t) = \sum_{t=1,2,3}^{N} \rho_t \tag{1}$$

where $\tilde{g}(\rho, t)$ denotes a deterministic discrete-time data sequence. Assume that $\tilde{g}(\rho, t)$ has finite energy, in general, the sequences possess a discrete-time Fourier transform (DTFT) defined as:

$$G(w) = \sum_{t=-\infty}^{\infty} g(t)\mathrm{e}^{-iwt}, \tag{2}$$

Furthermore, the corresponding inverse DTFT is then:

$$g(t) = \frac{1}{2\pi} \int_{-\pi}^{\pi} G(w) e^{iwt} \mathrm{d}w \tag{3}$$

where the angular frequency w is measured in radians per sampling interval. The conversion from w to the physical frequency variable $\bar{w} = \frac{w}{T_s}$ (rad/s). Let the corresponding energy spectral density be:

$$S(w) = |G(w)|^2. \tag{4}$$

The entropy [20] is a function that measures the regularity of the values in a signal or a system that quantifies the repeatability of the values belonging to a signal. This determines the conditional probability of the similarity between a data segment of a given duration and the next set of segments of the same duration.

For instance, if a signal provides repetitive patterns, then the signal is stable, and the approximate entropy is low. Entropy has been used in various disciplines as a nonlinear statistical method and is particularly useful in nonlinear dynamic signal analysis. A brief calculation algorithm of the approximative entropy is described here [21]:

First: Form $N - m + 1$ $N - m + 1$ vectors $X(1), \ldots, X(N - m + 1)$ defined by: $X(i) = [x(i), x(i + 1), \ldots, x(i + m - 1)]$ and $i = 1, \ldots, N - m + 1$. Fix m, an integer, and r, a positive real number. The value of m represents the window length of a compared run of data, and r specifies a filtering level.

Second: Define the distance $d[X(i), X(j)]$ between $X(i)$ and $X(j)$, as the maximum norm: $d[X(i), X(j)] = \max_{k=1,2,\ldots,m} |x(i + k - 1) - x(j + k - 1)|$. The variable d represents the distance between the vectors $x(i)$ and $x(j)$, given by the maximum difference in their respective scalar components. For a given $X(i)$, count the number $j (j = 1, \ldots, N - m + 1)$ so that $d[X(i), X(j)] \leq r$, denoted as $N^m(i)$. Then, for $i = 1 \ldots N - m + 1$, is $C_r^m(i) = \frac{N^m(i)}{(N-m+1)}$, where $C_r^m(i)$ measures, within a tolerance r, the frequency of patterns similar to a given window of length m.

Finally, compute the natural logarithm of each $C_r^m(i)$ and average it over i, $\phi^m(r) = \frac{1}{N-m+1} \sum_{i=1}^{N-m+1} \ln C_r^m(i)$, where $C_r^m(i)$ is the probability of vector $X_j^m X_j^m$ to lie within a distance r of the vector $X_i^m X_i^m$.

$$\mathrm{VecEn}\left(t + \frac{m}{2}\right) = \phi^m(r) \tag{5}$$

Increase the dimension to $m + 1$. Repeat steps to find the next $C_r^{m+1}(i)$ and $\phi^{m+1}(r)$.

ApEn is defined by: $\mathrm{ApEn}(m, r, N) = \phi^m(r) - \phi^{m+1}(r)$.

Poincare plot, considering a time series X of length n, $X = \{x_1, x_2, x_3, \ldots, x_{n-1}, x_n\}$, its Poincare plot vector $\vec{p}$ is the scatter plot representing the set of points [21]:

$$\vec{p} = (x_1, x_2), (x_3, x_4), (x_5, x_6), \ldots, (x_{n-1}, x_n) \tag{6}$$

The coarse-grained time series [22] are obtained using a non-overlapping moving average low-pass filter. The window length, s, determines the scale of the coarse-grained time series $\left\{\sum_s (j)\right\}$.

The elements of the coarse-grained time series for scale s are determined according to Equation:

$$\sum_s (j) = \frac{1}{s} \sum_{i=(j-i)s+1}^{j_s} x_i, \quad 1 \leq j \leq \frac{N}{s} \tag{7}$$

The traditional monochromatic Poincare plot can be enhanced by adding color to each of its data points to convey information about their normalized frequency of occurrence [23]. The probability density function can be estimated by employing the histogram technique. Specifically, we used the MATLAB, dscatter function to compute the smoothed normalized two-dimensional histogram of $\{(x_i, x_i + 1)\}$ [24].

2.3 Methods II

During the data capture of the eye or head movement, $g(t)$ can be represented as several points:

$$g(t) = [(x_1, y_1), (x_2, y_2), \ldots, (x_n, y_n)], \tag{8}$$

is a sequence of N scalar measurements obtained with a sampling rate T. Given such a series of observations, it is possible to reconstruct the dynamics of the original system.

All the datasets were processed by the same noise filtering method. The choice of filtering methods and their parameters were based on the review of the literature. The use of previously applied approaches for eye movement signals and solutions used in the analysis of other biological signals were assumed.

First, we apply the mean filter, with a window of size 7, this method replaces the central element of the defined window with the average value of the ordered list of elements of the window, and then we apply Savitzky–Golay (SG) filter [24], third grade filter, with window length of 15. The polynomial of the given order approximates the underlying signal within the window. The polynomial order was based on previously conducted research [25].

This method makes an adjustment of least squares of the consecutive data points to a polynomial, and the calculated central point of the adjusted polynomial curve is the new smoothed data point.

Subsequently, each time series, which represents the movement of the eye or head, was obtained by applying the standard procedure of two-point signal differentiation. They were subjected to an additional analysis that consisted of the following methodology:

- Calculating the power spectral density (with the method described above or any other method that calculates PSD.
- Vector entropy of the signal movement is calculated.
- The accumulative power of vector entropy is computed.
- Poincare plots and their descriptors are created.

All calculations were performed using the functions and packages available in MATLAB.

3 Results

The following figures describe the results obtained after the application of the analysis strategy mentioned above to the eye and head movements signals.

In Fig. 2, it can be seen the filtered data corresponding to the data series of the eye and head movements. The positive values correspond to the movement of the eye to the right side, while the negative ones to the saccadic movement of the left eye. The same direction corresponds to the movement of the head.

In Fig. 3, the power spectrum test revealed that the first session signal (Fig. 3, red line) presents a distribution of peaks in the higher frequency domain that seems to correspond to the quasi-periodic dynamic component superimposed on the chaotic one. For the second session, the evidence of the dynamic pattern seems to disappear being replaced by a distribution of the spectrum at slower and monotonous frequencies (Fig. 3, blue line).

In Fig. 4, the vector entropy test shows that the more regular the pattern of movement, the smaller the entropy. In this graph, we observe that the value of entropy during the second session for both eye and head movement was lower compared to the value of entropy for the first session, and this has generally been interpreted as less complicated.

To reinforce this assumption, we calculate the cumulative power associated with the entropy vector, and it is presented in Fig. 5.

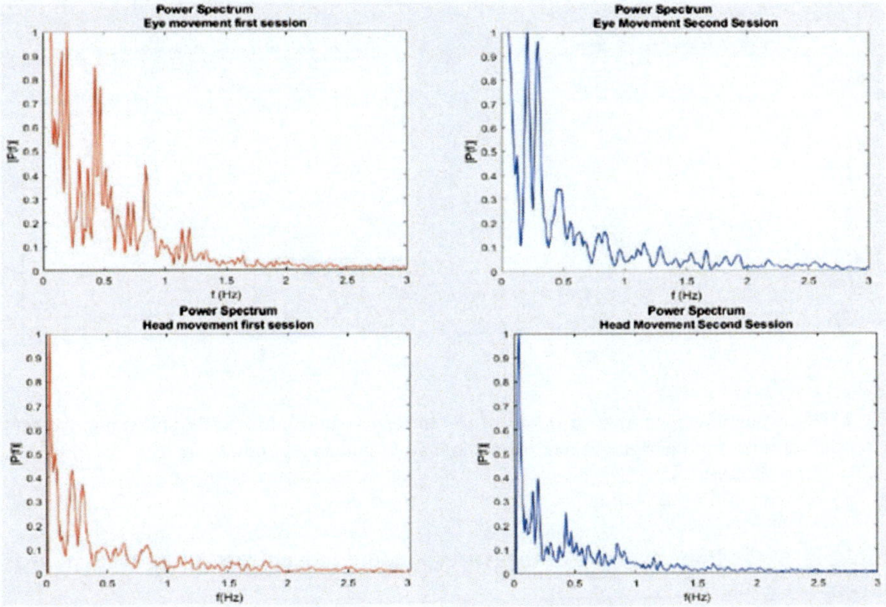

Fig. 2 Power spectrum, the first and second session, corresponding to the movement of the eyes and head for a pilot test

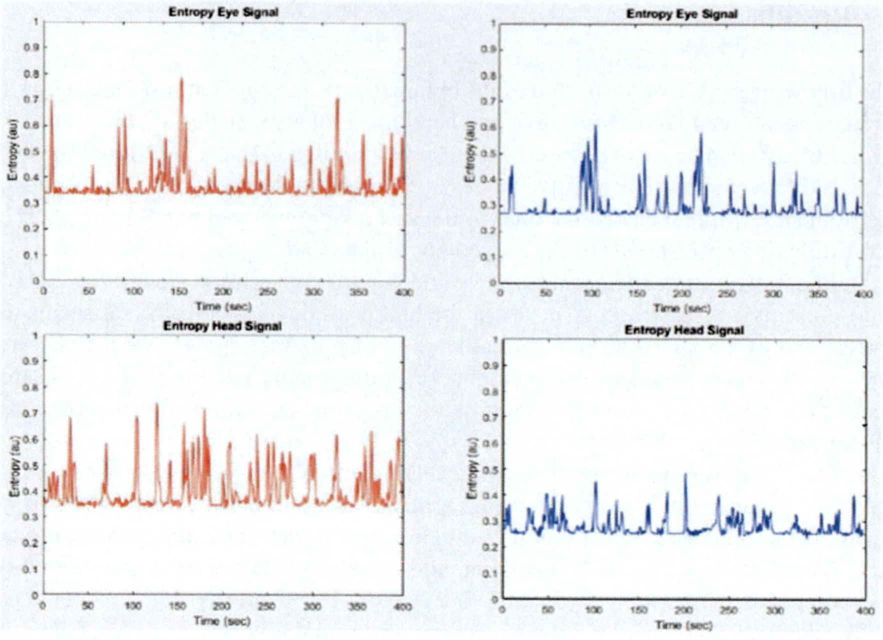

Fig. 3 Entropy vector showing the changes of entropy over time during the corresponding session for the pilot test

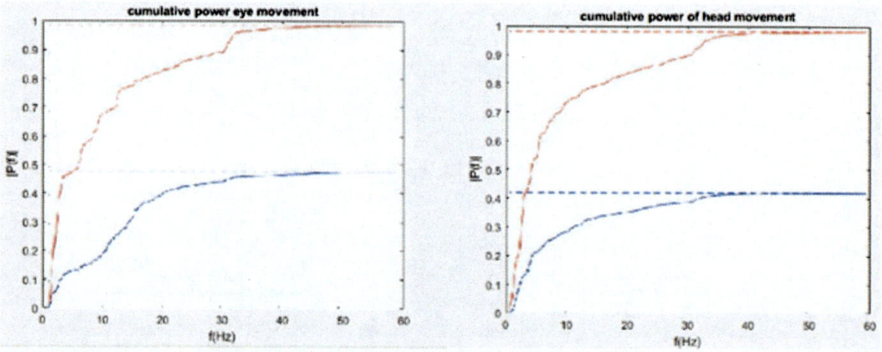

Fig. 4 Corresponding accumulated power of the entropy vector, in the left graph eye movement and head movement to the right, in red line is the first session and second sessions in the blue line, results of the pilot test

In Fig. 5, the Poincare map of the corresponding movements for both sessions is shown.

During all tests, we observed the decreasing change in the cumulative power of the spectrum and decrease of the entropy. Also, the Poincare indicator has a lower value when comparing the first and second sessions; Tables 1 and 2.

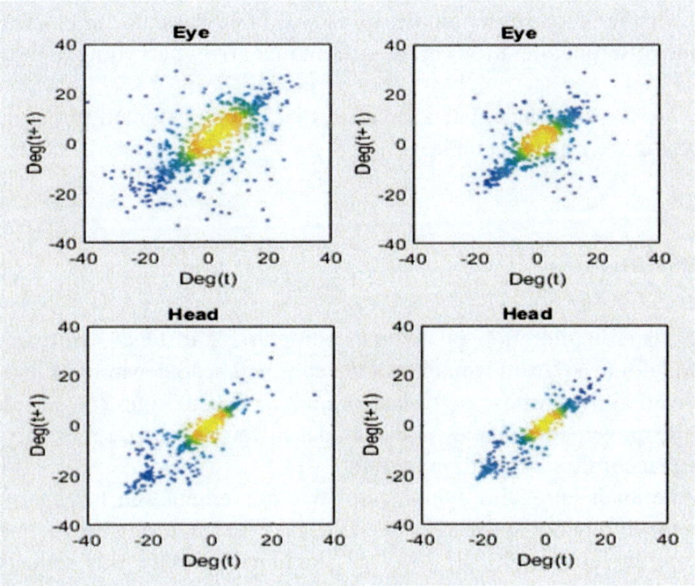

Fig. 5 Poincare map of head movements and eye movements for the pilot test corresponding to the two sessions

Table 1 Table eye dynamics results for all subjects (normalized power)

Parameter	First session	Second session
Energy PSD	0.845	0.426
Mean entropy	0.4985	0.280
Energy entropy	0.568	0.359
SD1	0.037	0.017
SD2	0.045	0.036
Ratio	0.834	0.466

Table 2 Head dynamics results for all subjects (normalized power)

Parameter	First session	Second session
Energy PSD	0.753	0.323
Mean entropy	0.4108	0.285
Energy entropy	0.756	0.286
SD1	0.037	0.016
SD2	0.087	0.097
Ratio	0.427	0.164

The loss or the increase in complexity is a reliable indicator in determining the impact of the different alterations in the system that arises according to the reference point.

The averaged results calculated for all subjects are presented in the next table, when analyzing the results in table.

4 Discussion

During the visual exploration of the environment, for instance, a driver during a driving simulator task, the movements of the eyes and head depend on the movement strategy as well as the purpose of the exploration of the environment. Besides, it was found that the movement of the eyes and head is different from one person to another and can vary according to multiple factors.

During the analysis of the dynamics of eye movement and head movement, a decrease in the number of spectral frequencies is revealed during the second session, and therefore, a decrease in the level of signal complexity was reflected correspondingly by a decrease either in the entropy values or Poincare's map descriptive parameters. Thus, random movements are more reduced in the second session.

Therefore, the value of the entropy and the values of the Poincare descriptors are of great potential as indicators to evaluate eye or head movements.

The dynamic analysis showed that it is feasible to compare the effects of training or the effects of external alterations of eye or head movements on the same subject.

It would also be valuable to repeat the test presented herein using other types of movement dynamics alterations, such as under the influence of drugs, degradation of the biomechanical system due to degenerative diseases or evaluation of neurological training.

5 Conclusions

The purpose of the study presented was to discover the dynamic characteristics of the eye and head movements to understand the dynamic behavior of the system based on the signals recorded for people during a virtual driving task.

When evaluating the characteristics of spectral frequency, entropy, and chaotic value, the results confirmed the viability of characterizing the dynamics of the system.

Based on the findings of this work, it is possible to begin comparative studies to provide a methodology to assess changes in movement dynamics. The differences in the results obtained when comparing two different sessions under similar conditions but with different movement dynamics provide arguments for future studies in this field.

Author Contributions M-R.S. designed and implemented the research method and conducted the data analysis. M-R.S was involved in preparing and carrying experiments. All authors took part in the paper preparation and edition.

Conflicts of Interest
The authors declare no conflict of interest. The founding sponsors had no role in the design of the study; in the collection, analyses, or interpretation of data; in the writing of the manuscript, and in the decision to publish the results.

Acknowledgments
We want to extend our gratitude to Jesse Michaels for providing the raw data collected during his Ph.D. research project, and all the participants who were involved in this study, as well as those directly involved in making this study possible. This research was partly funded by an NSERC Discovery grant and Essilor Industrial Research Chair (IRCPJ 305729-13), Research and development cooperative NSERC—Essilor Grant (CRDPJ 533187—2018), Prompt.

References

1. Ivanov et al (1996) Multifractality in human heartbeat dynamics. Naturev 383:323–327
2. Awrejcewicz J, Umberger DK (1989) Spatiotemporal dynamics in a dispersively coupled chain of nonlinear oscillators. Phys Rev A Gen Phys 1;39(9):4835–4842
3. Lai YC, Grebogi C (1999) Modeling of coupled chaotic oscillators. Phys Rev Lett 82:4803–4806
4. Shibata H (1998) Quantitative characterization of spatiotemporal chaos. Physica Av 252:428–449
5. Cross MC, Hohenberg PC (1993) Pattern formation outside of equilibrium. Rev Mod Phys 65:851–1112
6. Gollub JB, Langer JS (1999) Pattern formation in nonequilibrium physics. Rev Mod Phys 71:S396
7. Singer IM, Thorpe JA (1976) Lecture notes on elementary topology and geometry. Scott, Foresman & Company, Springer
8. Guckenheimer J, Holmes P (1983) Nonlinear oscillations, dynamical systems, and bifurcations of vector fields. Springer, Berlin Heidelberg New York
9. Lempel A, Ziv J (1976) On the complexity of finite sequences. IEEE Trans Inf Theory 22(1):75–81
10. Pincus SM (1991) Approximate entropy as a measure of system complexity. Proc Natl Acad Sci USA 88:2297–2301
11. Zhang X, Roy RJ (2001) EEG complexity as a measure of depth of anesthesia for patients. IEEE Trans Biomed Eng 48(12):1424–1433
12. Zhang X, Zhu Y, Thakor NV, Wang Z (1999) Detecting ventricular tachycardia and fibrillation by complexity measure. IEEE Trans Biomed Eng 46:548–555
13. Wiggins S, Strogatz S Nonlinear dynamics and chaos. Perseus. Takens, F (1981) Detecting strange attractors in turbulence
14. Lorenz EN Deterministic nonperiodic flow. J Atmos Sci Singh, 2009
15. Pincus SM, Cummins TR, Haddad GG (1993) Heart rate control in normal and aborted SIDS infants. Am J Physiol 264 (Regulatory & Integrative 33), R638–R646
16. VS500M car driving simulator used in this study (Virage Simulation Inc® 2005–2020)
17. Michaels J, Chaumillon R et al (2017) Driving simulator scenarios and measures to faithfully evaluate risk driving behavior: a comparative study of different driver age groups. PLoS ONE 12(10):e0185909

18. SensoMotoric Instruments and Noldus Information Technology combine eye tracking and video analysis. Noldus. Retrieved 2 April 2014
19. NaturalPoint, Motion capture systems—OptiTrack Webpage. [Online]. Available: optitrack.com. Accessed: 09-Jan-2017
20. Pincus SM, Goldberger AL (1994) Physiological time series analysis: what does regularity quantify? Am J Physiol (Heart Circ Physiol) 266:H1643–H1656
21. Abasolo D, Hornero R, Espino P (2009) Approximate entropy of EEG backgroung activity in Alzheimer's disease patients. Intell Autom Soft Comput 15(4):591–603
22. Yan R, Gao RX (2007) Approximate entropy as a diagnostic tool for machine health monitoring Mech. Syst Signal Process 21:824–839
23. Costa MD, Goldberger AL, Peng C-K (2002) Multiscale entropy analysis of complex physiologic time series. Phys Rev Lett 89(6):068102
24. Schafer RW (2011) What is a Savitzky-Golay filter. IEEE Signal Process Mag 28:111–117
25. Kasprowski P, Harężlak K, Stasch M (2014) Guidelines for the eye tracker calibration using points of regard. In: Piętka E, Kawa J, Wieclawek W (eds) Information technologies in biomedicine, vol 4. Springer International Publishing, Cham, Switzerland, pp 225–236

An Examination System to Classify the Breast Thermal Images into Early/Acute DCIS Class

Nilanjan Dey, V. Rajinikanth, and Aboul Ella Hassanien

Abstract The recent report by the World Health Organization (WHO) confirms that breast disease is one of the chief impacting cancers in women. The accessibility of the modern disease investigative arrangement and treatment process will support to enhance the survival rate of cancer-infected persons. This work considers the thermal imaging modality-based examination of the breast sections with the help of a computer-assisted analysis (CAA) scheme. The proposed work aims to implement a CAA to inspect and classify the breast thermal images (BTI) into early/acute ductal carcinoma in situ (DCIS) class. In this paper, initially, a multi-thresholding procedure with Shannon's entropy and firefly algorithm (FA) is performed to enhance the visibility of BTI. Later, essential texture features are extracted from the original and threshold BTIs. Later, the principal features selection is then implemented with statistical analysis, and these features are then considered to train and test the classifier systems. The classifiers, such as decision tree (DT), K-nearest neighbor (KNN), and support vector machine (SVM) are considered to classify the BTI, and the result confirmed that the proposed technique achieved a classification accuracy >89%.

Keywords Brest abnormality · Thermal imaging · DCIS · Texture feature · Classification

N. Dey
Department of Information Technology, Techno India College of Technology, Kolkata, West Bengal 700156, India
e-mail: neelanjan.dey@gmail.com

V. Rajinikanth (✉)
Department of Electronics and Instrumentation, St. Joseph's College of Engineering, Chennai, Tamil Nadu 600119, India
e-mail: v.rajinikanth@ieee.org

A. E. Hassanien
Faculty of Computers and Artificial Intelligence, Scientific Research Group in Egypt, Cairo University, Giza, Egypt

K. Ray et al. (eds.), *Proceedings of International Conference on Data Science and Applications*, Lecture Notes in Networks and Systems 148,
https://doi.org/10.1007/978-981-15-7561-7_17

1 Introduction

Cancer is one of the life-threatening sicknesses with a high death rate. The current statement by the World Health Organization (WHO) confirmed that in 2018 alone, 9.6 million predictable casualties worldwide because of cancer [1]. This report also confirms that breast and lung cancer is the leading cause of mortality compared to other cancers, and in 2018 alone, a registered death rate of 2.09 million is reached. The lung cancer (LC) is a common disease and affects the mankind irrespective of gender and race. The breast cancer (BC) affects largely the women, and in recent years, considerable awareness camps are conducted worldwide to present the BC and detect its occurrence in an early stage.

The early-phase diagnosis of BC will be done with a personal check followed by a clinical examination by an experienced doctor. At the clinical level, the BC can be examined and confirmed with the help of medical imaging procedures recorded with a range of modalities, such as mammograms, CT, MRI, ultrasound, and thermal imaging [2–10]. Thermal imaging (TI) is a recently adopted modality in clinics to identify the disease in internal body parts based by recording the infrared waves (IW) emitted by the body. In this process, a special camera is utilized to capture the IW emitted by the body, and the intensity of the IW will carry vital information regarding the body parts. By simply evaluating the intensity level in TI, it is possible to predict the abnormality in the internal body parts. Further, it is a non-invasive practice and helps to get the digital pictures directly with the help of a high resolution/pixel camera. Recently, the TI approach is considered to record and evaluate the abnormality in breast based on the breast TI (BTI) [11–13].

The premature phase of the BC can be predicted by examining ductal carcinoma. In situ (DCIS) a common disease arises due to the presence/growth of abnormal cells inside of the milk duct of the breast section. The early phase of the DCIS is non-invasive, and it can be cured easily with a prescribed treatment process. The acute stage is also treated with prescribed handling procedures, and the risk associated with the acute phase DCIS is more compared to the early phase [13, 14].

The proposed work aims to develop a computerized assessment system (CAS) to detect and classify the BTI into early/acute DCIS. The proposed CAS involves in: (i) collection and resizing (512×256 pixels) of the BTIs, (ii) multi-thresholding based on firefly algorithm and Shannon's entropy (FA + SE), (iii) extraction of Haralick texture features and entropies, (iv) selection of dominant features based on statistical measures, (v) training, testing, and validation of considered classifiers (DT, KNN and SVM) based on the selected features.

In the proposed work, the BTI database available in [15, 16] is considered for the examination. This dataset consists of the IW images of early- and acute-class DCIS. This is one of the benchmark BTI databases that consist of the RGB and grayscale versions of the BTIs. In this research work, the grayscale version of the BTIs is considered for the examination.

Initially, an image resizing is implemented to extract the section of interest from the raw test picture, and later, the necessary features are extracted, and these dominant features (eight features from each class of picture) are extracted from the images. These features are then considered to train and test the classifiers, and the performance of these classifiers are then computed based on the performance measures, such as accuracy (AC), precision (PR), sensitivity (SE), specificity (SP), and F1 score (F1S). Further, the additional measures, such as false-negative rate (FNR), false-positive rate (FRP), false detection rate (FDR), and false omission rate (FOR) are also computed. Based on these values, the performance of the classifier used in the developed CAS is verified. The remaining sections of this work are organized as; Sect. 2 illustrates the methodology, Sect. 3 presents the results and its discussions, and Sect. 4 presents the conclusion of the proposed work.

2 Methodology

The steps executed to develop the CAS are depicted in Fig. 1. Initially, the collected raw test images are preprocessed to maintain uniqueness in the test images. During this process, the essential BTI section is cropped and resized to 512×256 pixel sized images. Later, all the considered test images are enhanced with a multi-thresholding process based on the firefly algorithm and Shannon's entropy (FA + SE).

During this enhancement, the considered technique is implemented on these test images with bi-level, three-level, and four-level thresholds, and all these images are then considered during the next phase; called the feature extraction process. During feature extraction, the GLCM and entropy values are extracted from the original test pictures and the threshold pictures. Later, a dominant feature selection based on FA is then implemented, and the existing features are then ranked and selected along with student's *t*-test. This procedure helped to identify eight numbers of dominant texture features from each image class (8 features $\times$ 4 image class = 32 features). These features are then considered to train, test, and validate the classifiers, which classifies the BTI database into early/acute DCIS class.

2.1 BTI Database

The benchmark database was contributed by Silva et al. (2014) [16], and this dataset can be accessed from [15]. This database consists both the RGB scale and the grayscale pictures recorded with a controlled environment, and these images are similar to a high-quality clinical-grade database. In this work, the essential sections are initially cropped and resized for the evaluation purpose. The number of pictures considered for the assessment can be found in Table 1.

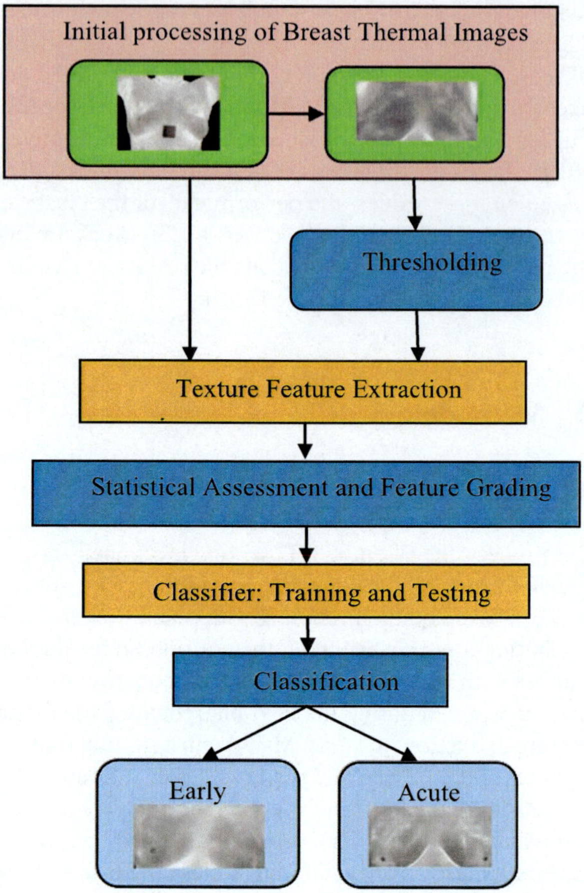

Fig. 1 Structure of the proposed CAS used to evaluate the BTI

Table 1 Number of BTI considered in the proposed work (512 × 256 pixels)

Image class	Number of BTI
Early	26
Acute	42
Total	68

2.2 *Image Enhancement*

During the medical image analysis, image enhancement schemes are widely implemented to remove the artifacts and also to improve the visibility of the test pictures. In this work, the visibility of the BTI is enhanced with Shannon's based multi-thresholding process. In order to attain optimal result: This thresholding procedure was implemented under the supervision of the firefly algorithm (FA), which helps to find the optimal threshold. Shannon's entropy (SE) for a picture of dimension M

* N with pixel group (A, B) is symbolized as $F(A, B)$, with $A \in \{1, 2, \ldots, p\}$ and $B \in \{1, 2, \ldots, p\}$.

Let T represents total pixels of image and location of every pixels {0, 1, 2, …, T − 1} can be denoted as U, as [17]:

$$F(A, B) \in U \ \forall (A, B) \in \text{picture} \tag{1}$$

Then, the regularized histogram is: $E = \{e_0, e_1, \ldots, e_{T-1}\}$.

For tri-level thresholding, it can be framed as:

$$E(\text{th}) = e_0(t_1) + e_1(t_2) + e_2(t_3) \tag{2}$$

where $\text{th} = \{t_1, t_2, \ldots, t_T\}$ are threshold values and th^* is the final threshold. Further information on SE can be found in [18–20].

In this work, maximization of Shannon's entropy based on FA is implemented and the enhanced image (with bi-, three-, and four-level threshold) is then considered for feature extraction process. The FA is responsible to arbitrarily vary the thresholds of the test picture, till the entropy value is maximized. The thresholding with the FA is largely available in the literature, and the outcome also confirms that FA technique will enhance the results.

The description of FA is as below:

$$P_{F_1}^{t+1} = P_{F_1}^{t} + \beta_0 e^{-\gamma d_{xy}^2}\left(P_{F_2}^{t} - P_{F_1}^{t}\right) + \alpha_1.\text{sign}\ (\text{rand} - 1/2) \oplus B(s) \tag{3}$$

where $P_{F_1}^{t+1}$ is modernized place of F_1, $P_{F_1}^{t}$ is opening background of firefly, $\beta_0 e^{-\gamma d_{xy}^2}\left(P_{F_2}^{t} - P_{F_1}^{t}\right)$ the striking energy among fireflies, $B(s) = A.|s|^{\alpha/2}$ is Brownian–Walk system, A is an random variable, β is spatial supporter, and α is temporal supporter. Related details on FA can be found in [21, 22].

In this paper, the FA parameters are assigned as: number of fireflies = 30, search dimension = threshold value (i.e., 2, 3, and 4), iteration level = 3000, and stopping criteria = maximized SE.

2.3 Texture Feature Extraction

The extraction of features during the image analysis is widely considered to develop an autonomous classifier system. In this work, the test pictures of the BTI datasets (normal and thresholded images) are considered for the feature extraction process. Initially, the texture features for the early- and acute-class DCIS images are extracted using the Haralick's approach [23]. This technique provides around 20 numbers of texture and shape features for each image [24–30]. Later, entropy values of these pictures are then extracted, and the entropies like Rényi, Max, Kapur, Vadja, Shannon, Yager, Tsallis, and Fuzzy are computed as discussed in [31]. The feature extraction process helped to attain around 28 (28 × 4 image class = 112 features) numbers of texture values, from which more dominant features are then identified with a feature selection process.

2.4 Feature Ranking and Selection

Principal texture features will demonstrate a maximal divergence between the early/acute class BTIs. In the literature, a substantial amount of primary feature selection techniques are proposed to identify and sort the accessible features based on its rank. In this work, 112 features were extracted from every image (normal, bi-level, three-level, and four-level threshold). To recognize the leading features, Student's *t*-test was employed, and these features were ranked based on the *p*- and *t*-values achieved during the statistical assessment practice. The particulars on *t*-test-based feature choice can be found in prior research work [32–35]. This process helps to select eight numbers of dominant features from each image class (8 × 4 = 32 features).

2.5 Classifier Implementation

This section presents the brief details on the considered classifiers, which are initially trained with the help of a feature set (32 features) for both the early and acute class BTIs. In this work, the well-known classifiers, such as decision tree (DT), K-nearest neighbor (KNN), support vector machine (SVM) with linear kernel (SVM1), and SVM with radial basis function (SVM2) are considered as the classifiers, and the essential information on these classifiers can be found in [31–33].

2.6 Validation

A relative inspection connecting the essential image class and identified class is then carried out to evaluate the superiority of proposed CAS with various classifiers. The performance of the classifiers, such as DT, KNN, SVM1, and SVM2, are assessed based on the values of the following performance measures [31, 36–38]:

$$\mathrm{AC} = \frac{T_{+\mathrm{ve}} + T_{-\mathrm{ve}}}{T_{+\mathrm{ve}} + T_{-\mathrm{ve}} + F_{+\mathrm{ve}} + F_{-\mathrm{ve}}} \tag{4}$$

$$\mathrm{PR} = \frac{T_{+\mathrm{ve}}}{T_{+\mathrm{ve}} + F_{+\mathrm{ve}}} \tag{5}$$

$$SE = \frac{T_{+ve}}{T_{+ve} + F_{-ve}} \tag{6}$$

$$SP = \frac{T_{-ve}}{T_{-ve} + F_{+ve}} \tag{7}$$

$$F1S = \frac{2T_{+ve}}{2T_{+ve} + F_{-ve} + F_{+ve}} \tag{8}$$

$$FNR = \frac{F_{-ve}}{F_{-ve} + T_{+ve}} \tag{9}$$

$$FPR = \frac{F_{+ve}}{F_{+ve} + T_{-ve}} \tag{10}$$

$$FDR = \frac{F_{+ve}}{F_{+ve} + T_{+ve}} \tag{11}$$

$$FOR = \frac{F_{-ve}}{F_{-ve} + T_{-ve}} \tag{12}$$

where T_{+ve}, T_{-ve}, F_{+ve}, and F_{-ve} denote true-positive, true-negative, false-positive, and false-negative, respectively.

3 Result and Discussion

This section present the investigational outcome acquired with the proposed CAS. All these results are attained using MATLAB7 software. During this study, the benchmark grayscale version of BTIs is (512 × 256 pixels) considered. Figure 2 depicts the sample test image of the considered dataset for both the early- and acute-class DCIS.

Fig. 2 Sample test BTIs of early and acute class

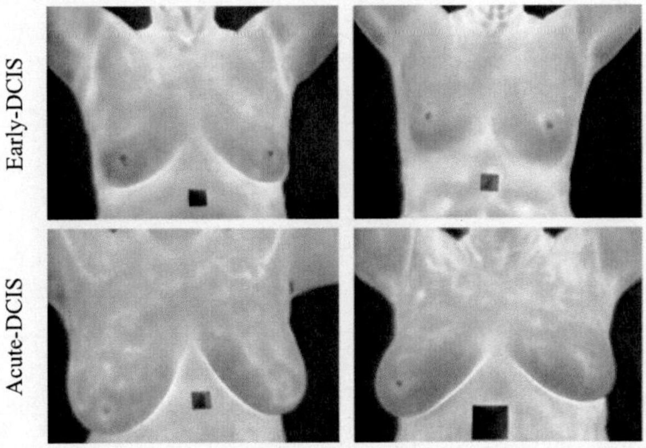

To attain a unique image, a cropping and image resizing process is then implemented on these test pictures.

Figure 3 depicts the regularized sample test pictures of early/acute class picture with a pixel dimension of 512 × 256. From this image, it can be noticed that, for the early class picture, the ductal abnormality value is less compared to the acute class picture. To examine the DCIS with the proposed CAS, it is necessary to extract the texture features. The earlier research work confirms that [31–33], to have better accuracy, it is necessary to have a more dominant feature, which can later be considered to train the classifier system. Further, the unprocessed test image may have artifacts or visibility issues, which also may decrease the classifier accuracy during the assessment. Hence, considered test images are then thresholded with SE + FA for a threshold value of 2, 3, and 4, and attained result's sample is depicted in Fig. 4 for both early and acute classes. In which, Fig. 4a presents a bi-level threshold outcome, and Fig. 4b, c presents the results of three- and four-level thresholds. From this, it can be noted that the threshold can be employed to improve the surface features in images.

Later, a feature extraction (Haralick and Entropy) process is implemented to mine the texture and shape features from these images and than a statistical analysis

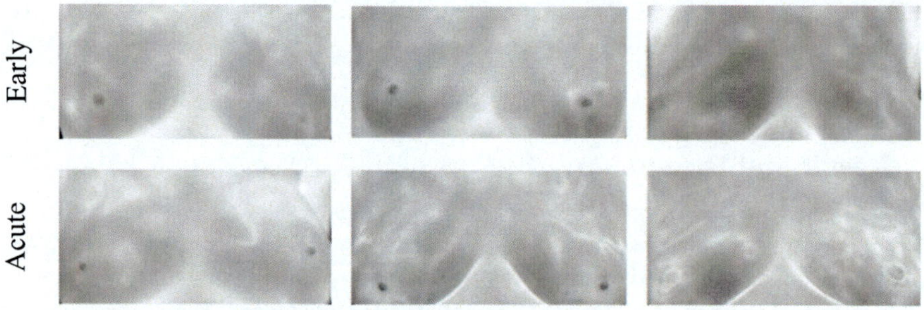

Fig. 3 Cropped and resized BTIs of size 512 × 256 pixels

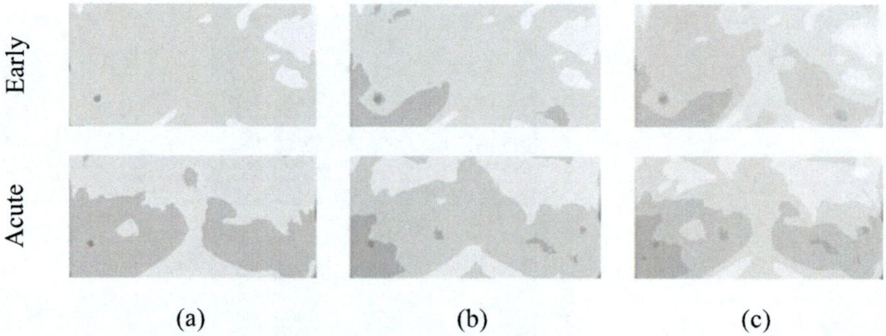

Fig. 4 Threshold outcome of the sample BTI. **a** Bi-level, **b** three-level, **c** four-level threshold

(Student's *t*-test) based ranking and selection of dominant features are then executed. This procedure helps to attain 32 dominant features (8 features from an image group × 4 classes = 32 features) from 112 primary features.

Table 2 presents the elected dominant features for the unprocessed test image of the BTI. This technique helped to attain eight features as given in table, which also presents the obtained statistical measures and the *p*- and *t*-values for these selected features. Similar features are also considered for the threshold pictures (bi-, three-, and four-level thresholds), and totally a 32 number of dominant features are then considered to train, test, and validate the classifier system.

Tables 3 and 4 presents the various performance measures attained with the considered classifier systems. Table 3 presents the AC, PR, SE, SP, and F1S attained with

Table 2 Elected dominant features of the raw test image of early/acute class

Selected features	Early				Acute			
	Mean	SD	Statistical measure		Mean	SD	Statistical measure	
			p-value	*t*-value			*p*-value	*t*-value
Rényi entropy	0.6527	0.0317	0.0000	28.0163	0.8218	0.0286	0.0000	30.1863
Max entropy	0.6373	0.0228	0.0000	27.0125	0.7706	0.0296	0.0000	29.1174
Kapur entropy	0.6218	0.0309	0.0000	26.9275	0.7626	0.0187	0.0000	28.9733
Vadja entropy	0.6231	0.0264	0.0000	26.6625	0.7581	0.0305	0.0000	28.6296
Shannon entropy	0.6152	0.0826	0.0000	24.0745	0.7429	0.0197	0.0000	26.9952
Yager entropy	0.6163	0.0173	0.0000	20.9773	0.7127	0.0076	0.0000	25.2871
Tsallis entropy	0.6054	0.0281	0.0000	20.2081	0.7015	0.0126	0.0000	23.2974
Energy	0.4183	0.0311	0.0000	15.1865	0.7186	0.0224	0.0000	16.2619

Table 3 Assessment of classifier performance based on performance measures

Method	TP	TN	FP	FN	AC (%)	PR (%)	SE (%)	SP (%)	F1S (%)
DT	23	38	3	4	89.71	88.46	85.19	92.68	86.79
KNN	25	37	1	5	91.18	***96.15***	83.33	***97.37***	89.29
SVM1	24	39	2	3	92.65	92.31	88.89	95.12	90.57
SVM2	24	41	2	1	***95.59***	92.31	***96.00***	95.35	***94.12***

Table 4 Assessment of classifier performance with predicted values of test images

Method	TP	TN	FP	FN	FNR	FPR	FDR	FOR
DT	23	38	3	4	0.1481	0.0732	0.1154	0.0952
KNN	25	37	1	5	0.1667	0.0263	0.0385	0.1190
SVM1	24	39	2	3	0.1111	0.0488	0.0769	0.0714
SVM2	24	41	2	1	0.0400	0.0465	0.0769	0.0238

the BTI database for the early/acute case. The outcome of this table confirms that the AC, SE, and F1S attained with SVM-RBF (SVM2) is better compared with DT, KNN, and SVM1. Further, the PR and SP attained with the KNN are superior to other techniques. The outcome of Table 4 confirms that the overall measure by SVM2 is superior compared with other classifiers. From these results, it can be confirmed that the proposed CAS with the SVM2 is better in classifying the BTIs with greater accuracy. The overall accuracy attained by the proposed CAS is >89%. In the future, the accuracy of the proposed CAS can be improved by considering the ensemble-based techniques during the feature selection. Further, the proposed CAS can also be tested on the RGB scale benchmark BTI database.

4 Conclusion

This paper proposes a computer-assisted system (CAS) to evaluate the BTI collected to form a benchmark database. In this work, BTIs of early and acute DCIS classes are considered for the evaluation, and the proposed system is considered to implement a two-class classifier based on the texture features. The considered test image is initially processed with a multi-thresholding technique based on the FA + SE with varied thresholds. Later, the texture features and the entropy features are then extracted from the raw and threshold images (112 features). A statistical evaluation based on the Student's *t*-test is then implemented to identify 32 numbers of principle features from 112 initial features based on the rank of *p*- and *t*-values. These 32 features are then considered to train, test, and validate the considered classifiers. In this work, well-known classifiers, such as DT, KNN, SVM1, and SVM2 are implemented, and the results of this study confirm that proposed CAS with SVM2 helps to attain better performance values compared to the alternatives. This CAS helps to achieve a classification accuracy of >89% on the considered database.

References

1. https://www.who.int/health-topics/cancer
2. Acharya UR, Ng EYK, Tan JH, Sree SV (2012) Thermography based breast cancer detection using texture features and support vector machine. J Med Syst 36(3):1503–1510
3. Sree SV, Ng EYK, Acharya UR, Faust O (2011) Breast imaging: a survey. World J Clin Oncol 2(4):171–178
4. Suganthi S, Ramakrishnan S (2014) Semiautomatic segmentation of breast thermograms using variational level set method. IFMBE Proc 43:231–234
5. Sayed GI, Soliman M, Hassanien AE (2016) Bio-inspired swarm techniques for thermogram breast cancer detection. Stud Comput Intell 651:487–506
6. Gonzlez FJ (2011) Non-invasive estimation of the metabolic heat production of breast tumors using digital infrared imaging. Quant Infrared Thermogr J 8:139–148
7. Raja NSM, Sukanya SA, Nikita Y (2015) Improved PSO based multi-level thresholding for cancer infected breast thermal images using Otsu. Procedia Comput Sci 48:524–529

8. Keatmanee C, Chaumrattanakul U, Kotani K, Makhanov SS (2019) Initialization of active contours for segmentation of breast cancer via fusion of ultrasound, Doppler, and elasticity images. Ultrasonics 94:438–453. https://doi.org/10.1016/j.ultras.2017.12.008
9. Rodtook A, Kirimasthong K, Lohitvisate W, Makhanov SS (2018) Automatic initialization of active contours and level set method in ultrasound images of breast abnormalities. Pattern Recogn 79:172–182
10. Rajinikanth V, Raja NSM, Satapathy SC, Dey N, Devadhas GG (2018) Thermogram assisted detection and analysis of ductal carcinoma in situ (DCIS). In: Proceedings international conference intelligence computer instrument control technology. IEEE, pp 1641–1646. https://doi.org/10.1109/icicict1.2017.8342817
11. Cheriguene S, Azizi N, Zemmal N, Dey N, Djellali H, Farah N (2016) Optimized tumor breast cancer classification using combining random subspace and static classifiers selection paradigms. Appl Intell Optim Biol Med 96:289–307. https://doi.org/10.1007/978-3-319-21212-8_13
12. Fernandes SL, Rajinikanth V, Kadry S (2019) A hybrid framework to evaluate breast abnormality using infrared thermal images. IEEE Consum Electron Mag 8(5):31–36. https://doi.org/10.1109/mce.2019.2923926
13. Bejnordi BE, Balkenhol M, Litjens G, Holland R, Bult P, Karssemeijer N, Laak JAWMVD (2016) Automated detection of DCIS in whole-slide H&E stained breast histopathology images. IEEE Trans Med Imaging 35(9):2141–2150
14. Ng EYK (2009) A review of thermography as promising non-invasive detection modality for breast tumor. Int J Therm Sci 48(5):849–859
15. http://visual.ic.uff.br/dmi/
16. Silva LF et al (2014) A new database for breast research with infrared image. J Med Imaging Health Inf 4(1):92–100. https://doi.org/10.1166/jmihi.2014.1226
17. Kannappan PL (1972) On Shannon's entropy directed divergence and inaccuracy. Probab Theory Rel Fields 22:95–100. https://doi.org/10.1016/S0019-9958(73)90246-5
18. Raja NSM, Arunmozhi S, Lin H, Dey N, Rajinikanth V (2019) A study on segmentation of Leukocyte image with Shannon's entropy. In: Histopathological image analysis in medical decision making 1–27. https://doi.org/10.4018/978-1-5225-6316-7.ch001
19. Raj SPS, Raja NSM, Madhumitha MR, Rajinikanth V (2018) Examination of digital mammogram using Otsu's function and watershed segmentation. In: Fourth international conference on biosignals, images and instrumentation (ICBSII), IEEE, pp 206–212. https://doi.org/10.1109/icbsii.2018.8524794
20. Nair MV et al (2018) Investigation of breast melanoma using hybrid image-processing-tool. In: International conference on recent trends in advance computing (ICRTAC). IEEE, pp 174–179. https://doi.org/10.1109/ICRTAC.2018.8679193
21. Yang XS (2010) Firefly algorithmstochastic test functions and design optimization. Int J Bio-inspired Comput 2(2):78–84
22. Yang XS (2011) Nature-inspired metaheuristic algorithms, 2nd edn. Luniver Press, Frome, UK
23. Haralick RM, Shanmugam K, Dinstein I (1973) Textural features for Image classification. IEEE Trans Syst Man Cybern 3(6):610–621
24. Samanta S, Ahmed SkS, Salem MA-MM, Nath SS, Dey N, Chowdhury SS (2014) Haralick features based automated glaucoma classification using back propagation neural network. Adv Intell Syst Comput 327:351–358
25. Zayed N, Elnemr HA (2015) Statistical analysis of haralick texture features to discriminate lung abnormalities. Int J Biomed Imaging 2015:7
26. Virmani J, Dey N, Kumar V (2016) PCA-PNN and PCA-SVM based CAD systems for breast density classification. Appl Intell Optim Biol Med 96:159–180. https://doi.org/10.1007/978-3-319-21212-8_7
27. Chaki J, Dey N (2019) A beginner's guide to image shape feature extraction techniques. CRC Press
28. Ali MNY, Sarowar MG, Rahman ML, Chaki J, Dey N, Ravares JMRS (2019) Adam deep learning with SOM for human sentiment classification. Int J Ambient Comput Intell (IJACI) 10(3):92–116. https://doi.org/10.4018/IJACI.2019070106

29. Shi F et al (2019) Texture features based microscopic image classification of liver cellular granuloma using artificial neural networks. In. 8th Joint international information technology and artificial intelligence conference (ITAIC), IEEE, pp 432–439. https://doi.org/10.1109/itaic.2019.8785563
30. Wang Y et al Morphological segmentation analysis and texture-based support vector machines classification on mice liver fibrosis microscopic images. Curr Bioinf 14(4):282–294. https://doi.org/10.2174/1574893614666190304125221
31. Acharya UR et al (2019) Automated detection of Alzheimer's disease using brain MRI images—a study with various feature extraction techniques. J Med Syst 43:302. https://doi.org/10.1007/s10916-019-1428-9
32. Dey N et al (2019) Social-group-optimization based tumor evaluation tool for clinical brain MRI of Flair/diffusion-weighted modality. Biocybern Biomed Eng 39(3):843–856. https://doi.org/10.1016/j.bbe.2019.07.005
33. Chen Y, Chen G, Wang Y, Dey N, Sherratt RS, Shi F (2019) A distance regularized level-set evolution model based MRI dataset segmentation of brain's caudate nucleus. IEEE Access 7:124128–124140
34. Chaki J, Dey N, Moraru L, Shi F (2019) Fragmented plant leaf recognition: bag-of-features, fuzzy-color and edge-texture histogram descriptors with multi-layer perceptron. Optik 181:639–650. https://doi.org/10.1016/j.ijleo.2018.12.107
35. Zemmal N, Azizi N, Dey N, Sellami M (2016) Adaptive semi supervised support vector machine semi supervised learning with features cooperation for breast cancer classification. J Med Imaging Health Inf 6(1):53–62
36. Rajinikanth V, Thanaraj KP, Satapathy SC, Fernandes SL, Dey N (2019) Shannon's entropy and watershed algorithm based technique to inspect ischemic stroke wound. Smart Intell Comput Appl 105:23–31. https://doi.org/10.1007/978-981-13-1927-3_3
37. Lakehal A, Alti A, Laborie S, Roose P (2020) A semantic agile approach for reconfigurable distributed applications in pervasive environments. Int J Ambient Comput Intell (IJACI) 11(2):48–67. https://doi.org/10.4018/IJACI.2020040103
38. Chandrakar P (2019) A secure remote user authentication protocol for healthcare monitoring using wireless medical sensor networks. Int J Ambient Comput Intell (IJACI) 10(1):96–116. https://doi.org/10.4018/IJACI.2019010106

Implementation of Hybrid Wind–Solar Energy Conversion Systems

Pooja Joshi and K. C. Roy

Abstract Along with the growing issues regarding global warming and reduction of reserves of fossil fuels, researchers all around the world are attempting to look for alternate resourced required for saving the planet for the coming future. For our requirement, more and more plants are set up using the wind, hydro and solar generation of power. It is known that wind power generation can provide power to some huge extent; however this resource is also not predictable in terms of its presence in the atmosphere. In a similar manner power through sun cannot be dependent on, it can be present for whole day but atmospheric conditions can be never reliable and unexpected clouds, rains, trees can be factors of limitations related to such power generation. Most basic signs of climate especially in sun and wind systems are due to their irregular natures of appearance which makes there a very unreliable source of power. Although using the two resources in combination, by using algorithms such as (MPPT): "Maximum Power Point Tracking", an effective transfer of power can be done to enhance the reliability and effectiveness of the power generation system.

Keywords Solar power · Wind power · MPPT

1 Introduction

Implementation of Hybrid Wind-Solar Energy Conversion Systems, Solar power and wind power are considered clean, unobtrusive, limitless and environmentally friendly. Such features have attracted the potential field to utilize large-scale renewable potential sources. Although almost every renewable source of energy has its concede limitations. Solar and wind sources depend upon factors that are unpredictable like the changing conditions of the climate or the weather.

P. Joshi (✉)
Department of Electrical Engineering, School of Engineering, Rai University, Ahmadabad, India
e-mail: puji.joshi@gmail.com

K. C. Roy
Kautilya Institute of Technology and Engineering, Jaipur, India

K. Ray et al. (eds.), *Proceedings of International Conference on Data Science and Applications*, Lecture Notes in Networks and Systems 148,
https://doi.org/10.1007/978-981-15-7561-7_18

Both sources because of their ever-changing aspects can, however, remove the issue by removing the limitation of one using other resource's power. Such an idea can give rise to the concept of "Hybrid Solar–wind power plant." Such hybrid stations of producing energy are considered as most beneficial in reducing fossil fuel depletion rates, along with being used in some small areas, with less or no power supply with no harm done to the natural environment.

The first is AC-to-AC, the second is AC-to-DC, the third is DC-to-AC, and finally, DC-to-DC. In the matter of functional description, modern power electronics fixed plan performs one or more of the following conversion functions. An optimistic outlook for renewable potential sources over the past few years so as to meet the argument of increased viability and reduce environmental issues. Renewable sources Fancy solar, wind, hydro have enough capabilities to compensate for increased demand. Commercial wind turbine generators located in wind farms are capable of generating large amounts of MW alone, but the presence of wind is a very unreliable factor because they can be very high at that point, safe operating point, especially in fancy tornadoes. Or at times it is too low, the cut-wind required to start the windmill is less than the wind speed.

2 Literature Review

Micro-hybrid power systems—A feasibility study [1], with information of designs on wind turbine and solar PV hybrid power generation, with specific plans to power 100 homes as well as elementary school and health clinics model communities. This particular study initiated by examining the potential of wind and solar sources in the interest area [2].

The Kyoto Protocol has established goals for participating countries to minimize greenhouse gas emissions by a minimum of five%. The 1990 amount is reduced in the commitment time via 2008 to 2012 [3]. US based on the Potency Information Administration, the planet's electric usage, is going to increase through 12,833 TWh in 1999 to 22,230 TWh throughout 2020, primarily driven by developing nations, wherever 2 billion folks continue to be with no electricity usage [4].

Wind power forecast as integration into the power grid, to evaluate alternatives for F-grid electrification in Bhutan Kingdom's rural villages [5]. This study was conducted in the country's 4 distinct locations. Moreover, only communications and lighting services consider the demand load. This paper mainly emphasized on the hybrid power generating units' optimization [6].

The wind/battery fixed plan is to be implemented in the Yangtze site. This particular session summarizes the presentation: "Digital Signal Processing for Green Power Systems and Delivery". There is a significant increase in the electrical systems that are accessed through wind power [7].

The mini-grid category introduces multipurpose electrical power service for communities with a population of Saw to Indelible El Thousand (probably 50 to

500 homes or more). With an overall strength lead from the most extraordinary daily of the day to the vaguely aloes KWH [8].

Because of the wind power's disruptive characteristics the wind power development faces a primary issue of grid integration. The power fixed plan security and stability issues help in accomplishing the wind farm's unpredictable power generation [9] (Table 1).

3 Research Methodology

3.1 Objectives Methodology

Renewable sources such as hydro, fancy solar as well as wind have more ability to recompense for the increased potential. Commercial wind turbine generators are basically situated in wind farms and they have the capability to produce high amount of MW. However, due to wind factor, it can be very cold at times, a safe operating fact, especially in fancy method or for a short time it is much less than the cut-wind speed required starting a windmill. This is the main reason for solar power, throughout the day it is present in scattered form, but the level of irradiation changes based on natural situations like shadow creates by trees, objects as well as clouds and so on. And solar power makes them an incredible source of energy. However, these 2 sources combined and MPPT algorithm was implemented, by this system power reliability maybe enhance. If one of the sources is not present or might be insufficient to fulfill the demands of loads, other energy may be able to recompense power difference. MPPT regulates by boost converter topology for PV panel.

3.2 Boundaries of Methodology

Few drawbacks were found during this research period, and assumptions were also made:

(a) In this particular research work, load application has been made as the actual capacity consumption as well as residents' income in work. Therefore, in order to accurate load quality data, better option is to take a survey with the power needs of the villagers.
(b) The imaginary method R of the houses taken is the old method in the particular area, though current method R was considered to calculate the actual electricity load applications of society.
(c) All techniques have applied to obtain the maximum accurate cost of power specific plan installation along with components costs.

Table 1 Parameter analysis of different methods

Title of the research papers	Methodology used	Disadvantage of research paper	Research gap	Advantage of my paper
Simulation of solar hybrid systems using PSIM	Integrated into PSIM software	System integrated, small wind into PCM software, the turbine is selected for maximum power to 2 kW	System Integrated into MATLAB software, a maximum 5 MW power is used for wind turbine	MATLAB is a fast one. Use a selection software at the system as well as circuit level. Having a private user interface. Mainly utilized in the control power converters and circuits' simulation
Micro-hybrid power systems	Electricity is used to avail the power which is a stand-alone direct power inverter system	Systems are currently using solar energy, unable to find more work than using windows. Energy	MPPT, the sun's maxim potential power is extracted from the system that is utilized by the boost converter system whenever it is available	Solar is a hybrid system integrated into MATLAB software
Wind-Solar Hybrid Power System for Annual Applications	Electricity and solar resources combine specific potential in electricity generation as a static stand-alone or grid-connected hybrid power system	1.5 CW is the hot season and 1.0 kW is the rainy season	Only grid-connected MPPT controller, maximum power of 5 kW	Having a private user interface. Mainly utilized in the control power converters and circuits' simulation
A Hybrid Wind-Solar Energy System: New Rectifier Stage Top OL Logic	Depending on the availability of front-end rectifier stage, two sources, the OUR source	There are additional input filters Do not fill the high-freedom harmonics	High-frequency harmonics are filtered with help of extra input filters: 2) All renewable resources are up/down (huge support) wind input as well as PV ranges. MPPT is realistic	The high-performance DC-DC converter is better in comparison with NeNeParent. PWM technology offers good harmonic decrease

(continued)

Table 1 (continued)

Title of the research papers	Methodology used	Disadvantage of research paper	Research gap	Advantage of my paper
Village Power Hybrid Systems Development	Study case. no methodology is developed	Adv. the developing world will experience Rapid growth in the application of electricity_ to meet community economic and social needs over the next decade	Modeling	Village power hybrid system development
Modeling and Control for Smart Grid Integration of Solar/Wind Energy Conversion System	Synchronous (induction) generator, P&O algorithm, 28.8 kw solar power system	The PV system available power greatly depends on solar radiation	A maximum of 5 MW power is used for wind turbine. MPPT controller is used	Every source realized through MPPT. The DC-DC converter having higher efficiency of nearly more than 90
Optimization of Hybrid Pv/Wind Power System for Remote Telecom	Telecommunication networks, In Hswps Sizing and analyzing. pragmatic applications are found by the proposed system	The telecom load that is controlled by these PV systems is 750 W	The sun's maximum possible power is extracted with the help of designed boost converter that uses MPPT	Direct mode is used for work, by the proposed hybrid system. Also, in case, the generated power by the solar and wind energy is greater than load, and then battery directly stores the extra power
Optimization of A Combined Wind and Solar Power Plant	Optimized active areas of a photovoltaic conversion system. Battery storage units	Very small unit of solar and wind different and then energy is stored in battery then connected in parallel	Direct mode is used for work by the proposed hybrid system. Also, in case, the generated power by the solar and wind energy is greater than load, and then battery directly stores the extra power	The sun's maximum possible power is extracted with the help of designed boost converter that uses MPPT

(continued)

Table 1 (continued)

Title of the research papers	Methodology used	Disadvantage of research paper	Research gap	Advantage of my paper
A Simple Sizing Algorithm for Stand-Alone Pv/Wind/Battery Hybrid Microgrid	This algorithm determines the photovoltaic array and wind turbine generator's generating units that are needed	Observation-based algorithm. hybrid micro grid is stand-alone	MPPT algorithm and system developed	Direct grid-connected system using MPPT controller and DC to DC booster
Wind Energy Conversion Systems	Wind energy case study	No solar part included in hybrid system of wind and solar	Modeling of hybrid system	Wind energy conversion systems

4 Experiment Simulations on MATLAB

See Tables 2 and 3.

4.1 MATLAB/Simulink Model Implementation

The simulation model was created primarily for the opportunity analysis of the use of renewable potential equipment, for their management in the design phase, and to study the problems that might be caused by the solution adopted. The solution adopted concerns the management solutions that are adopted, but also the supervision, regulation and mandate of renewable energy and the consumer.

The studies performed considered various configurations and the availability of renewable energy. The availability of renewable power sources was taken in the range of 0.8–3 kWh/m^2 for solar source, wind speed is between 2 and 20 m/s and water

Table 2 Parameters for wind turbines

Wind turbine parameter	
Nominal output power	5 kW
Base wind speed	12 m/s
Base rotational speed	10 m/s
Initial rotational speed	0.8 rpm
Moment of inertia	1 m kg m^2
Torque flag	0
Master/slave flag	1

Table 3 Parameters for the permanent magnet synchronous machine

Permanent magnet synchronous machine	
R_s (stator resistance)	1 m ohm
L_d (d-axis ind.)	1 m II
L_q (q-axis ind.)	1 mH. The d–q coordinate is defined such that the d-axis passes through the center of the magnet, and the q-axis is in the middle between two magnets. The q-axis is leading the d-axis
Vpk/krpm [peak line-to-line back emf constant, in V/krpm (mechanical speed)]	7112
No. of poles	P 30
Moment of inertia	100 m kg m^2
Master/slave flag	0

flow is between 30 and 100 l/s and it varies from 50 m level. Hydro Resource. Linear and nonlinear consumers (pi = 33 kW) were also considered.

4.2 *Result and Discussion*

4.2.1 Discussion

Thesis load application comes from a combination of PV arrays, wind turbines and batteries. An inverter is used to convert the output from solar and wind systems to AC power output. The circuit breaker is used to connect an additional load of 5 kW in a given time. This hybrid is controlled to provide maximum output power under all operating conditions to meet a fixed plan load. Either wind or solar is supported by batteries to meet specific plan loads. Also, simultaneous operation of wind and solar specific scheme is supported by battery for uniform load.

4.2.2 PV Modules

The power output of the solar cell is given by $P = V * I$. For simulation, solar irradiance and wind speed are used. The data is the input of the PV and the wind energy system. PV and wind energy are shown as waveforms of the fixed scheme and also the grid voltage and flow waveform. Grid voltage and current are analyzed by Fourier transform, and both harmonics grid voltage and current are calculated (Fig. 1).

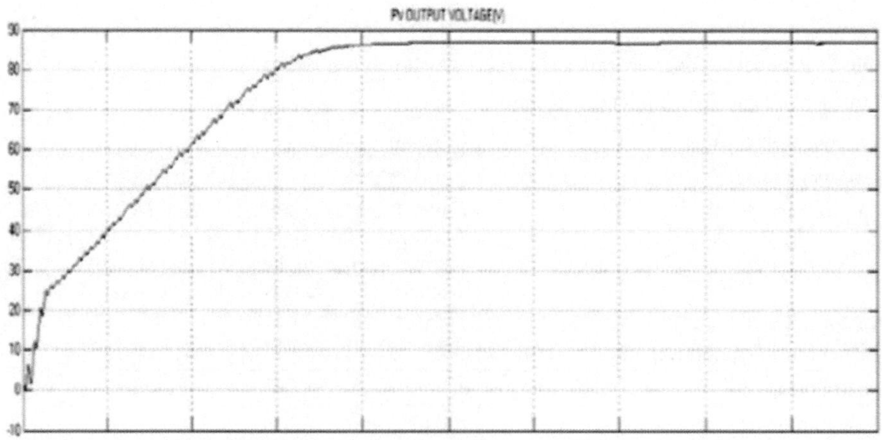

Fig. 1 Output voltage of PV system

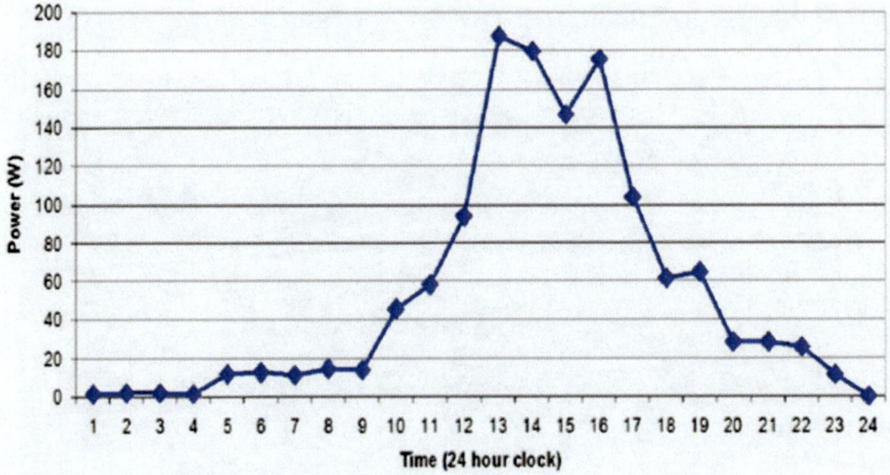

Fig. 2 Availability of power per assemblage area during a day

It is clear that the Earth receives maximum power from the sun between 11:00 am and 03:00 am. The designed battery bank is 48 V, 200 Ah. We also considered a DOD of 50%. Thus, the total capacity to charge 5000 drained batteries is 4800 Wh. For ease of design, let us assume that the power of the sun is available for at least 10 h a day. Thus, ideally a 480 W rated solar panel can produce 4800 Wh in 10 h. But in the practical case, this does not happen because of the variations in the availability of the sun's radiation. The panel rating of 175% is on the safe side of the design. Thus, the panel rating is 480 x 1.75 = 840 W. The easy options interning at 1 kW panel, considering the converter effect is 90%. This designed rating was only suitable for charging half-drained batteries at full charge (Fig. 2).

4.2.3 Wind Modular

Setting Gen Model = 3 allows users to write user-defined generator models in Fortran. Nevertheless, we have chosen to develop generator models using Simulink's more visual block diagram representation. For this purpose, we could not set VSContrl = 0 because FAST requires input from Simulink. Setting VSControl = 1 will use a fast-built, simple, variable-speed generator model: While setting VSContrl = 2 allows users to write their own variable-speed generator model in a Fortran. None of these options was applicable. Setting VSControl = 3 allows input from Simulink (Figs. 3 and 4).

The results show that from a significant step in wind speed, from 12 m/s to 15 m/m, HSS speed made very small (less than 1%) changes (128–128.5 rad/s) adeptly, the output torque and power showed major changes. There was no pitch regulation at the venue. The oscillations in the volume were the end of the torque shadow effect, which

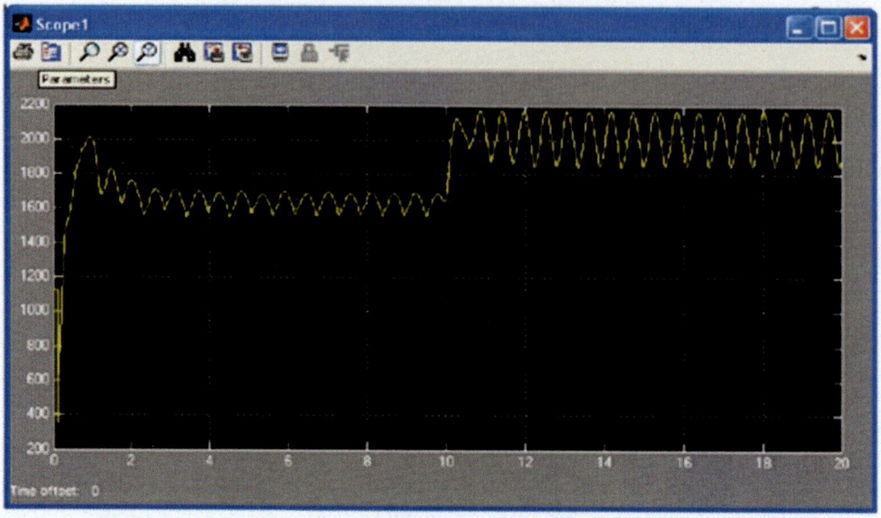

Fig. 3 Example of a MATLAB scope output during run time

Fig. 4 Completion of wind power plant in MATLAB

caused by changing the medium-torque from disturbing the tower's wind, causing the wind to flow in a downwind turbine. This model does not have the ability to convert electrical stimulation into a generator, as it was effectively connected to a fully fixed-voltage, fault-proof, "infinite" bus. Thus, electrical defects and their effects cannot be pretended to the extent associated with the turbine execution (Figs. 5 and 6).

This model does not have the ability to convert electrical stimulation into a generator, as it was effectively connected to a fully fixed-voltage, fault-proof, "infinite" bus.

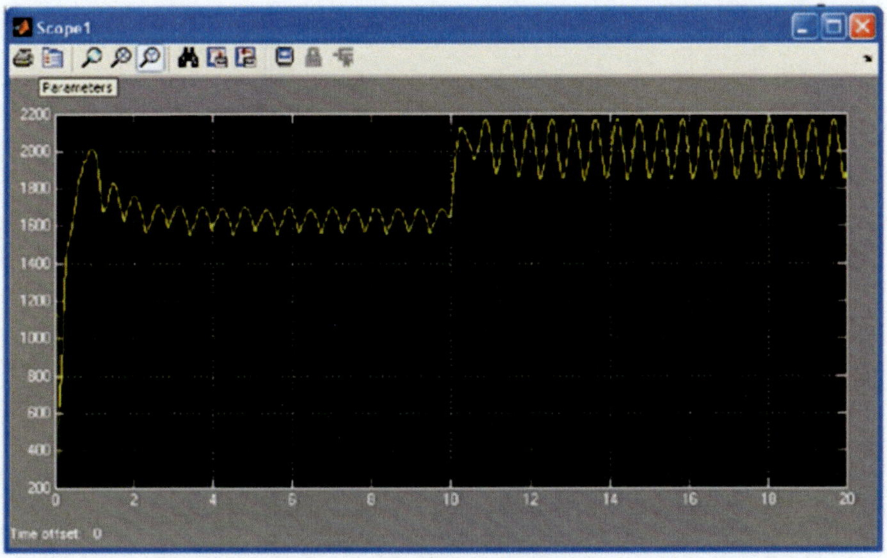

Fig. 5 Example of a MATLAB scope output during run time

Fig. 6 Completion of wind power plant in MATLAB

Thus, electrical defects and their effects cannot be pretended to the extent associated with the turbine execution.

The physical diagram and power-speed characteristic of the type 3 WTG is shown in Fig. 7a, b. This shows some results from the simulation of the single-phase sag. Figure 7a shows the torque, speed and power on the high-speed shaft. This signal

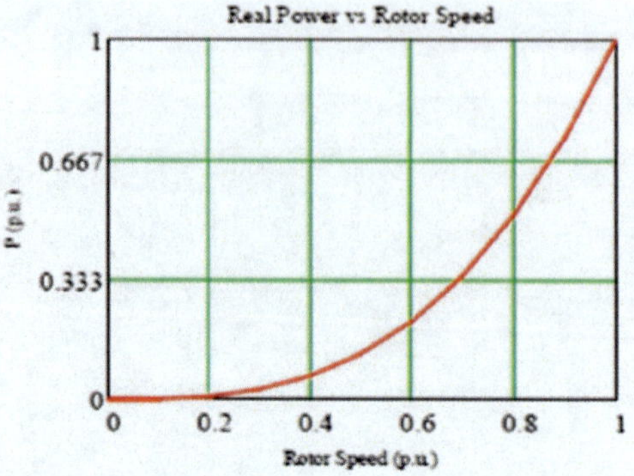

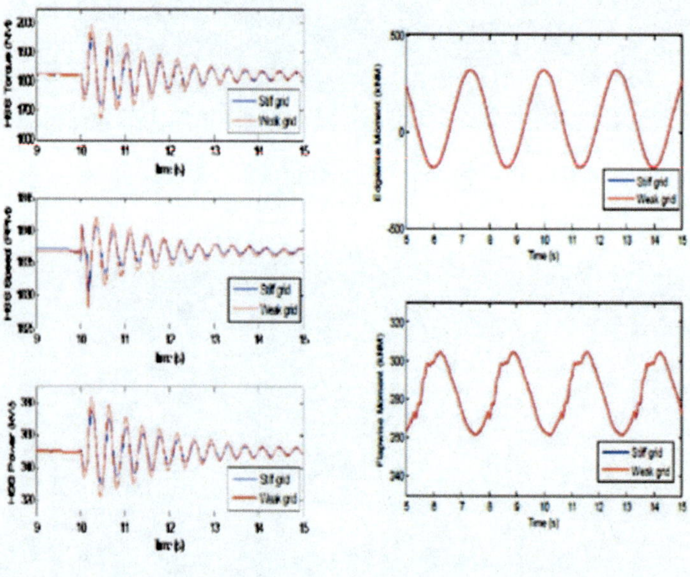

Fig. 7 Simulation results showing the impact of a single-phase voltage sag for a type 3 WTG on **a** high-speed shaft torque, speed and power and **b** edgewise and flap-wise blade moments at the blade root

causes the fault to be about 2.5-Hz oscillations that last long after the fault is cleared, indicating that some mechanical oscillations in the drive were fancy excited.

4.2.4 Three-Phase Regulatory Rectifiers

Three-phase regulated rectifiers were completed using the MATLAB Simulation. The results of the simulation are as follows. The rectifier diagram shows the output

Fig. 8 Output waveform of three-phase full wave regulated rectifier with R load

waveform of a three-phase regulated rectifier, a pulse generator of six IGBTs, a model file in the MATLAB/Simulink, respectively (Figs. 8 and 9).

5 Buck Boost Convener

The plot shows a lower output voltage compared to the reference voltage. It shows the changing load current and the power of the PWM cycle average of the two MOSFETs (Figs. 10, 11 and 12).

6 Wind Power MPPT

Time Simulation The hybrid power is carried out for a specific scheme with constant load under sufficient wind and senses. Turbine output power characteristics and wind power output characteristics are analyzed (Figs. 13 and 14).

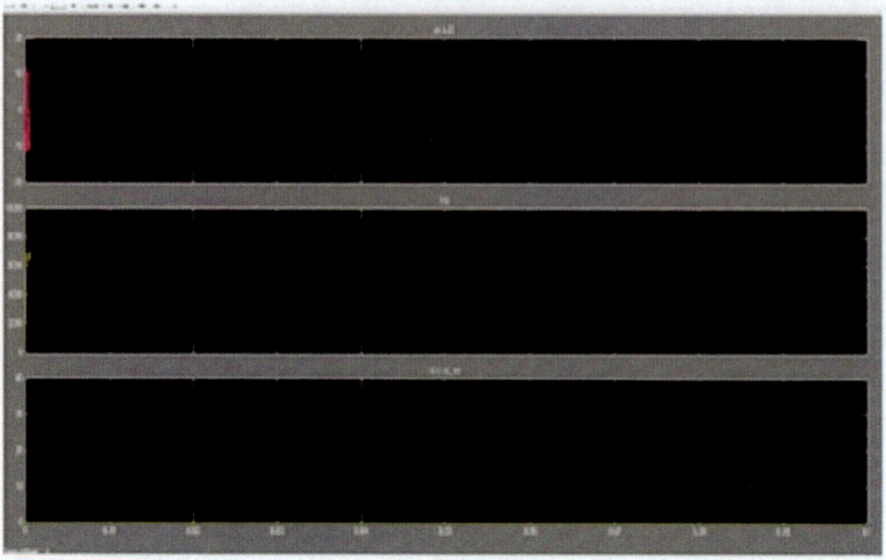

Fig. 9 Result of rectifier

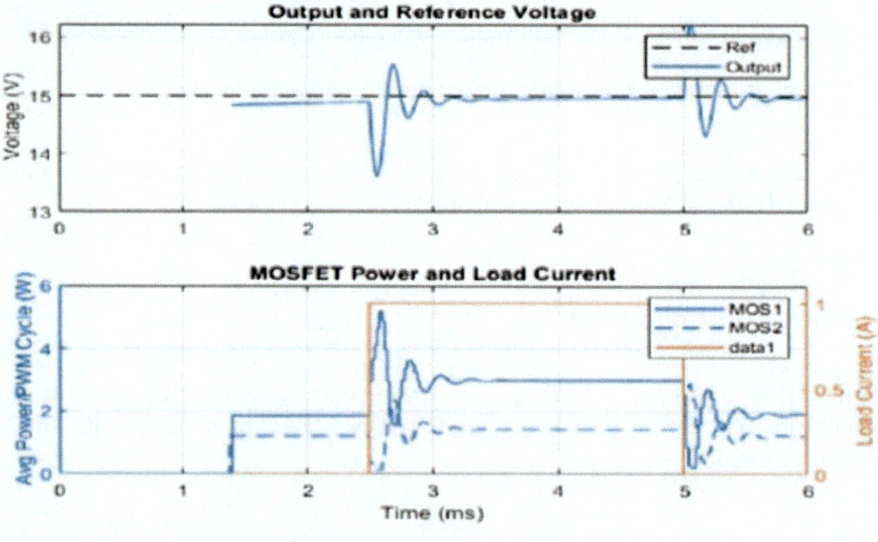

Fig. 10 The plots show different implementations of the PI Regulator

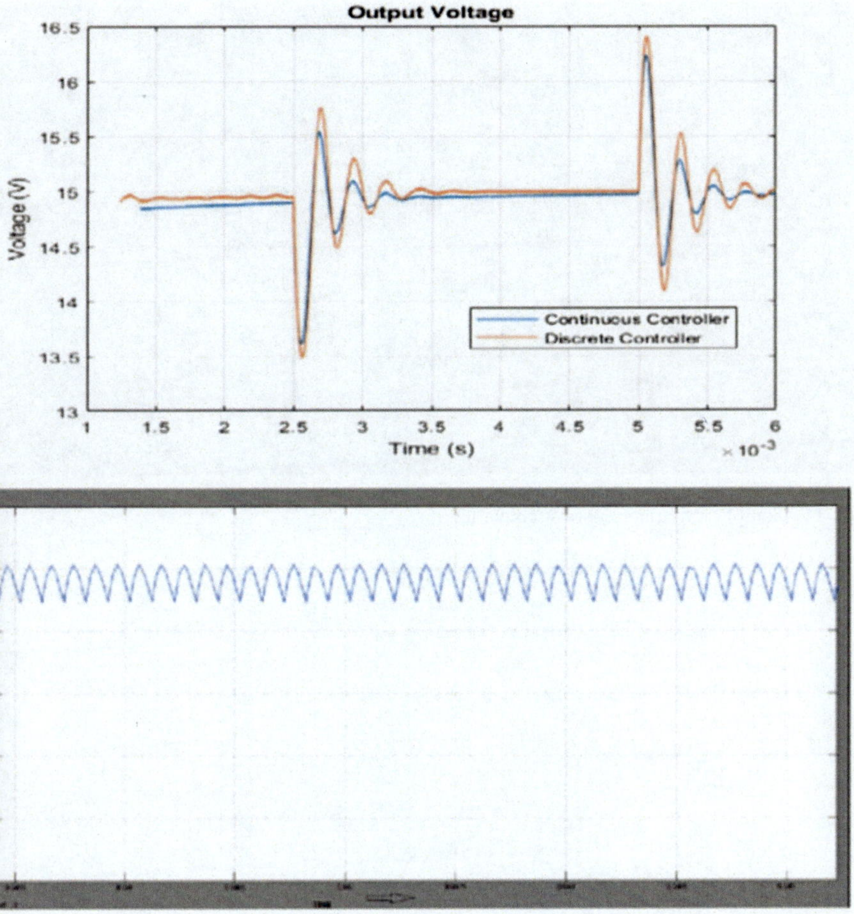

Fig. 11 Simulation completion of DC-DC converter (voltage versus time)

The consultation algorithm is tested under different conditions of the rapidly changing wind speed as shown in Fig. Wind turbines find the maximum power point of the wind speed. The reference input regulator MPPT algorithm was developed accordingly to track current injector. An effective tracking feature of the consultant MPPT algorithm, separated by power characteristics, is shown in Fig.

The power coefficient of the particular scheme shown in Fig. By observing the CP, the change in wind capacity proves the maximum power point tracking capacity of the fixed plan. This completion proves the work of the wind potency conversion fixed plan at maximum power point.

Fig. 12 Completion of buck–boost converter

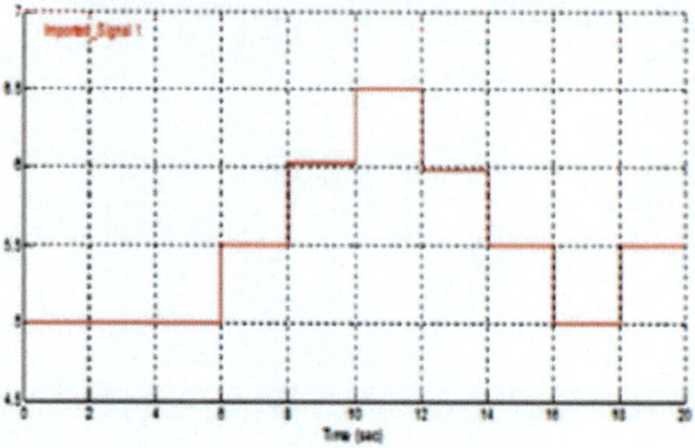

Fig. 13 Wind velocity variation

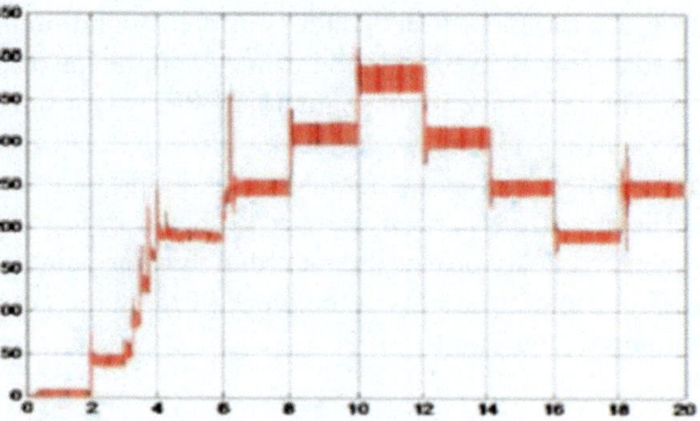

Fig. 14 Power variation with change in wind velocity

7 Grid Connected Hybrid System

The p–v characteristics of the solar cell are taken for the purpose of simulation. DC-DC Boost Converter Topology is used for stutter performance. Boost converter topology is also used in additional carrying method for high efficiency. An LC filter is added at the end of the boost stage to remove the high frequency ripple from the output voltage waveform (Fig. 15).

Wind velocity is not so high, but sufficient to generate electric power, so the first and important step is to increase the swell area. One priority area is 3.5 m^2. In addition, the power engender by considering the its limit is shown in the figure. Taking a turbine's cut in speed of 3.5 m/s, it is clear from the figure that a total power of more than 1 kW (assuming that the generator effects are at least 90%) can ignite the wind during the daytime (Fig. 16).

As the wind potentials are aerodynamically comparable to solar energy, we will reduce the capacity of the solar panel connected to the circuit. The radiation of the sun as it approaches the project site is a variable amount. The specific incident strength of the assemblageria against the time of day is shown in Fig. The design rating of the 1 kW panel is shown in the figure (Figs. 17 and 18).

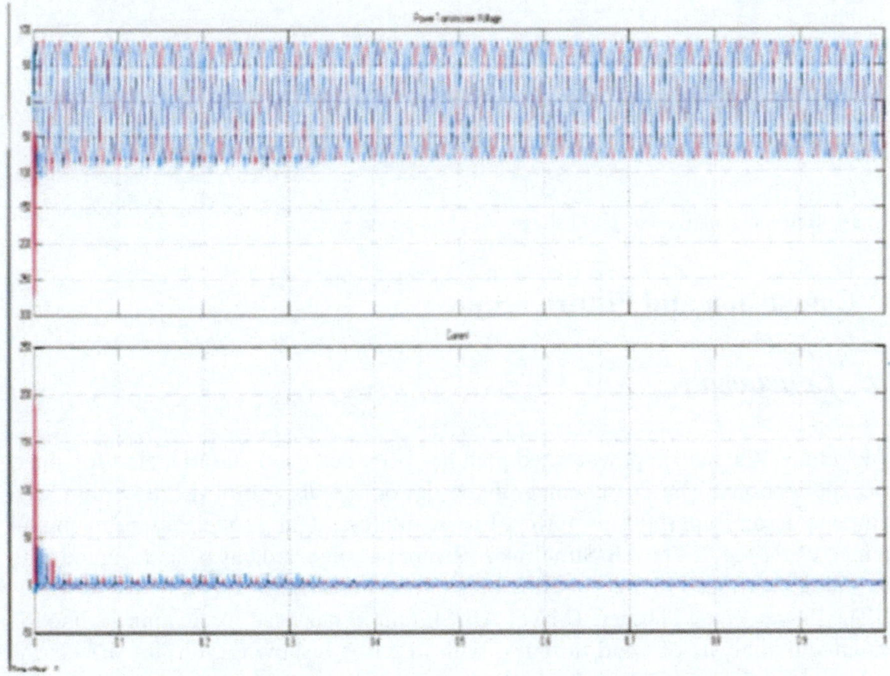

Fig. 15 Completion of the hybrid specific scheme shown in the Simulink above

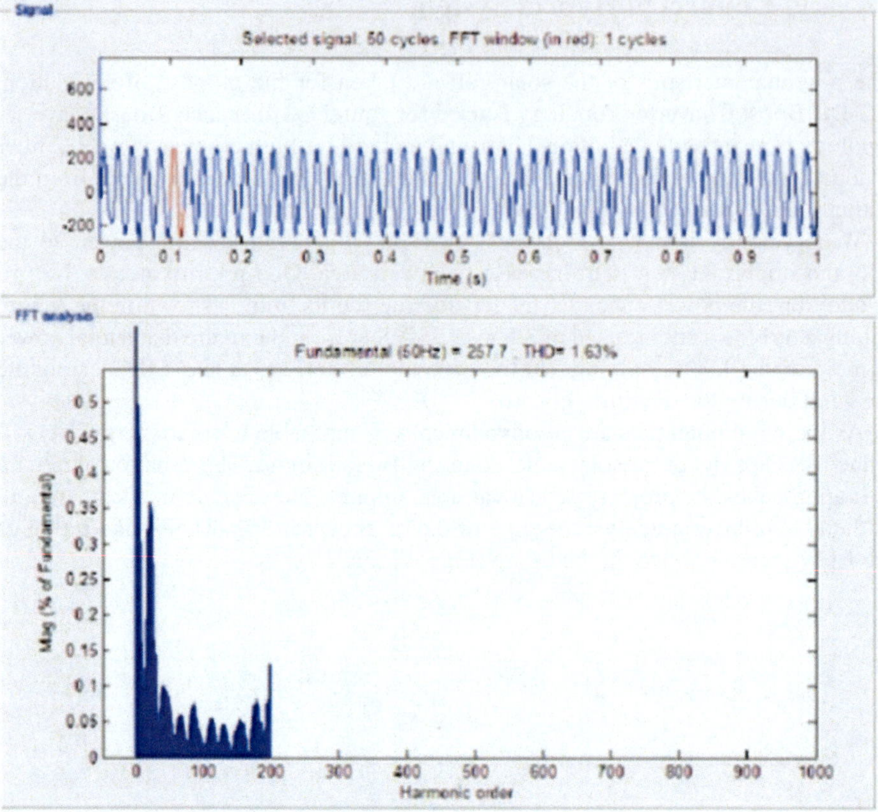

Fig. 16 Harmonic analysis of grid voltage of hybrid power S) stem

8 Conclusion and Future Scope

8.1 Conclusion

A novel PV/WT hybrid power fixed plan has been designed and modeled for smart grid applications. The co-schemes of the developed algorithm make certain plan components and appropriate power flow regulators. This model has been implemented using the MATLAB/Simulink software package and has been designed with the dialog buffs used in the Simulink blockchain libraries.

The Power Fixed Planners (MATLAB/Simulink) has used for an approach to the design and analysis of wind turbines, both in terms of how the turbine affects the grid and how the grid affects the turbine.

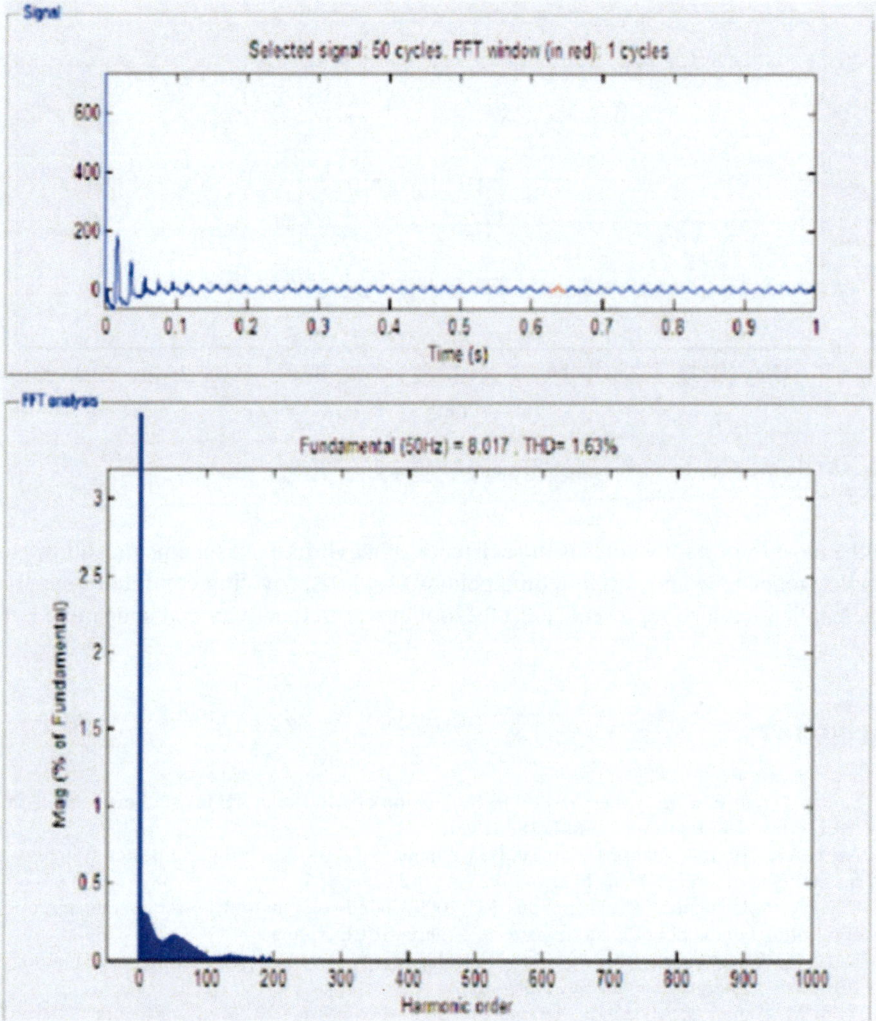

Fig. 17 Harmonic analysis of grid current

8.2 Discussions

Modelling of solar–wind–hydroelectric hybrid fixed scheme in the Simulink environment. The application is useful for analyzing and simulating the actual hybrid solar–wind–hydroelectric precision scheme connected to the public grid. Each component is built on an application modular architecture for easy study of module performance. Blocks fancy wind model, solar model, hydroelectric model, potency conversion and load are applied and the results of the simulation are also presented. For example, one

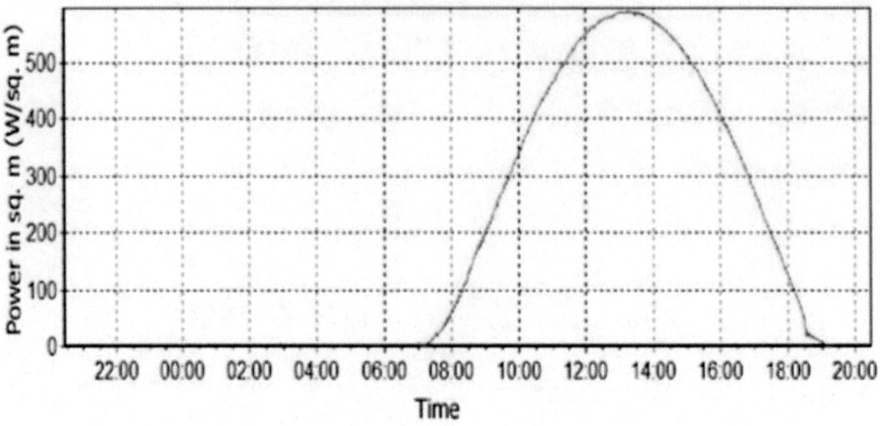

Fig. 18 Hourly average wind powers harnessed during July 2018

of the most important studies is the existence of a hybrid fixed scheme that allows to employ renewable and variable time potassiors while providing continuous supply. The application also represents a useful tool in research activity and education.

References

1. Patil AP, Vatti RA, Morankar AS (1999) Simulation of wind solar hybrid systems using Psim. IEEE Proc-Elect Pow Appl 146(2):193–199
2. Arjun AK, Athul S, Ayub M, Ramesh N, Krishnan A (2005) Micro-hybrid power systems—a feasibility study. IEEE Trans Energy Convers 20(2):398–405
3. Cajethan N,Uchenna UC, Theophilus M (2009) Wind-solar hybrid power system for rural applications in the South Eastern States of Nigeria. IEEE Xplore
4. Hui J, Bakhshai A, Jain PK A hybrid wind-solar energy system: a new rectifier stage topology. IEEE Trans Energy Convers 20(2):398–405
5. Laidi M, Hanini S, Abbad B, Merzoukand NK, Abbas M (1999) Study of a solar Pv-wind-battery hybrid power system for a remotely located region in The Southern Algerian Sahara. IEEE Proc-Elect Pow Appl 146(2):193–199
6. Flowers L, Green J, Bergey M, Lilley A, Mott L (2007) Village power hybrid systems development in The United States. IEEE Xplore
7. Patel PR, Singh NK (2012) Modelling and control for smart grid integration of solar/wind energy conversion system. IEEE Xplore
8. Paudel S, Shrestha MS, Neto JN, Ferreira FJ, Adhikari JAF (2003) Optimization of hybrid Pv/wind power system for remote telecom station. 18(4):493–502
9. Chang L(2001) Nb wind energy conversion systems. University of N. Brunswick

Accident Prediction Modeling for Yamuna Expressway

Parveen Kumar and Jinendra Kumar Jain

Abstract Expressways, which are being made at very fast pace in recent times in India, are very unsafe for road users. In this study, accident data of 165.5 km of Yamuna expressway was analyzed, and a negative binomial model was developed for prediction of accidents on the expressway. Percentage of cars in traffic volume, horizontal curvature in degree per km, length of horizontal curve per km, number of vertical curves per km, and traffic in terms of 1000 PCU per day are taken as explanatory variables. Three explanatory variables, percentage of cars in traffic volume, number of vertical curves per km, and traffic in terms of 1000 PCU per day have significant effect on accident frequency for negative binomial model. According to the model, the accidents increase by 125% with increase of 1000 PCU per day in traffic and 5.4% with 1% increase in percentage of cars. However, accidents decrease by 3.3% with increase of one vertical curve per km.

Keywords Highway safety · Accident prediction model · Negative binomial model

1 Introduction

To travel long distances by roads in lesser time period controlled access expressways are being made in India. Approximately 1581 km of expressway is presently operational in India. However, The National Highways Development Project by Government of India aims to expand the highway network and plans to add an additional 18,637 km of expressways by 2022.

The rapid growth of numbers of vehicles and good road network is required for the fast development of the country, but it should not be at the cost of safety of road users. Unfortunately, in India, safety of road users is not on priority. Consequently, road

P. Kumar (✉) · J. K. Jain
Civil Engineering Department, Malaviya National Institute of Technology, Jaipur, India
e-mail: parveen.rh.kumar@gmail.com

J. K. Jain
e-mail: jkjain05@gmail.com

K. Ray et al. (eds.), *Proceedings of International Conference on Data Science and Applications*, Lecture Notes in Networks and Systems 148,
https://doi.org/10.1007/978-981-15-7561-7_19

traffic injuries are one of the leading causes of death, disabilities, and hospitalization in India. In 2017 only, 1, 47,913 persons have lost their lives and 4, 70,975 persons have injured in road accidents in India. While the absolute numbers for road accident, fatal accident, fatality, and injury are on decline in 2017 as compared to previous years, howerver, accident severity, i.e., number of persons killed per 100 accidents, is continuously kept on rising in India [1]. Fatal road accident victims largely constitute young people in the productive age groups. The socio-economic cost of road accidents is estimated about 3% of GDP of Indian economy, whereas the grief and pain of families of victim of accidents cannot be calculated in monetary terms.

Fatality rate on national highways is 0.67 deaths per km per year. However, fatality rate on expressway is 1.8 deaths per km per year [2]. This high rate of fatalities on expressways needs to be critically examined to find out causal factors of these accidents and to improve safety on the expressways. No study has been conducted to quantify the effects of various factors responsible for accidents on its frequency on expressway in India. Keeping this in mind, this study examines road safety on Yamuna expressway, which is India's first longest expressway (165.537 km) connecting the national capital of India, Delhi to star tourist attraction Taj Mahal at Agra.

2 Objective of the Study

The main objective of the study was to quantify the effect of identified causal factors on accidents by using best predictive model.

3 Literature Review

Singh et al. [3] reviewed the problems associated with the accident prediction modeling in Indian conditions and reported that under reporting, poor quality of available data, over dispersion, low sample mean and small sample size, under dispersion, heterogeneity of traffic, and fixed parameters are main problems associated with the accident prediction modeling.

La Torre et al. [4] developed negative binomial (NB) model to improve road safety along Italian freeway network analyzing accident data collected on 884 km over five years using highway safety manual approach. Base line models and crash reduction factors were developed for horizontal curvature, inside and outside shoulder width, proportion of the segment with a barrier in the median and roadside, proportion of AADT during hours where volume exceeds 1000 veh/hr/lane and percentage of heavy good vehicles. The R^2 achieved was 0.271. The model developed in this study was used to estimate fatal and injury single or multiple vehicles accidents per year in one direction of travel.

Ma et al. [5] developed random effect negative binomial (RENB) model to investigate the relationship between accident frequency and potential influencing factors on a 50 km long expressway in china by taking 567 accidents during 2006 to 2008. Degree of curvature, curve length ratio, curvature change rate, weighted curvature, longitudinal grade, grade differences, super-elevation, road width (m), ratio of longitudinal grade and curve radius, special segment, and AADT (passenger cars per day) were taken as explanatory variables in this study. Three explanatory variables, longitudinal grade, road width, and AADT were found having significant effect on accident frequency for RENB model. Longitudinal grade may have a decreasing impact on accident frequency. Road width and AADT both have an increasing impact on accident frequency.

Ture Kibar et al. [6] developed NB model for analyzing truck accidents on interurban roads in Turkey. Section length, vertical grade, curvature per km, lane width, number of lanes, shoulder width, median width, number of junctions, AADT/lane/1000, truck percentage, and average speed of trucks were taken as explanatory variables in this study. AADT per lane, truck percentage, and lane width were found significant. Increased AADT per lane and truck percentage results in more truck accidents while increase in lane width results in less truck accidents.

The literature review suggests that no study has been reported on Yamuna expressway which tried to find out statistical correlation between accidents on the expressway and the possible risk factors.

4 Methodology

Road accident data, traffic volume data, and road features data like horizontal and vertical alignment, radius of horizontal curve, intersections, flyovers, underpasses, acceleration, and deceleration lanes were received from Yamuna Expressway Industrial Development Authority. To perform accident prediction modeling, the whole stretch of Yamuna expressway was divided into 16 equal sections of 10.3125 km each. Accident data of each year was inserted with each section of the expressway. Thus, for six years of accident data, 96 entries were obtained. Out of 96 data point, four year data from 2012–16 forming 64 data points was taken as training dataset, and remaining two year dataset pertaining to 2017–18 forming 32 data points was used as test dataset. The explanatory variables considered in the prediction model were traffic in terms of 1000 PCU per day (T), traffic composition, i.e., percentage of cars in traffic volume (P Car), horizontal curvature in degree per km (HCDKM), length of horizontal curve per km (LCKM), and number of vertical curves per km (VCKM). Accident frequency in terms of accident per year (A) was considered as dependent variable. The descriptive statistics of these variables are given in Table 1.

Poisson model and negative binomial model were applied on training data using SPSS software. As the data of accident frequency showed over dispersion (variance > mean; Table 1 column 7 and 8), negative binomial model was selected for modeling as suggested in the literature [7]. Both the models were compared by goodness of fit statistics, and the model that has lesser AIC (Akaike's information criterion) and BIC (Bayesian information criterion) was finally selected. Hypothesis testing was

Table 1 Descriptive statistics of model variables

Type of variable	Variable name	Abbreviation	No. of data point	Min.	Max.	Mean	Std. deviation
Dependent	Accident frequency per year	A	96	15	126	54.66	22.544
Independent	Percentage of cars in traffic volume	P Car	96	72.14	77.83	74.287	2.034
	Horizontal curvature in deg. per km	HCDKM	96	0.587	26.863	12.721	7.178
	Length of horizontal curve per km	LCKM	96	0.079	1	0.589	0.250
	Number of vertical curves per km	VCKM	96	7.144	21.300	17.27	3.279
	Traffic in terms of 1000 PCU per day	T	96	8.160	9.526	8.939	0.350

done in SPSS software to check the significance of effect of different independent variables taken into the model on the numbers of accident. Taking 95% confidence interval, independent variable that had significance value less than 0.05 was kept in the model, and those having significance value more than 0.05 were excluded from the model. Using the final model, accident frequencies were predicted for both train and test data, and *R* square values were calculated for both test and train dataset.

5 Data Analysis

5.1 Exploratory Data Analysis for Predictive Modeling

To understand relationship of accidents with various independent variables, scatter plots were made as shown in Fig. 1. Fig. 1 shows a possibility of positive association of accidents with traffic volume and an indication of a negative association between accidents and vertical curves. Accidents first decrease as percentage of car increases but again increases as percentage of car increase, although the rise seems not very

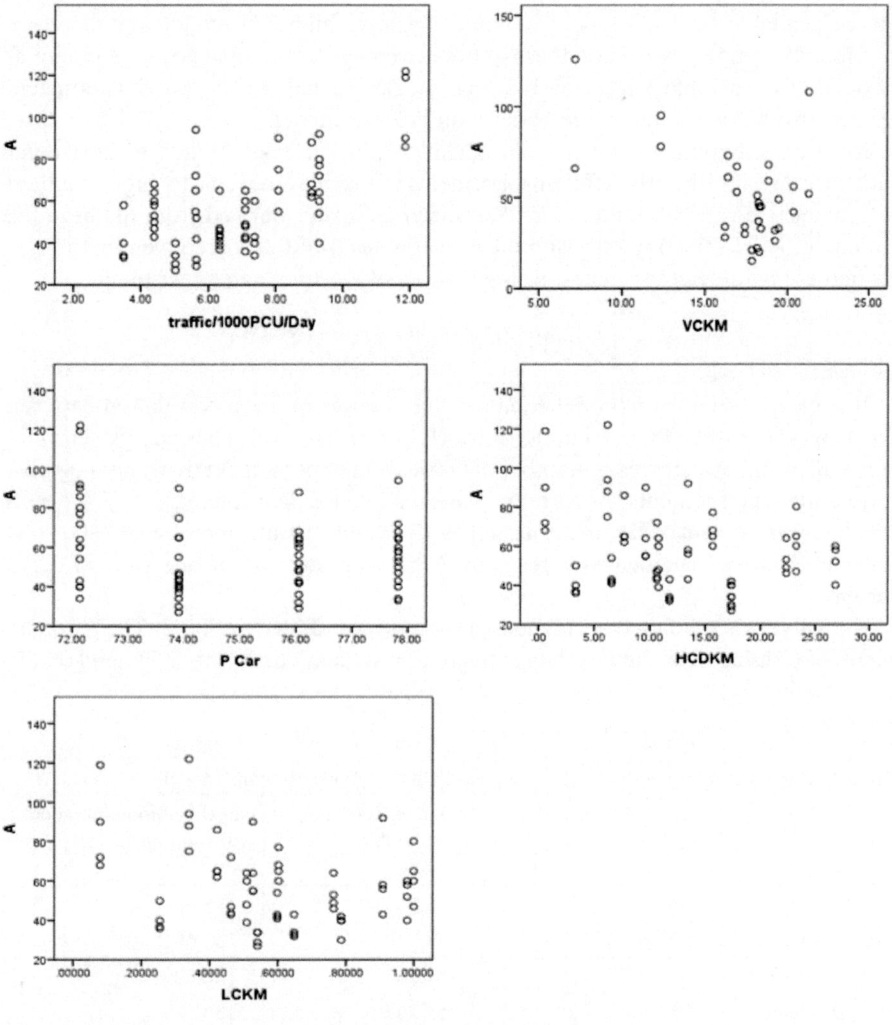

Fig. 1 Variation of accident with various independent variables

prominent. There seems a decreasing trend in accidents with increase in horizontal curvature in degree per km and also a reduction in accidents with increase in length of horizontal curve per km length of expressway.

5.2 Accident Prediction Modeling

SPSS software was used to develop prediction model for accident frequency. As already given in Table 1, the accident data being over dispersed, and Poisson model was not suitable for such type of dataset. Negative binomial model was therefore developed. Comparison of goodness of fit derived by Poisson and negative binomial model is given in Table 2. It is clearly seen from Table 2 that goodness of fit parameters for negative binomial model are better than Poisson model.

Negative binomial model did not consider two variables related to horizontal curvature as significantly affecting accidents. This may be due to high standard of geometric design of expressway. Parameter estimates derived from the negative binomial model selecting only significant variables ($p < 0.05$) are given in Table 3. From these parameter estimates, following model equation can be written:

$$A = 0.001(\mathrm{T})^{0.811} e^{[0.053(\mathrm{P\ Car})-0.034(\mathrm{VCKM})]}$$

It is clearly seen from model equation that there is positive correlation between accidents and traffic (T). Percentage of car (P Car) in traffic also has positive correlation with accidents. However, number of vertical curve per km (VCKM) has negative correlation with accidents. According to this model, accidents increase by 5.4% with 1% change in percent of cars in traffic and by 125% with a traffic increase of 1000 PCU per day. However, accidents decrease by 3.3% with increase of one vertical curve per km.

Actual versus predicted accidents scatter plots are shown in Fig. 2. The R square values of training data and test data extracted from these curves are 0.577 and 0.385, respectively.

Table 2 Comparison of goodness of fit by poisson and negative binomial model

Description	Poisson model with all variables	Negative binomial model with all variables	Negative binomial model with significant variables only
Deviance	160.047	148.940	135.681
Scaled deviance	160.047	148.940	135.681
Person chi-square	162.467	151.263	136.629
Scaled person chi-square	162.467	151.263	136.629
Log likelihood	−266.344	−268.628	−260.940
AIC[a]	544.682	551.256	531.880
Finite sample corrected	546.162	553.256	532.915
BIC[b]	557.642	566.368	542.675
Consistent AIC	563.642	573.368	547.675

[a]Akaike's information criterion
[b]Bayesian information criterion

Table 3 Parameter estimate

Parameter	B	Std. Error	95% Wald confidence interval		Hypothesis test			Exp(B)	95% Wald confidence interval for Exp(B)	
			Lower	Upper	Wald Chi-square	df	Sig.		Lower	Upper
(Intercept)	−6.503	2.440	−11.287	−1.720	7.100	1	0.008	0.001	0.00001	0.179
P Car	0.053	0.018	0.016	0.090	7.847	1	0.005	1.054	1.016	1.094
VCKM	−0.034	0.007	−0.048	−0.019	20.162	1	0.000	0.967	0.953	0.981
T	0.811	0.138	0.540	1.083	34.235	1	0.000	2.251	1.715	2.954
Over-dispersion parameter				0.005[a]						

[a]Hessian matrix singularity is caused by the scale or negative binomial parameter

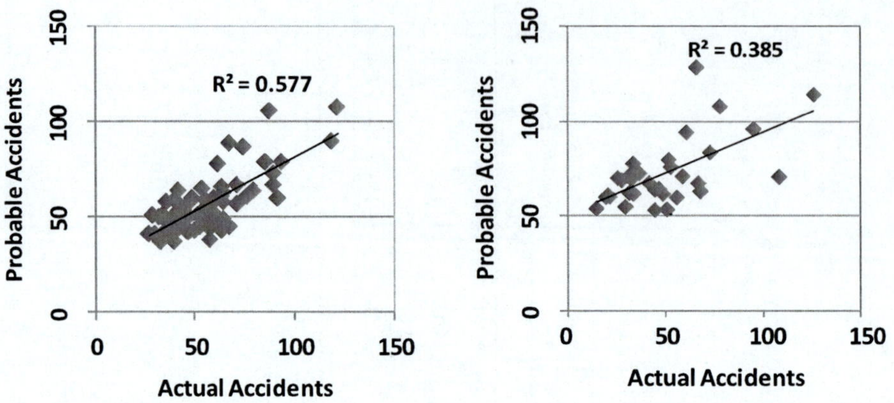

Fig. 2 Scatter plot of actual versus predicted accidents with training and test datasets

6 Conclusions

This study was conducted on Yamuna expressway, and six year data of accidents comprising a total of 5247 accidents involving 9013 persons was analyzed. Three explanatory variables, percentage of cars in traffic volume, no. of vertical curves per km, and traffic in terms of 1000 PCU per day have significant effect on accident frequency for negative binomial model.

The study suggests that the accidents increase by 125% with increase of 1000 PCU per day in traffic and by 5.4% with 1% increase in percent of cars. However, accidents decrease by 3.3% with increase of one vertical curve per km.

References

1. Ministry of Road Transport and Highways, Government of India, New Delhi. Road Accidents in India 2016. https://morth.nic.in/road-accident-in-india
2. Mohan D, Tiwari G, Bhalla K (2016) Road safety in India: status report 2016. http://tripp.iitd.ernet.in/publication/report
3. Singh G, Sachdeva SN, Pal M (2017) Predictive modelling of road accidents in India: a review. Ind High 45(6):29–38
4. La Torre F, Meocci M, Domenichini L, Branzi V, Paliotto A (2019) Development of an accident prediction model for Italian freeways. Acc Anal Prev 124:1–11
5. Ma Z, Zhang H, Steven I, Chien J, Wang J, Dong C (2017) Predicting expressway crash frequency using a random effect negative binomial model: a case study in China. Acc Anal Prev 98:214–222
6. Ture Kibar F, Celik F, Wegman F (2017) Analyzing truck accident data on the interurban road Ankara–Aksaray–Eregli in Turkey: comparing the performances of negative binomial regression and the artificial neural networks models. J Trans Safe Secu 11(2):129–149
7. Sharma AK, Landge VS (2013) Zero inflated negative binomial for modeling heavy vehicle crash rate on indian rural highway. Inter J Adv Eng Tech 5(2):292

Optical Image Encryption Algorithm Based on Chaotic Tinker Bell Map with Random Phase Masks in Fourier Domain

Sachin, Archana, and Phool Singh

Abstract Double random phase encoding is a well-received optical encryption scheme for images by the scientific community. However, with an increase in computational power, double random phase encoding has been proven to be vulnerable by an intruder to some basic attacks like known plaintext, chosen plaintext and chosen ciphertext attacks. Some researcher applied a chaotic map with double random phase mask to endure these basic attacks. In this paper, a novel image encryption scheme in which Tinker bell map is used along with a double random phase mask in the frequency domain. Tinker bell map is a chaotic map which is relatively less explored. It has four parameters and two initial values which are highly sensitive to their values. Simulations are carried on Matlab on grayscale images. Statistical attacks based on histogram, entropy, 3-D plot and Correlation distribution analysis are performed on the scheme. The scheme is also evaluated for its efficacy against the noise and occlusion attack. The results show that the proposed scheme is highly secure and resist basic attacks.

Keywords Chaos · Double random phase encoding · Chaotic map · Tinker bell map · Image encryption · Phase mask

1 Introduction

As a rapid growth has been witnessed in the use of computer and internet, mostly the information stored and transmitted in the form of images and videos. During the transmission and storage of information, security remains the primary goal. The information may be social media account details, bank account details, one-time password, biometric details, medical reports, police identification procedure, government

Sachin (✉) · Archana
Department of Mathematics, Central University of Haryana, Mahendergarh 123031, India
e-mail: sachin191219@cuh.ac.in

P. Singh
Department of Mathematics, SOET, Central University of Haryana, Mahendergarh 123031, India

K. Ray et al. (eds.), *Proceedings of International Conference on Data Science and Applications*, Lecture Notes in Networks and Systems 148,
https://doi.org/10.1007/978-981-15-7561-7_20

secrets, confidential video conferences, and many more. If information is not secured then it can be easily hacked by an intruder. The unauthorized use of information is very dangerous. There are many conventional ways to protect the information from an intruder. Encryption of the original information of an image or a video is the most popular way to secure the information. The conventional image encryption methods like data encryption standard (DES), advance encryption standard (AES) are a popular method for the security of data but their complex computation and complexity, make them not easy to perform in real-time. To send the information in a secure way, an encryption algorithm is used and its private keys are required at both ends, i.e., the sender's as well receiver's end. Keys play an important role in the secure transmission of information. If keyspace is small then an intruder can easily apply brute force attack and retrieve the original key. Many cryptographic theories rely on mathematics. Fundamentals of cryptography rely on probability and number theory. In number theory, modular function provide a large keyspace from a small set of numbers.

In the last two decade, a wide infrastructure of cryptographic technologies has been developed. But still, there is a huge demand for advanced encryption algorithms. Therefore, many institutions provide a huge amount of money for developing cryptographic technologies. For image encryption, recently digital, optical and hybrid, i.e., optoelectronic scheme have been developed [1–20]. Most popular optical encryption scheme is double random phase encoding (DRPE) has been proposed by Refregier and Javidi in 1995 [21]. In DRPE, two random phase mask is applied in spatial and frequency domain. The encrypted image resembles a white noise stationary image in the spatial domain. Several optical and optoelectronic encryption methods were also proposed in the light of DRPE. However, attackers proved that DRPE is vulnerable under some basic cryptographic attacks like known plaintext, chosen plaintext and chosen ciphertext attack [15, 22, 23]. Therefore, many encryption schemes are proposed to endure these basic attacks.

Chaotic maps and cryptosystems are closely related to each other. Chaotic maps have a property like uncertain to predict, sensitive to their parameter and initial value, random behavior and many more which make it a perfect part of cryptosystem. Initial values of their parameter are the main key in order to study the behavior of the chaotic system. If the investigator doesn't know the exact initial value of parameter then chaotic system shows abrupt behavior. Due to these properties of chaotic system, chaos-based cryptosystem is secure for image encryption.

There are many chaotic dynamical maps exist. In this paper we present a novel image encryption scheme in which a chaotic map Tinker bell is used along with double random phase encoding. Many chaotic maps like a logistic map, cubic map, baker map, Gauss map, sine map and many more exist but we choose Tinker bell map because it has more parameters thus enlarge keyspace. Tinker bell map is less explored as compared to the other chaotic maps. Since Tinker bell map is very sensitive to their parameters and initial values, so it provided a large keyspace which makes it very difficult to apply brute force attack on it.

2 The Principle

2.1 Tinker Bell Map

Tinker bell map [24, 7] is a two-dimensional discrete-time map having four parameters which is used for generating a random sequence. The graphical picture of map is like a bell as shown in Fig. 1. Tinker bell map produces a periodic but divergent sequence which is used as a key in chaos-based cryptosystem. Mathematically, Tinker bell is given by

$$x_{n+1} = x_n^2 - y_n^2 + a_n x_n + b_n y_n$$

$$y_{n+1} = 2x_n y_n + c_n x_n + d_n y_n$$

Here x_0 and y_0 are the initial value which works as the main key in sequence generation, a_n, b_n, c_n, d_n are parameters. Initial value of parameter $a_0 = 0.9, b_0 = -0.6013, c_0 = 2, d_0 = 0.5$ is considered in this paper. For the purpose of getting

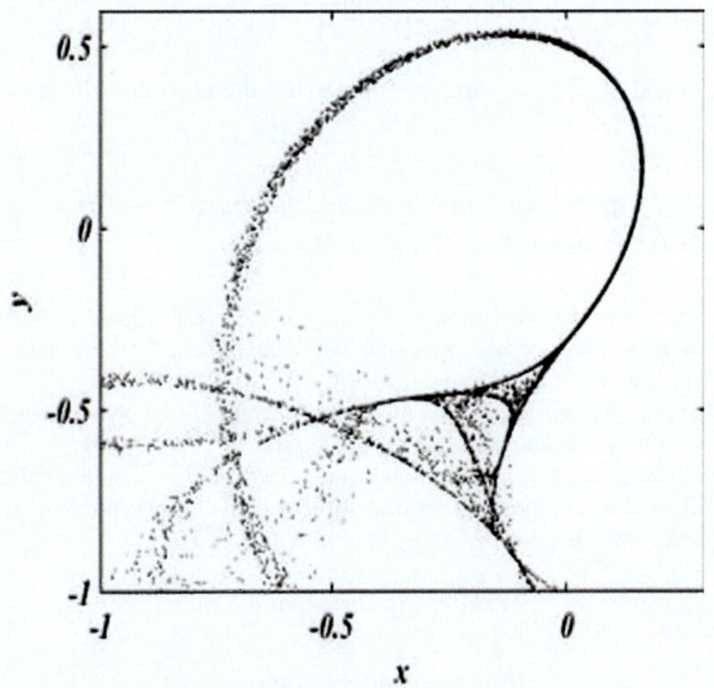

Fig. 1 Bifurcation diagram of Tinker bell map

bifurcation diagram of the map, parameter b varies in interval $(-0.6013, -0.54791)$ and initial values $x_0 = 0.1576$, $y_0 = 0.123$ are considered.

In this paper, we have taken 55,000 iterations to generate a random sequence using Tinker bell map. Tinker bell map shows a periodic property. The bifurcation diagram of Tinker bell map is depicted in Fig. 1.

2.2 Fourier Transform

In single variable, Fourier transform (FT) [25] of a function $f(x)$ is defined by

$$F\{f(x)\} = \int_{-\infty}^{\infty} f(x)\mathrm{e}^{-2\pi i x u}\mathrm{d}x.$$

In two variables, Fourier transform of a function $f(x, y)$ is defined by

$$F\{f(x, y)\} = \int_{-\infty}^{\infty}\int_{-\infty}^{\infty} f(x, y)\mathrm{e}^{-2\pi i(xu+yv)}\mathrm{d}x\mathrm{d}y.$$

where (x, y) and (u, v) are coordinates in spatial and Fourier domain, respectively.

2.3 Basic Concepts of Double Random Phase Encoding (DRPE) Technique in Fourier Domain

In double random phase encryption technique (DRPE), an image is encrypted using two random phase mask of the same size that of the image. Second random phase mask work as the key of the scheme [26–30].

Input image f(x, y) is pixel-wise multiplying with random phase mask (RPM1) (RPM1) then Fourier transform is performed on it.

Resulting image from the previous step bonded with a second random phase mask (RPM2) followed by another Fourier transform to give the encrypted image.

RPM1 and RPM2 is given by

$$\text{RPM1} = \exp(2\pi i \times m(x, y))$$

$$\text{RPM2} = \exp(2\pi i \times n(x, y))$$

where $m(x, y)$ and $n(x, y)$ are random matrix of same size that of input image $f(x, y)$.

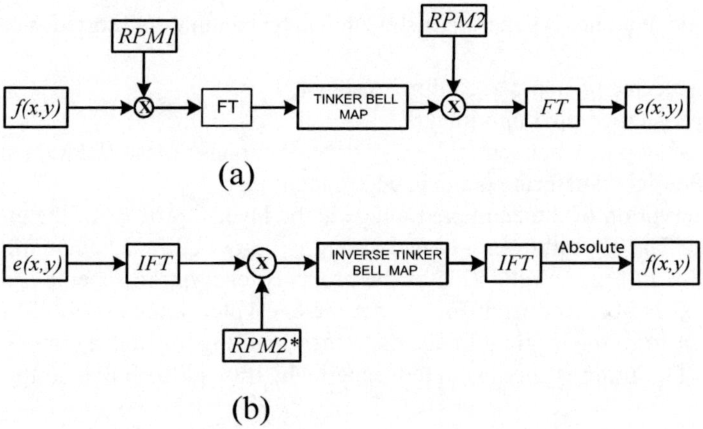

Fig. 2 Schematic diagram of encryption and decryption process of double random phase encoding

The whole process of image encryption is mathematically described as follow:

$$e(x, y) = \mathrm{FT}(\mathrm{FT}(f(x, y) * \mathrm{RPM1}) * \mathrm{RPM2})$$

To decrypt the encrypted image, inverse Fourier transform is applied on encrypted image $e(x, y)$ and bonded with a conjugate of the second random phase mask ($\mathrm{RPM2}^*$). Again inverse Fourier transforms applied on resulting image followed by absolute operation on the image which results in the decrypted image. The whole process of decryption is mathematically described as follows:

$$f(x, y) = \mathrm{abs}\big(\mathrm{IFT}\big(\mathrm{IFT}(e(x, y)) * \mathrm{RPM2}^*\big)\big)$$

Here $\mathrm{RPM2}^*$ stands for conjugate RPM2, IFT stands for inverse Fourier transform and abs stand for absolute of function. The complete double random phase encoding process is explained by using a schematic flowchart in Fig. 2.

2.4 Proposed Scheme and Validation

In the proposed scheme, chaotic Tinker bell map is used along with DRPE in the following manner:

1. Without loss of generality, input image is assumed of the size $M \times N$.
2. Image is bounded with a random phase mask (RPM1) and apply Fourier transform.
3. Divide the image into smaller blocks and convert each block in a vector by appending column after column.

4. A chaotic sequence is generated using the Tinker bell map and sorted in increasing order.
5. Sort the vector of step 3 according to step 4.
6. Reshape vector step 5 into image.
7. Image of step 6 is bonded with second random phase mask (RPM2) and apply again Fourier transform results in encrypted image.
8. The decryption of the encrypted image is the inverse process of the encryption scheme. The algorithm of a cryptosystem is described in schematic flow chart described in Fig. 3. Validation of the proposed encryption scheme is shown in Fig. 4. It is observed from Fig. 4 that the encrypted image (Fig. 4b) is totally different from the original image of Cameraman Fig. 4a and resembles a noisy image. Figure 4c is the decrypted image which is quite similar to the original image.

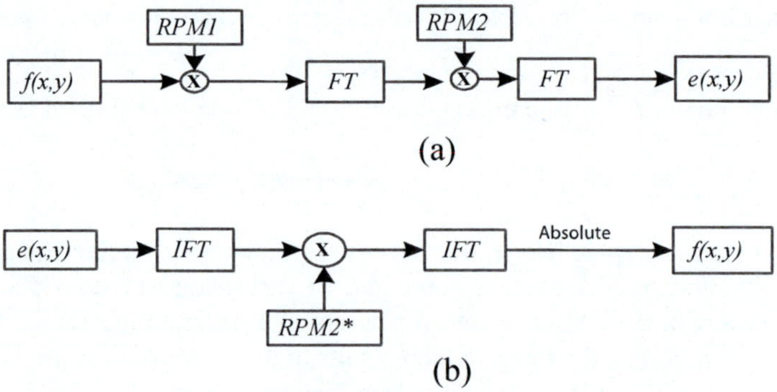

Fig. 3 Schematic diagram of proposed **a** encryption; **b** decryption process

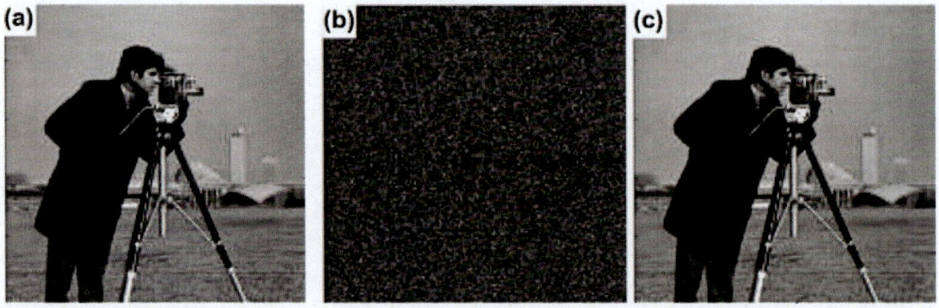

Fig. 4 **a** Input image; **b** encrypted image; **c** recovered image of Cameraman by the proposed scheme

3 Results and Discussion

The proposed scheme has been validated by using well-acknowledged grayscale image of Cameraman in literature by performing simulation on Matlab. In our simulation, the initial parameters of Tinker bell map $a = 0.9, b = -0.6013, c = 2, d = 0.5$ with initial condition $x_0 = 0.1787$, $y_0 = 0.178$ are considered.

In the next subsections, we will discuss the statistical attacks analysis on the scheme such as histogram and 3-D plots analysis, Information entropy and Correlation distribution analysis. Thereafter, scheme sensitivity to the parameters will be discussed followed by attack analysis.

Various statistical information such as mean squared error (MSE), correlation coefficient (CC), peak signal-noise ratio (PSNR) is computed for grayscale images.

Mathematically mean squared error is given by the following expression:

$$\text{MSE} = \frac{1}{MN}\sum_{x=0}^{M}\sum_{y=0}^{N}[I_0(x, y) - I_r(x, y)]^2$$

Here $I_0(x, y)$ and $I_r(x, y)$ denotes the pixel value of original image and recovered image.

Correlation coefficient [31] is given by the following expression:

$$\text{CC} = \frac{\text{cov}(I_0(x, y), I_r(x, y))}{\sigma(I_0(x, y), I_r(x, y))}$$

Here, cov stand for covariance and σ stand for standard deviation.

PSNR is given by the following expression:

$$\text{PSNR} = 10 \times \log_{10}\frac{255^2}{\text{MSE}}$$

The mean squared error between the original and recovered images of the proposed scheme for gray scale Cameraman image is $4.9654\,\text{e}^{-27}$, PSNR is 311.1712 and CC = 1. All these statistical measures suggest that a faithful recovery of the original imagc is achieved by the proposed scheme.

3.1 Histogram and 3-D Plots Analysis

The goodness of encryption algorithm can be verified by knowing the histogram and 3-D plot diagram of the plaintext image and the ciphertext image. If the histogram and 3-D plot of ciphertext image are completely different from histogram and 3-D plot diagram of the plaintext image then the algorithm of image encryption known as good encryption scheme. From Fig. 5a–c, it is clearly visible that histogram of

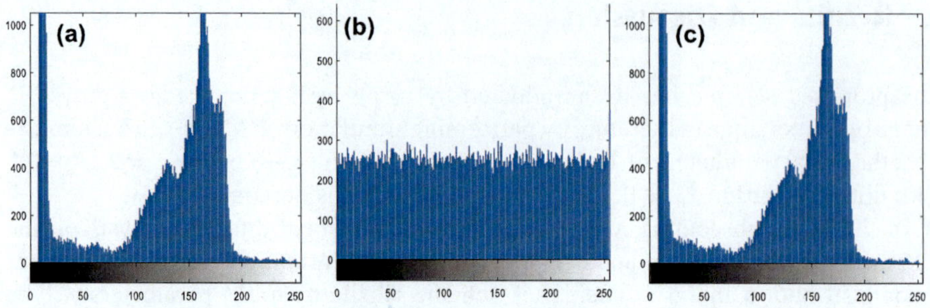

Fig. 5 Histogram of **a** plaintext; **b** ciphertext; **c** decrypted image of Cameraman

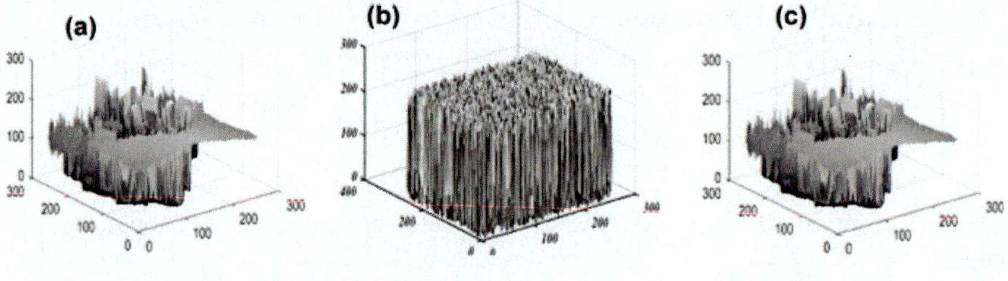

Fig. 6 3-D plot of **a** plaintext image; **b** ciphertext image; **c** recovered image of Cameraman

plaintext and ciphertext is completely different. Histogram of ciphertext image shows that pixels are equally distributed in the ciphertext image and thereafter no statistical information can be obtained through it. From the histograms, we can observe the goodness of an encryption algorithm.

From Fig. 6a, b, one can observe that 3-D plot of ciphertext image is completely different from the 3-D plot of plaintext image as that of a histogram. In Fig. 6b, the 3-D plot of ciphertext image shows that pixels are equally distributed. Therefore, it not so easy to find out any information about the plaintext image. Figures 6a, c show that 3-D plots of plaintext image and the recovered image are same.

3.2 *Information Entropy*

Information entropy [32] is defined as the average rate at which information is produced by source of data. If a source of data produces low probability values then it contains more information as compare to that source of data which contain high probability values.

Information entropy $E(m)$ of source m is defined as

$$E(m) = \sum_{k=1}^{256} P(m_k)\log_2 \frac{1}{P(m_k)}$$

here $P(m_k)$ is probability of m_k. Entropy of grayscale image lies between 0 and 8. Entropy of grayscale Cameraman image of plaintext and ciphertext for the proposed scheme is 7.0097 and 7.9954, respectively. The result shows that as randomness increase in ciphertext and so information entropy also increases. Entropy of ciphertext image of Cameraman for the proposed scheme is 7.9954 which is very close to the maximum value of information entropy of grayscale image.

3.3 Correlation Distribution Analysis

Another way to show the goodness of the encryption algorithm is based on Correlation distribution analysis of the neighbouring pixels in the plaintext and ciphertext images. Here we randomly select 5000 pixels of plaintext image in diagonal direction. Its Correlation distribution is ploted as a scattered diagram in Fig. 7a whereas the Correlation distribution of ciphertext is shown in Fig. 7b. In Correlation distribution figure that they are highly related to each other while the Correlation distribution Fig. 7b of ciphertext is random and showing that pixels are not related to each of plaintext image, pixels are distributed along a straight line stating others.

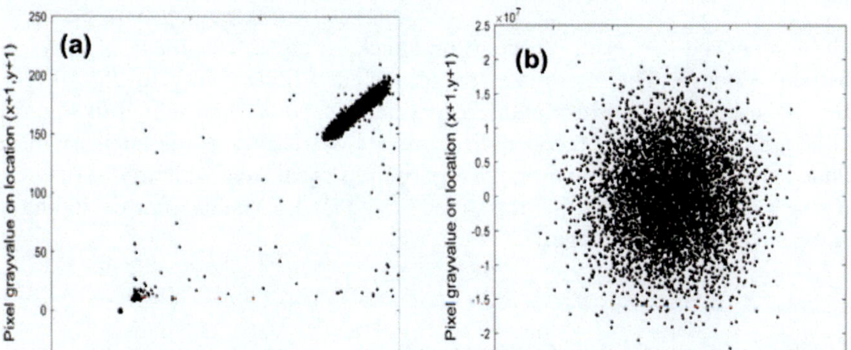

Fig. 7 Correlation distribution of **a** plaintext image; **b** ciphertext image

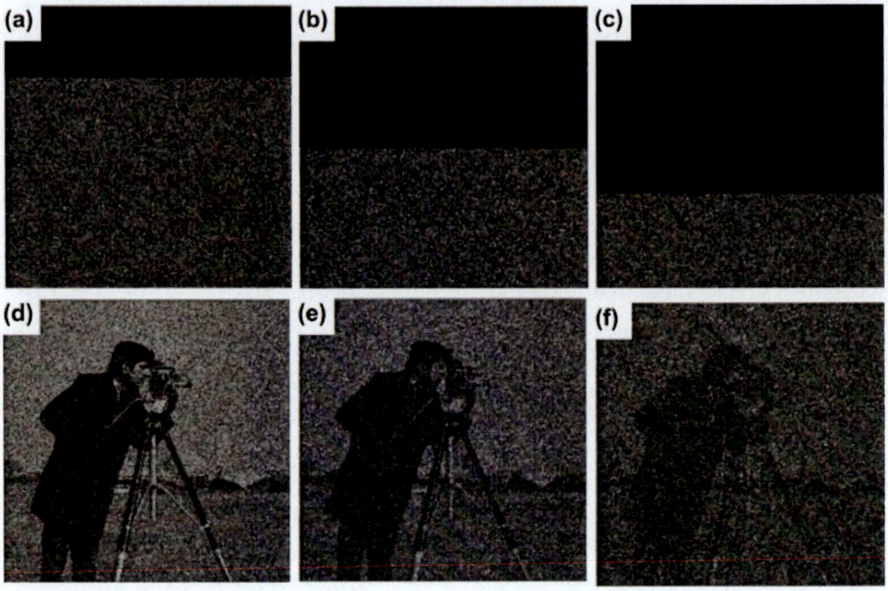

Fig. 8 Encrypted image of Cameraman with occlusion **a** 25%; **b** 50%; **c** 70%; and corresponding decrypted images; **d–f** vertically

3.4 Occlusion Attack Analysis

We have analyzed the result of occlusion attack on ciphertext image of grayscale Cameraman image. The encrypted image occluded horizontally to 25, 50, 70% in Fig. 8a–c and their corresponding decrypted images are presented in Fig. 8d–f. Although, the quality of recovered images are decreasing as the intensity of the occluded area increased. However, the algorithm is robust to a wider range of occlusion attack. The data shows that recovered image is also visible after occluding up to 70% area of plaintext image.

3.5 Sensitivity of Secrete Key Analysis

In the success of any encryption algorithm, secret key plays a very important role. To check the strength of the secret key of the proposed scheme, we analyze the sensitivity of the key of Tinker bell map. The key sensitivity results are shown in Fig. 9a–f. Figure 9 shows that decrypted images when the slightly wrong value of Tinker bell map are used. From Fig. 8, one can see that the decrypted image is completely unrecognizable even if we make a slight change in a parameter. The keys are sensitive up to 15 decimal places for each parameter and initial value. Thus, keyspace is large enough to resist any brute force attack for a reasonable time.

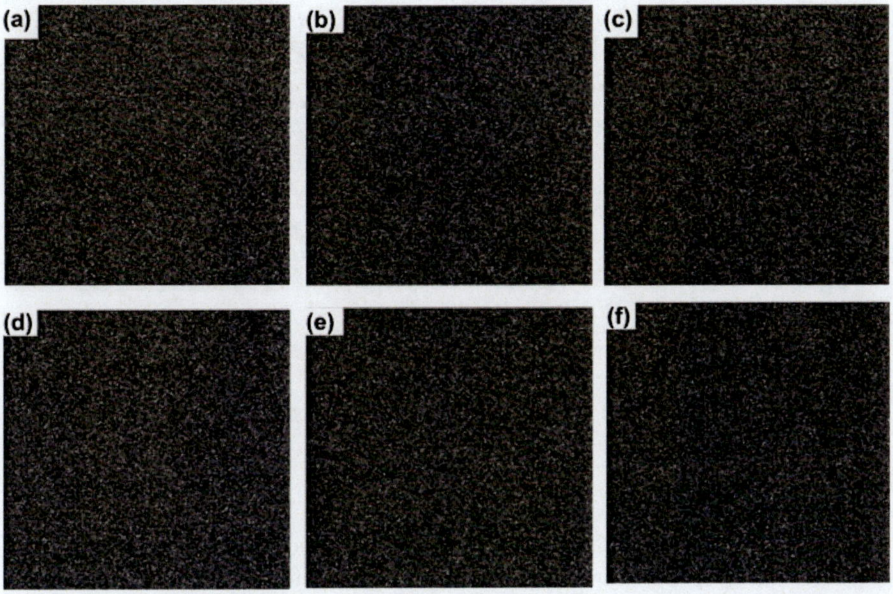

Fig. 9 Recovered image of Cameraman when the wrong parameter is used in Tinker bell map: **a** wrong parameter $\boldsymbol{a} = 0.8999999999999999$ used in the place of $\boldsymbol{a} = 0.9$; **b** wrong parameter $\boldsymbol{b} = -0.601299999999999$ used in place of $\boldsymbol{b} = -0.6013$; **c** wrong parameter $\boldsymbol{c} = 1.999999999999999$ used in the place of $\boldsymbol{c} = 2$; **d** wrong parameter $\boldsymbol{d} = 0.4999999999999999$ used in place of $\boldsymbol{d} = 0.5$; **e** wrong initial value $\boldsymbol{x} = 0.178699999999999999$ used in place of $\boldsymbol{x} = 0.1787$; **f** wrong initial value $\boldsymbol{y} = 0.177999$ is used in the place of $\boldsymbol{y} = 0.178$

3.6 Noise Attack Analysis

Scheme is also analyzed for its resistance to noise attack. Gaussian noise of strength k is added using relation

$$EN = e(x, y) * N$$

where $N = 1 + kG$, G is Gaussian noise function with mean 0 and variance 1. Figure 10 shows the results of noise attack where Gaussian noise of strength $k = 2$, 4 and 6 is applied on the encrypted image of Cameraman. Results show that as the strength of noise is increasing, the quality of recovered image decreases but, image is still recognizable.

Fig. 10 **a** Recovered image with noise intensity $\boldsymbol{k} = 2$; **b** recovered image with noise intensity $\boldsymbol{k} = 4$; **c** recovered image with noise intensity $\boldsymbol{k} = 6$

4 Conclusion

In the paper, an encryption scheme for grayscale images is presented. The scheme uses the Tinker bell chaotic map for generating a random sequence and used for pixel scrambling. Double random phase encoding technique along with Tinker bell map in frequency domain is used in the proposed encryption scheme. The scheme is simulated in Matlab environment. Sensitivity of the Tinker bell parameters has been carried out. Efficacy and performance of the scheme is also established through statistical attacks via histogram and 3-D plots. In the simulation of the proposed scheme, we recovered faithful recovery of input image such as CC = 1, PSNR = 311.1712 and MSE = 4.9654 e^{-27}. The results indicate that the proposed scheme is highly sensitive to the encryption keys. The results indicate that the scheme provides a large keyspace and offers a high level of security and resistivity towards brute-force, occlusion and noise attacks.

References

1. Singh P, Yadav AK, Singh K, Saini I (2019) Asymmetric watermarking scheme in fractional Hartley domain using modified equal modulus decomposition. J Opt 21:484–491
2. Yadav AK, Singh P, Singh K (2018) Cryptosystem based on devil's vortex Fresnel lens in the fractional Hartley domain. J Opt 47(2):208–219. https://doi.org/10.1007/s12596-017-0435-9
3. Rakheja P, Vig R, Singh P (2019) Optical asymmetric watermarking using 4D hyperchaotic system and modified equal modulus decomposition in hybrid multi resolution wavelet domain. Optik 176:425–437. https://doi.org/10.1016/j.ijleo.2018.09.088
4. Hanchinamani G, Kulakarni L (2014) A novel approach for image encryption based on parametric mixing chaotic system. Int J Comput Appl 96(11):29–37. https://doi.org/10.5120/16839-6690
5. Anees A (2015) An image encryption scheme based on lorenz system for low profile applications. 3D Res 6(3):24. https://doi.org/10.1007/s13319-015-0059-2
6. Gao T, Chen Z (2008) Image encryption based on a new total shuffling algorithm. Chaos, Solitons Fractals 38(1):213–220. https://doi.org/10.1016/j.chaos.2006.11.009

7. Zhu A, Li L (2010) Improving for chaotic image encryption algorithm based on logistic map. In: 2010 The 2nd conference on environmental science and information application technology, Wuhan, China, pp 211–214. https://doi.org/10.1109/esiat.2010.5568374
8. Alvarez G, Li S (2006) Some basic frypyographic requirments for chaose-based cryptosystem. Int J Bifurc Chaos 16(08):2129–2151. https://doi.org/10.1142/S0218127406015970
9. Rakheja P, Vig R, Singh P (2019) Asymmetric hybrid encryption scheme based on modified equal modulus decomposition in hybrid multi-resolution wavelet domain. J Mod Opt 66(7):799–811. https://doi.org/10.1080/09500340.2019.1574037
10. Chen W, Javidi B, Chen X (2014) Advances in optical security systems. Adv Opt Photonics 6(2):120. https://doi.org/10.1364/AOP.6.000120
11. Elshamy AM et al (2013) Optical image encryption based on chaotic baker map and double random phase encoding. J Light Technol 31(15):2533–2539. https://doi.org/10.1109/JLT.2013.2267891
12. Faragallah OS, Afifi A (2017) Optical color image cryptosystem using chaotic baker mapping based-double random phase encoding. Opt Quantum Electron 49(3):89. https://doi.org/10.1007/s11082-017-0909-7
13. Cuche E, Bevilacqua F, Depeursinge C (1999) Digital holography for quantitative phase-contrast imaging. Opt Lett 24(5):291. https://doi.org/10.1364/OL.24.000291
14. Kumar J, Singh P, Yadav AK, Kumar A (2018) Asymmetric cryptosystem for phase images in fractional Fourier domain using LU-decomposition and Arnold transform. Procedia Comput Sci 132:1570–1577. https://doi.org/10.1016/j.procs.2018.05.121
15. Singh P, Yadav AK, Singh K (2019) Known-plaintext attack on cryptosystem based on fractional hartley transform using particle swarm optimization algorithm. In: Ray K, Sharan SN, Rawat S, Jain SK, Srivastava S, Bandyopadhyay A (eds) Engineering vibration, communication and information processing, vol 478. Springer Singapore, Singapore, pp 317–327
16. Rakheja P, Vig R, Singh P (2019) An asymmetric hybrid cryptosystem using hyperchaotic system and random decomposition in hybrid multi resolution wavelet domain. Multimed Tools Appl 78(15):20809–20834. https://doi.org/10.1007/s11042-019-7406-x
17. Rakheja P, Vig R, Singh P, Kumar R (2019) An iris biometric protection scheme using 4D hyperchaotic system and modified equal modulus decomposition in hybrid multi resolution wavelet domain. Opt Quantum Electron 51(6):204. https://doi.org/10.1007/s11082-019-1921-x
18. Singh P, Yadav AK, Singh K (2017) Phase image encryption in the fractional Hartley domain using Arnold transform and singular value decomposition. Opt Lasers Eng 91:187–195. https://doi.org/10.1016/j.optlaseng.2016.11.022
19. Kumar J, Singh P, Yadav AK, Kumar A (2019) Asymmetric image encryption using Gyrator transform with singular value decomposition. In: Ray K, Sharan SN, Rawat S, Jain SK, Srivastava S, Bandyopadhyay A (eds) Engineering vibration, communication and information processing, vol 478. Springer Singapore, Singapore, pp 375–383
20. Rakheja P, Vig R, Singh P (2019) A hybrid multiresolution wavelet transform based encryption scheme. Presented at the emerging trands in mathematics and its applications: proceedings of the 3rd international conference on recent advances in mathematical sciences and its applications (RAMSA-2019), Noida, India, p 020008. https://doi.org/10.1063/1.5086630
21. Refregier P, Javidi B (1995) Optical image encryption based on input plane and Fourier plane random encoding. Opt Lett 20(7):767. https://doi.org/10.1364/OL.20.000767
22. Biryukov A (2005) The boomerang attack on 5 and 6-round reduced AES. In: Dobbertin H, Rijmen, V, Sowa A (eds) Advanced Encryption Standard—AES, vol 3373. Springer, Berlin Heidelberg, pp 11–15
23. Hasib AA, Md. AA, Haque M (2008) A comparative study of the performance and security issues of AES and RSA cryptography. In: 2008 third international conference on convergence and hybrid information technology, Busan, Korea, pp 505–510. https://doi.org/10.1109/iccit.2008.179
24. Ding K, Xu X (2016) Chaotic synchronization of modified discrete-time Tinkerbell systems. Discrete Dyn Nat Soc 2016:1–7. https://doi.org/10.1155/2016/5218080
25. Pratt WK (2002) Digital image processing, 3rd edn. Wiley-Liss, Hoboken, NJ

26. Abd-El-Hafiz SK, AbdElHaleem SH, Radwan AG (2016) Novel permutation measures for image encryption algorithms. Opt Lasers Eng 85:72–83. https://doi.org/10.1016/j.optlaseng.2016.04.023
27. Sharma N, Saini I, Yadav A, Singh P (2017) Phase image encryption based on 3D-Lorenz chaotic system and double random phase encoding. 3D Res 8(4):39. https://doi.org/10.1007/s13319-017-0149-4
28. Frauel Y, Castro A, Naughton TJ, Javidi B (2005) Security analysis of optical encryption. Presented at the European Symposium on Optics and Photonics for Defence and Security, Bruges, Belgium, p 598603. https://doi.org/10.1117/12.633677
29. Rakheja P, Vig R, Singh P (2019) An asymmetric hybrid watermarking mechanism using hyperchaotic system and random decomposition in 2D Non-separable linear canonical domain. Proc Indian Natl Sci Acad. https://doi.org/10.16943/ptinsa/2019/49590
30. Rakheja P, Vig R, Singh P (2020) Double image encryption using 3D Lorenz chaotic system, 2D non-separable linear canonical transform and QR decomposition. Opt Quantum Electron 52(2):103. https://doi.org/10.1007/s11082-020-2219-8
31. Lai CS et al (2019) A robust correlation analysis framework for imbalanced and dichotomous data with uncertainty. Inf Sci 470:58–77. https://doi.org/10.1016/j.ins.2018.08.017
32. Lyda R, Hamrock J (2007) Using entropy analysis to find encrypted and packed malwaresss-susing. IEEE Secur Priv Mag 5(2):40–45. https://doi.org/10.1109/MSP.2007.48

Fiber Optics Near-Infrared Wavelengths Analysis to Detect the Presence of Liquefied Petroleum Gas

H. H. Cerecedo-Núñez, Rosa Ma Rodríguez-Méndez, P. Padilla-Sosa, and J. E. Lugo-Arce

Abstract This paper is based on the use of near-infrared wavelength (NIR) spectrum to identify the presence of liquid petroleum gas (LPG) in high concentrations. We tested a fixed proportion of standard propane/methane mix, analyzing variations of the surface area under transmitted spectra curves. We tested spectra ranging from 900 up to 1200 nm, through an optical fiber coupling array in a closed container; we observed transmission changes with the gas presence at ambient conditions. Second, we selected four specific wavelengths (920, 980, 1020, and 1064 nm) inside the mentioned range. We calculated the absorbance and compared the obtained results. Finally, we showed an analysis of the signal-to-noise ratio (SNR) as a complement to see the viability of our research. We mention these results in the aim of studying the feasibility of using fiber optics technology to develop in the future an accessible and flexible LPG detection system in the near-infrared band. Today, the use of liquefied petroleum gas is very common in many industrial and domestic uses. Therefore, it is necessary its detection for security uses.

Keywords Optical fiber sensors · Spectrometry measurements · NIR spectrometry · LPG sensors · Gas sensors

1 Introduction

Daily, millions of people use liquefied petroleum gas (LPG), and they depend on it for lots of uses: third sector, industry, transport, agriculture, energy power generation, or even cooking, which comprises heating and entertainment [1, 2]. LPG is colorless

H. H. Cerecedo-Núñez (✉) · R. M. Rodríguez-Méndez · P. Padilla-Sosa
Laboratorio de Óptica Aplicada, Facultad de Física, Universidad Veracruzana, Xalapa, Veracruz, Mexico
e-mail: hcerecedo@uv.mx

J. E. Lugo-Arce
Faubert Lab, School of Optometry, University of Montreal, Montreal, QC H3C3J7, Canada
e-mail: jesus-eduardo.lugo.arce@umontreal.ca

K. Ray et al. (eds.), *Proceedings of International Conference on Data Science and Applications*, Lecture Notes in Networks and Systems 148,
https://doi.org/10.1007/978-981-15-7561-7_21

and odorless, its emanation is only detected when it receives a small quantity of scented agent, as itself, is neither toxic nor corrosive, but in higher concentrations, it can cause asphyxia because replaces oxygen and it is heavier than air [1–5]. Also, it is excessively cold because while liquefying the gas, the temperature is below 0 °C.

According to the Health and Human Services department (HSS, USA), the liquid petroleum gas is a standard domestic and industrial fuel, highly inflammable. The permissible exposition limit is of 1000 ppm for eight working hours. Higher concentrations than this cause dangerous effects in our respiratory and central nervous system. Here, propane and butane usually constitute the LPG, and it is used as combustible, calefaction, and as a raw material in chemistry [1, 2].

Most electronic sensors used in gas measurements are electrochemical, they have to be exposed directly to a gas, and usually, for shorts periods [6–8]. Moreover, current sensing technology is capable of measuring gas remotely, for instance, optical sensors based on optical fibers. Optical fiber coupling can work in places without power supply and locate far away from the city. Attaching them to infrared absorbance cells is potentially a flexible and efficient way to control the number of harmful and dangerous gases in closed spaces, such as tunnels and mines [7, 8].

A commercial method to detect LPG is the use of TGS (822, 3870, 4160 y 2600) sensors, from FIGARO USA, Inc. These are semiconductors with small dimensions, and plenty of them have a sensor element for two types of gases: butane and propane, as pure gases or as a mixture. There are also new sensor proposals, and some of these consume a low quantity of energy [9–13].

In 1998, Ryan et al. [6], used a novel sensing methodology using a spectrum optical analyzer Anritsu MS9702B and a halogen lamp to detect propane, butane, 2-methylpropane, and a commercial mixture of LPG in the wavelength range between 900 and 1200 nm. Their normalized transmittance findings have shown that there are two sensitive wavelengths at 920 and 1020 nm, where the optical transmission is low. Besides, recent studies support even more the viability to use near-infrared wavelengths to detect LPG [7, 8, 14] and using some alternative analysis [6–9, 15].

In this paper, we detected different high concentration levels of gas LP in the near-infrared wavelengths. We showed the feasibility of a methodology to detect gas using a fiber optic probe that would be more convenient and easily accessible than some previous reports.

2 Materials and Methods

We used a commercial 275 gr., LP gas cartridge with a mixture of 75% butane and 25% propane. Our research was developed with an experimental setup, displayed in Figs. 1 and 2, principally composed of these elements: tungsten lamp LS-1 (Ocean Optics), two optical fibers NT 57-751 (Edmund Optics), Polyvinylchloride (PVC) cell, a valve, and an AQ6317B (ANDO) spectrum analyzer. We coupled two optical fibers inside the cell. One fiber receives the light from the tungsten lamp, and the second fiber inputs the spectrum analyzer.

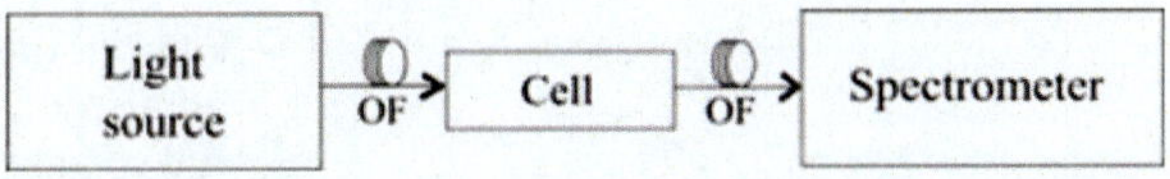

Fig. 1 General scheme used for the testing

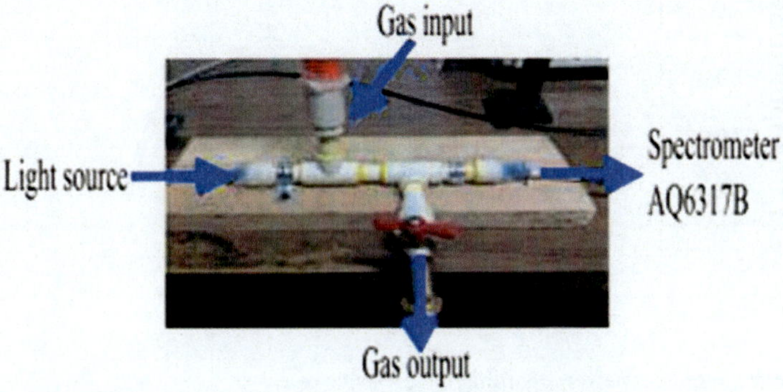

Fig. 2 Gas cell prototype

In our experimental setup, the cylindrical cell has a gas input at the top of it, while the gas exit is located in one of the cell sides. Its dimensions are: a diameter of 1/2″, and a length of 8.26″, this should imply a maximum volume gas of 1.62 in^3. For a better understanding, we are going to consider it in metric unities, 26.58 cm^3.

In that configuration, we got a reference that transmitted power spectrum without gas, volume = 0 cm^3, and then, we observed that the transmitted power spectrum continuously after the gas was introduced (see Fig. 3).

Once the spectra were taken, we considered first studying global changes instead of specific local changes in all transmitted spectra. We integrated all the optical spectra to obtain their area under the curve and plot the area value versus time. In this way, the time series can show visible changes that may occur due to the gas thermodinamical changes inside the cell. The results will be described in the following section. Second, we studied specific local changes in all transmitted spectra within the experimental optical bandwidth (from 900 to 1200 nm). This is achieved by selecting four wavelengths to develop our analysis, as it is depicted in Fig. 4. The procedure will be explained in the next section.

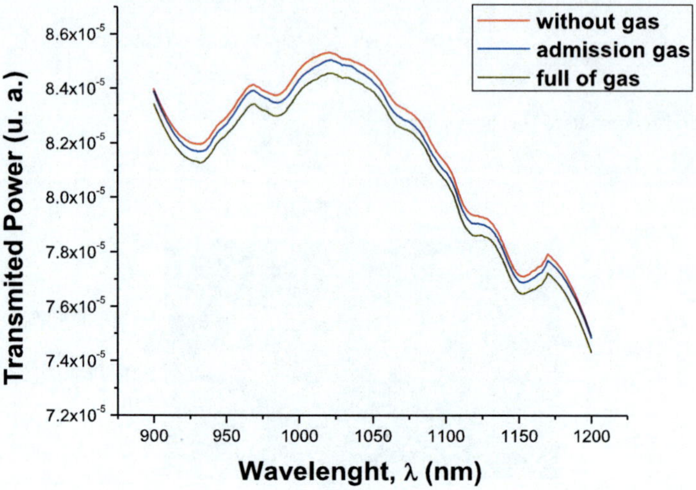

Fig. 3 Changes in optical power regarding the presence of gas

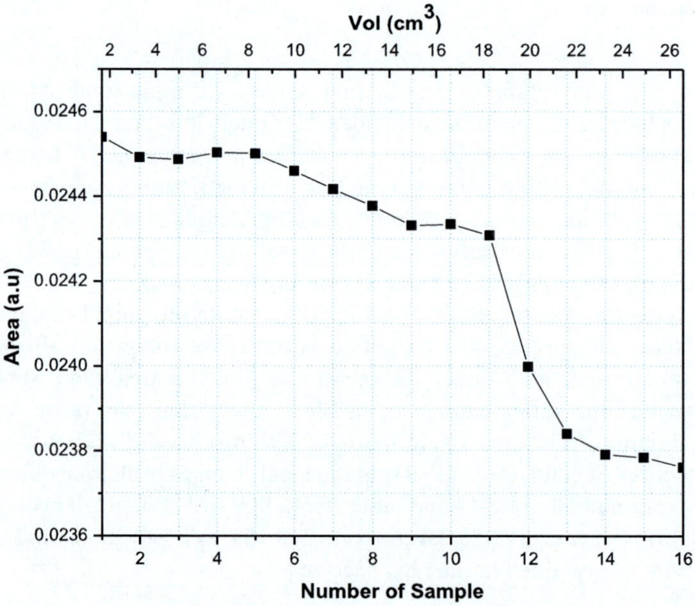

Fig. 4 Areas under transmitted spectra curves. We can notice an abrupt change from sample 11 to 16 probably due to a gas–liquid phase transition whereby liquid condensation is created. Consequently, liquid droplets can scatter the light, and the transmission amplitude decreases

3 Analysis and Discussions

3.1 *Transmission Changes Assessed by the Area Under Spectra Curves*

In the first analysis, we propose to observe optical transmission changes using the area under the transmitted spectra curves. This is achieved by mathematical integration of the spectra. An example of a typical transmission optical change, at different gas volumes inside the chamber, is plotted in Fig. 4; upper scale represents the expected volume (of the fixed concentration) of mixed gas, inside the tester chamber. From here, we can infer the different physical changes inside the chamber due to the continuous presence of mixed gas. Sample 1 corresponds to the area value at the beginning of the test (volume = 0), at an initial time ($t = 0$). Subsequently, times were increased in steps of 3 min for each sample, and the gas (volume) was accumulative. From point 1 to point 11, we observed the gradual decreasing value, due to the increase of absorption and scattering by the gas mix. Then, from point 11 to 16, we can observe an abrupt change, and this may be due to a thermodynamic gas–liquid phase transition inside the chamber. This would raise significantly light scattering by liquid droplets formation. Using the same transmission spectroscopy technique, similar results were found for ethanol sensing [16], where abrupt transmission changes were observed when ethanol change phase from liquid to gas. Small fluctuations in sample points 4–6 and 10–11 can be attributed to little turbulence inside the chamber. That is, gas turbulence domains could modulate light scattering adding noise to the optical measurements. This behavior was consistent in all the tests we did.

The dynamics within the tester chamber is well known. Initially, the LPG is under high pressure inside the cartridge (over atmospheric pressure) at room temperature. Our environmental conditions where media temperature, 18.6 °C, atmospheric pressure, 865.9 mmHg, and relative humidity, 66%. When LPG is left out, it inputs the tester chamber at high velocity, thus pressure and temperature drop-off. The chamber cools down, and the gas mixture condenses. The gas mixture high velocity inside the tester chamber causes the formation of turbulence domains as the gas is filling the chamber. As we mentioned before, in such conditions, transmitted light through the tester chamber should suffer absorption and scattering. Thus, the dynamic gas process inside the chamber also should produce dynamic light variations of transmittance.

3.2 *Changes by Transmitted Wavelength*

Most flammable gases have a good absorption capacity in the infrared spectral region [6, 7, 15]. If the spectrum of the light source covers a range of wavelengths, which contains one or more lines of absorption of the gas, then attenuation occurs in those lines of absorption. An alternative method to express the attenuation of electromagnetic radiation is the absorbance. This parameter is the most common to express the attenuation of radiation, and it can be represented by Eq. (1) also known as Beer–Lambert Law [7, 17]. This equation does not consider light scattering, and it is written as

$$A = -\log \frac{P_T}{P_0} \tag{1}$$

where P_0 and P_T are the initial and subsequent light power, respectively.

Once target wavelengths were selected from measured spectrums, their associated absorbance is calculated using Eq. (1). Wavelengths (at 920 and 1020 nm) were selected based on previous research [14], in which the absorbance happened in this specific spectrum range. The other two wavelengths, at 980 and 1064 nm, were chosen accordingly to the wavelength of available commercial laser diodes.

For each selected wavelength (920, 980, 1020, and 1064 nm), we plot fifteen points, corresponding to different gas volumes. Each point represents an average of thirty spectrums, and they are separated in intervals of three minutes. This interval was optimized experimentally, and we looked for the best time window capable of capturing noticeable changes on the optical transmission (see Fig. 5).

In Fig. 6, we show the calculated absorbance values at different selected wavelengths. Remembering that the number of samples is equivalent to an accumulative volume into the tester chamber. Due to gas mixture thermodynamics inside the chamber, which was previously mentioned, the presence of fluctuations on the optical spectra is inevitable; this is indicated by the high standard errors depicted in Fig. 5. We also noticed that error value decreased significantly for the wavelength of 1064 nm, unlike the others. Even so, the averaged absorbance values increased

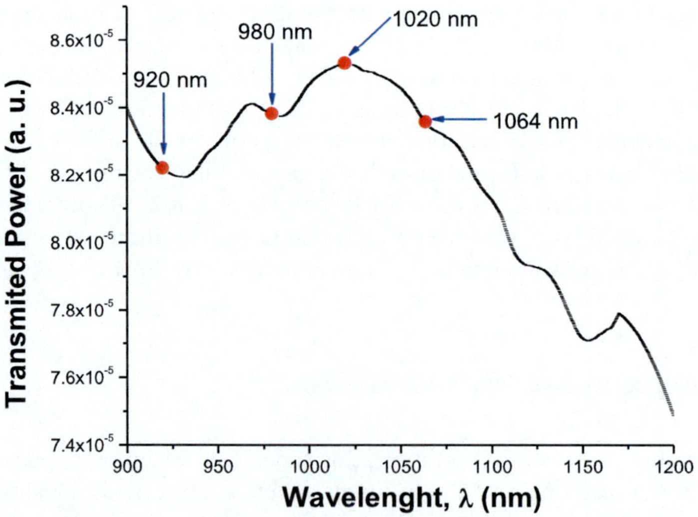

Fig. 5 Example of selected spectral points for analysis

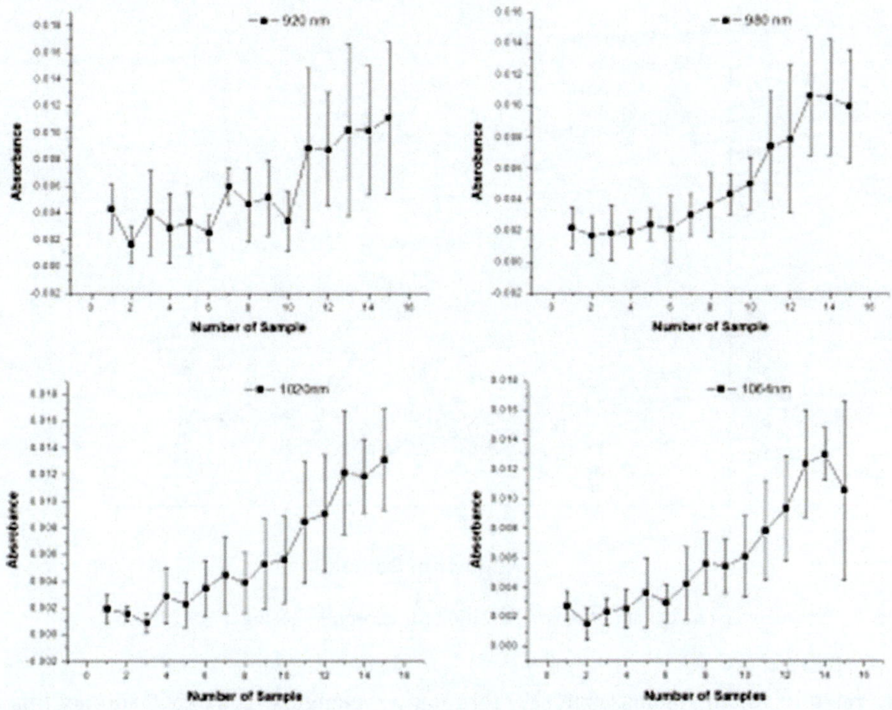

Fig. 6 Absorbance results from different wavelengths. However, the high standard errors indicate the presence of fluctuations on the optical spectra, we also noticed that error value decreased significantly for the wavelength of 1064 nm, unlike the others. Even so, the averaged absorbance values increased with time. This means that as the gas mixture volume increases with time inside the chamber, the average optical absorbance value increases. This is a significant result that shows the feasibility using the selected near-infrared wavelengths for future sensing applications

with time. This means that as the gas mixture volume increases with time inside the chamber, the average optical absorbance value increases. This is a significant result that shows the feasibility using the selected near-infrared wavelengths for future sensing applications. One drawback of the sensing technique presented here is that it is not fast. On average, it took 15 min up to 21 min to observe a distinguishable increment in the average absorbance value. This result is inherent to the experimental setup design, and by enhancing it, we should expect much fewer fluctuations caused by temperature, pressure, phase changes, and liquid condensation. Nevertheless, the primary goal of our work was showing the potential use of near-infrared wavelengths along with fiber optics technologies; such goal merely is demonstrated by Figs. 4 and 6 we believe.

Even more, we can explore how thermodynamic fluctuations impact the absorption inferred signal by plotting the signal-to-noise ratio (SNR) between the average absorbance values and their standard deviation, and for each wavelength, Fig. 7 shows this result. Here, we can observe a tendency to increase the SNR when the

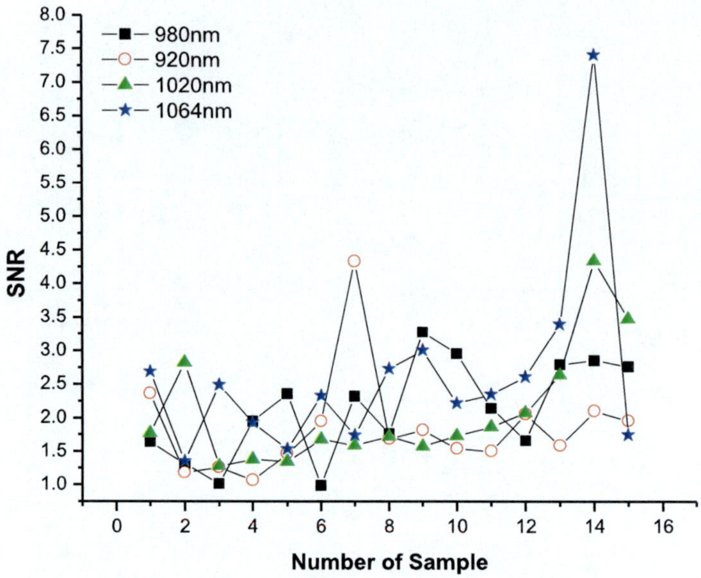

Fig. 7 SNR between media and standard deviation for chosen wavelengths

gas mixture volume increases mostly for long wavelengths. The explanation of this trend is twofold; first, as the gas mixture concentration increases, there is more light absorption; second, signal fluctuations due to light scattering by liquid droplets should decrease as the light wavelength increases.

4 Conclusions

In order to investigate the feasibility to detect LP gas with fiber optics technology, a methodology that senses the presence of LP gas in the range from 900 to 1200 nm was presented. In one hand, we determine that changes in the area under transmission spectra curves should be consistent with the presence or absence of gas, which would be suitable to quantify gas concentration in the future. In the other hand, in the case of wavelength analysis, the conclusion is that for our particular experimental setup, long wavelengths, specifically the wavelength of 1064 nm, could have a better response to gas presence, which could be a suitable option for sensing. The analysis of SNR also shows a trend to corroborate the use of 1064 nm to correlation with gas quantities.

The presented study is a first approximation to a next step, the quantification, toward a volume gas detector. The proposed experimental setup is simple and easy to implement. At this point in our research, the detection limit is considering by the maximum volume inside the tester chamber, but it could be increased using a more suitable test chamber. The time response in this research is still low. From the results in Fig. 6, we can observe a good linear response in all range we tested. However, we always have to consider better conditions for the measurements and their repeatability, to avoid noise sources such as light scattering, for example. Thus, if we look forward to quantifying gas concentrations, we need to control and measure pressure and temperature values inside the cell. Finally, the experimental information, provided from the proposed setup and the time-resolved spectroscopy we used, gave us some insights regarding LP gas dynamics, but systematization of the present methodology and technique should require additional research that it was out of the scope of this work for the moment.

Acknowledgements Rosa Ma. Rodríguez-Méndez is grateful to CONACyT, México, for scholarship support.

References

1. What is LPG?, World LPG Association (WLPGA) Glossary. https://www.wlpga.org/about-lpg/what-is-lpg/, 10 Dec 2018
2. ELGAS, LPG Gas Blog. https://www.elgas.com.au/blog/453-the-science-a-properties-of-lpg, 10 Dec 2018
3. Raslavičius L et al (2014) Renew Sustain Energy Rev 32:513–525
4. Setiyo M et al (2017) Int J Technol 1:112–121
5. Drews AW (ed) (1998) Manual on hydrocarbon analysis, 6th edn. American Society for Testing and Materials, p 16
6. Ryan JD et al (1998) Proc SPIE 3540:58–65
7. Wen-qing W et al (2013) Procedia Eng 52:401–407
8. Wang Y et al (2018) IEEE Sens J 18:(20)
9. Morsi I (2008) SAS 2008—IEEE sensors applications symposium Atlanta, GA, February, pp 12–14
10. DS Dhawale et al (2010) Sens Actuators, B Chem 147(2):488–494
11. Rey JM et al (2014) Appl Phys B 117:935–939
12. Shimpi NG et al (2016) Appl Surf Sci 390:17–24
13. Sonawane NB et al (2017) Mater Chem Phys 191:168–172
14. Kluczynski P et al (2012) Appl Phys B 108:183–188
15. Stewart G et al (2010) Proc of SPIE 7675:1–9
16. Jiménez MDR et al (2018) Materials 11:894
17. Harvey D (2000) Modern analytic chemistry. Edit Mc Graw-Hill, p 556

A Novel Approach to Optimize SLAM Using GP-GPU

Rohit Mittal, Vibhakar Pathak, and Amit Mithal

Abstract Mapping is the process to scale down and represent actual geo-allied information whereas localization is to identify and accommodate itself in the environment. Various papers had been published in this respect which sees SLAM algorithms, but a few had considered and examined for identification of object through SONAR and IR sensors. Apart from identification of object, and error due to curve path is also not considered in fictional steering system. The paper examines the above leftover and produces experimental results which are motivating.

Keywords SLAM · Fictional steering · SEP · Error · GP-GPU

1 Introduction

Mapping is affect which predefined scaled down geophysical information of a given area on which designed robot voyages but if we talk about localization, it predefines to distinguish the robot itself in a given map; the identification of object is based with propositional location in the map. In this respect, some researchers had done some work in optimization of Simultaneous Localization and Mapping (SLAM) algorithm in respect of precision and computation. Hence, localization and mapping are crucial parameters in self-directed robotics. Most of mapping parameters are in focus of precision in prediction by loading more sensory information into the

R. Mittal (✉) · V. Pathak
Arya College of Engineering & I.T., Jaipur, India
e-mail: rohit18mittal@yahoo.com

V. Pathak
e-mail: vibhakarp@rediffmail.com

A. Mithal
Jaipur Engineering College & Research Centre, Jaipur, India
e-mail: amit.mithal81@gmail.com

K. Ray et al. (eds.), *Proceedings of International Conference on Data Science and Applications*, Lecture Notes in Networks and Systems 148,
https://doi.org/10.1007/978-981-15-7561-7_22

prediction system (Extended Kalman Filter, Random Walk) or reducing computing complexity by either distributed processing or mathematical optimization.

A few researches had been done in optimization of error in navigation viz when the robot is in move in general and specific when it is steering itself in a curvature path. The objective of this paper is to present effect of errors in navigation when robot with fiction steering is moving in curve path. The errors during navigation of robot are presented in terms of Sectorial Error Probable (SEP). The SEP observed was based on Extended Kalamn Filter (EKF)-based SLAM algorithms, which is carefully selected among various popular SLAM algorithms viz: iSAM, DPG SLAM, FAST SLAM, Mono SLAM, Full SLAM, etc. [1].

To optimize computation of SLAM algorithm based on Extended Kalman Filter, it was proposed to use GP-GPU optimization by exploring parallel computation facility of GPU that is under control of any graphics processor which is loosely coupled on Graphics Core Next (GCN). To localize itself among various objects, an object identification algorithm is also an employed case.

In this paper, the researchers' categories the work in such a way that next section describes about the previous work done by the researchers in allied area. After that, the following section discusses an analysis on sectorial error probability when robot navigates under control of different SLAM algorithms. After then, researchers discuss the optimization done on chosen SLAM algorithm and finally researchers talked on result set and discussion.

2 Previous Work

Some research had been reported on localized and mapped parameters which are based on particle filter; it has seen that particle filter is based on hypothesis of probability of particle distribution which spread over of space in a given map. The particle filter-based SLAM algorithm had inherited challenges like it will not consider noise and computation is exponential to map area. Previously, researchers had done work on IoT-based sensors for voyage control of robot on many types of surfaces. Others done work on experimented localized and mapped parameters. Some works are focused on SLAM which is based on physical parameters and move on galze tile, boulder tile, etc. [1–3].

Majority of the previous work on simultaneous localized and mapped parameters techniques like iSAM represents affiliation of data with exponential growth in it in both online or offline modes to solve SLAM problem using smoothing technique [4]. Some work had been reported on FastSLAM algorithm in which researchers analyze various landmarks (point location) which are independent and give robot's pose. Simultaneous localized and mapped parameters based on Real Time Appearance based Mapping (RTAB-MAP) which uses OpenCV to extract the features from image to obtain visual arc, directly from kinect camera [4, 5].

Most of the previous work will not discuss sectorial error probable analysis. The paper presents upgraded SLAM based on Extended Kalman Filter (EKF) with Sectorial Error Probable check and its optimization based on GP-GPU computing.

3 Analysis on SEP Using Standard SLAM Algorithms

In this paper, authors evaluate and anlyze various experiments to analyze SEP in navigation of robot, i.e., probability of finding a robot when it moved in basic position within its boundaries. The analysis will be done using EKF, mSLAM, fSLAM.

In Table 1, it is extended that the Extended Kalman Filter, SLAM algorithm computes better SEP (less error) than other SLAM algorithms.

For (0°–15°) small sector in EKF, the error probability alters from 3.2 to 4.8 cm, whereas for mSLAM algorithm, variation from 2.2 to 6.5 cm and whereas for fSLAM algorithm, it is 3.2–7.8 cm.

Table 1 Analysis on standard SLAM algorithms

Turn angle	EKF	mSLAM	fSLAM
0°–15°	3.2 to 4.8 cm	2.2 to 6.5 cm	3.2 to 7.8 cm
15°–30°	2.5 to 4.6	2.4 to 4.4	3.6 to 4.2
30°–45°	2.7 to 4.4	2.3 to 4.0	3.0 to 4.2
45°–60°	2.3 to 4.5	2.2 to 3.9	3.8 to 5.0
60°–75°	2.2 to 4.3	2.1 to 4.8	4.4 to 5.2
75°–90°	2.1 to 3.3	1.8to 5.0	4.2 to 5.1
90°–105°	1.4 to 3.3	3.3 to 4.0	4.4 to 5.8
105°–120°	1.5 to 3.6	3.4 to 4.8	1.3 to 3.4
120°–135°	1.6 to 3.3	3.1 to 4.6	1.5 to 3.4
135°–150°	0.9 to 2.8	1.3 to 2.8	1.9 to 5.0
150°–165°	1 to 2.8	1.8 to 2.5	2.1 to 4.2
165°–180°	1.1 to 3.0	1.1 to 2.4	3.0 to 4.5
Average	1.87 to 3.72	2.25 to 4.14	3.03 to 4.81

For (15°–30°) small sector EKF, the error probability alters from 2.5 to 4.6 cm, for mSLAM algorithm, variation from 2.4 to 4.4 cm, whereas for fSLAM algorithm, it is 3.6–4.2 cm.

Hence, it has been concluded from abovementioned results, that as distance among boundaries in which robot move gets decreases, hence, EKF-based localized and mapped parameters give accurate outcomes with minimized SEP.

4 Algorithm Optimization Along with Pre-pinned Buffers and Shaders

Predicted learning algorithm is used along with GCN in shaders of AMD Radeon to optimize Extended Kalman Filter-based SLAM algorithm. The shader instructions gets decode, fetches after then execution of instructions are for group of inputs; the AMD architecture is best in rendering; the architecture has better parallel programmability than Nvidia architecture as computation floating point computation is better addressed in AMD architecture due to shaders and pre pinned buffers, which is used by OpenCL. When instructions are encoded in compute shader, it manifolds the performance of thread group size that is used in GPU architecture. When instructions get transferred from memory to GPU [6, 7], the pages gets locked in memory, due to which cost of GPU computation gets increased and are directly proportional to size of memory [8–10].

These buffers gets pinned (variable pinned) at the time of creation of variables and can be used to achieve peak interconnect bandwidth; it can create a cache copy within the device. As instruction decreases, parallelism in GPU architecture increases and increases in accuracy with speed. SONAR and IR sensors get attached in robot and it can be controlled through Bluetooth at 9600 baud rate for serial communication. Port monitor is needed to check port is opened or not and results got received [11, 12].

5 Results Set and Discussion

After the successful runs of robot along with execution of code, it can be seen that there is deviation in the findings of aforesaid circuits during the movement of robot and it found some fixed obstacles which are in rectangular shape and seen that obstacle will not so profound (Table 2) on 6.5 cm but when it move a little bit then at 12.0 cm, which is shown in Table 2. After some more successful runs (20), it can be

Table 2 Result set for obstacle in rectangular shape

Number of runs	Actual distance (cm)	Observed obstacle distance (cm)	Obstacle detects
1.	6.5	6.7	No
2.	6.9	7.2	Yes
3.	7.4	8.3	Yes
4.	9.4	10.9	Yes
5.	12.0	14.1	Yes
6.	14.0	18.2	No
7.	19.1	18.4	No
8.	19.4	21.0	No
9.	23.2	23.4	No
10.	25.4	24.0	No
11.	27.0	25.8	No
12.	29.2	27.9	No
13.	31.1	28.8	No
14.	33.4	30.4	No
15.	34.9	34.1	No
16.	36.1	36.2	No
17.	38.8	37.0	No
18.	40.1	38.9	No
19.	43.0	42.0	No
20.	44.5	44.7	No

seen that IR and SONAR sensor values for every distance that covered by robot as shown in Table 2 are different. Sensor value changes as there is increase in distance.

It has been concluded that after movement on robot on path with 20 successful runs using EKF, its data is shown in Table 3 which tells that obstacle is not significant when measured distance is in range, but after navigating some distance, the obstacle can easily be detected; if the case of cylindrical shape obstacle, but after that the actual distance has steep increase (17.1–37.1 cm) with infrared sensor and SONAR sensor bound on robot but IR sensor is not so sensitive in detection. Result is only 3 times obstacle presence is shown.

If conical shape is as shown in Table 4, infrared sensor is very sensitive but SONAR is not so sensitive. After executing the robot 20 successful runs, it is found that in Table 4 ,one-time sensor found the object. Hence, obstacle detection is best in case, if it is of rectangular shape.

Table 3 Result set if obstacles are in cylindrical shape

Number of runs	Actual distance	Observed obstacle distance (cm)	Obstacle detects
1.	5.5	6.6	No
2.	6.8	7.1	No
3.	7.9	8.2	Yes
4.	9.5	10.4	Yes
5.	12.1	13.6	Yes
6.	17.2	17.4	No
7.	17.5	18.2	No
8.	18.5	19.3	No
9.	22.2	23.4	No
10.	25.3	26.1	No
11.	25.4	26.2	No
12.	26.5	26.8	No
13.	27.8	28.1	No
14.	28.9	29.2	No
15.	29.1	29.9	No
16.	31.4	32.0	No
17.	33.4	32.8	No
18.	34.9	34.1	No
19.	35.4	35.6	No
20.	37.1	36.8	No

6 Conclusion and Future Scope

The algorithm and its experiment present have advantage of path and pose prediction with obstacle detection with least possible detection like IR and SONAR, which is readily available in low cost. Therefore, it has been concluded that EKF and sensory system (SONAR & IR) are used for acquiring sensor data of robot which focuses on obstacle detection in different physical environment. For further research, different types of sensory system can be used to optimize the objects with different physical parameters like IR dispersive, SONIC reflective, etc.

Table 4 Result set if obstacles are in conical shape

Number of runs	Actual distance	Observed obstacle distance	Obstacle detects
1.	5.6	6.6	No
2.	6.9	7.1	No
3.	7.8	8.1	No
4.	9.8	10.2	No
5.	13.1	14.3	No
6.	15.4	17.4	No
7.	17.1	17.6	Yes
8.	19.8	21.0	No
9.	22.6	23.0	No
10.	21.1	24.8	No
11.	25.2	25.6	No
12.	27.6	28.1	No
13.	28.9	30.2	No
14.	31.2	31.9	No
15.	32.3	33.0	No
16.	34.8	34.9	No
17.	36.1	36.8	No
18.	36.8	37.9	No
19.	38.1	40.4	No
20.	41.2	44.3	No

References

1. Frese U, Hirzinger G (2001) Simultaneous localization and mapping—a discussion
2. Mittal R, Pathak V et al (2018) A review of robotics through cloud computing. J Adv Robot
3. Apriaskar E, Nugraha YP et al (2017) Simulation of simultaneous localization and mapping using hexacopter and RGBD camera. In: 2017 2nd international conference on automation, cognitive science, optics, micro electro-mechanical system, and information technology (ICACOMIT)
4. Dhiman NK, Deodhare D et al (2012) Where am I? Creating spatial awareness in unmanned ground robots using SLAM: a survey
5. Balasuriya BLEA, Chathuranga BAH et al (2016) Outdoor robot navigation using Gmapping based SLAM algorithm. In: 2016 Moratuwa engineering research conference
6. http://haifux.org/lectures/267/Introduction-to-GPUs.pdf
7. https://www.boston.co.uk/info/nvidia-kepler/what-is-gpu-computing.aspx
8. Fatahalian K (2012) How GPU works
9. AMD accelerated parallel processing guide.pdf (2013)

10. Najam S, Ahmed J et al (2019) Run-time resource management controller for power efficiency of GP-GPU architecture. IEEE Access 7. 22 Feb 2019
11. Mittal R, Pathak V, Goyal S, Mithal A (2020) Chapter 17 a novel approach to localized a robot in a given map with optimization using GP-GPU. Springer Science and Business Media LLC, 2020
12. https://armkeil.blob.core.windows.net/developer/.../pdf/.../Mali_GPU_Architecture

Making of Streptavidin Conjugated Crypto-Nanobot: An Advanced Resonance Drug for Cancer Cell Membrane Specificity

Anup Singhania, Pathik Sahoo, Kanad Ray, Anirban Bandyopadhyay, and Subrata Ghosh

Abstract Due to various reasons, the drug cargo delivery technology in nanomedicine is not advancing toward front-end applications. One of the prime reasons is that the uncontrolled drug releases from the payload due to various adverse factors like proteolytic enzymes' intervention, pH imbalance, etc. We need a different approach to succeed. So, we can develop a stand-alone nanobot drug that activates chemically bonded payload by a resonance trigger. Herein we aim to develop a 'crypto-nanobot' resonance drug that would be body-friendly and also properly shielded to protect the drug from the human proteolytic enzymes. We have described the general method of preparation of such advanced crypto-nanobot resonance drugs. This will help the resonance drug to travel through the body fluids and reach the target tissues avoiding premature degradation.

Keywords Nanomedicine · Crypto-nanobot · Resonance drugs · Dendrimer

1 Introduction

Nanobot development for practical application is a prerogative to solve the problems related to the nanoscale environment. Thus, the development of nanobot has

A. Singhania · S. Ghosh (✉)
Chemical Science & Technology Division, CSIR-North East Institute of Science & Technology, Jorhat, Assam 785006, India
e-mail: ocsgin@gmail.com

Academy of Scientific and Innovative Research (AcSIR), CSIR-NEIST Campus, Jorhat, Assam 785006, India

P. Sahoo · A. Bandyopadhyay
International Center for Materials and Nanoarchitectronics (MANA), Research Center for Advanced Measurement and Characterization (RCAMC), National Institute for Materials Science, 1-2-1 Sengen, Tsukuba, Ibaraki 3050047, Japan

K. Ray
Amity School of Applied Sciences, Amity University, Jaipur, Rajasthan, India

K. Ray et al. (eds.), *Proceedings of International Conference on Data Science and Applications*, Lecture Notes in Networks and Systems 148,
https://doi.org/10.1007/978-981-15-7561-7_23

also become an important topic of research for its huge potential application in nanomedicine [1, 2]. In order to tackle the challenge of inducing suitable workability in the nanobot for working in the nanosize domain, there are two unique approaches. The top-down approach follows miniaturization of macroscopic robots [3, 4], whereas the bottom-up approach prefers the process of self-assembly or nano-organization of microscopic molecules [5, 6]. The former approach has its advantages because the model can be suitably designed from macromaterials, and the activity can be standardized in the macrodomain and that eventually replicated in micro- and nanoscale. So, an opportunity is there to optimize the structure–activity relationship of a nanobot beforehand. Whereas in the other approach molecular engineering of design is more decisive because structure–activity relationship maintenance is the bigger challenge [7, 8]. Now to investigate the potential of the unconventional nanobot resonance drugs, we propose to follow the later method. In the present case, we are dealing with the making of a resonance drug that is surrounded by a protective shell of protein. In this development process, we keep in our mind that in the communication of the nanobot drug to the outer shell local environment via the resonance energy transfer processes should not be disrupted by the protein shell and at the same time the protein shell does not marginalize the efficiency of the nanobot drugs [9, 10].

Thus, in the current context, we are purposefully modifying our previously developed PCMS nanobot resonance drug by replacing certain components to make it more advanced and more cell membrane specific for cancer cells. It will increase the druggability of the tissues. At the same time, we are taking care of drug lifetime by protecting it from outside till it could reach the target [11, 12]. For doing that, we have selected biotin molecule which is well-known to form selective and robust conjugation with streptavidin. If we know the molecular properties of the molecules at the individual levels, it seems easy to entangle them in a complex system to work in a network of desired performance [13, 14]. However, in practice, it is not so easy because the engineering of entanglement is a critical challenge and often fails for many unknown reasons [15, 16]. The molecular building blocks are sequentially incorporated in a single platform with the help of chemical interaction via hydrogen bond and chemical bond formation. Herein we report the proton NMR spectra of a novel crypto nanobot PCBM. The NMR spectra show the stepwise structure formation of the PCBM. As per the concept of our nanobot drugs, besides the activity of biotin-streptavidin conjugate as the membrane protein sensor, it also has to play a role in the active triangular communication pathway [17–19] (between sensors, controller, and motors), which we could further observe in the CEES spectroscopy.

2 Result and Discussion

Since premature drug degradation is an obstruction toward successful drug delivery, therefore, the protection of a potential drug from proteolytic enzymes is essentially required. So, in the new nanobot design, we have introduced new functional part biotin which is available at the dendrimer terminals. The selection of biotin was done based on its unique structural motif and ease of conjugation with the amine groups of the nanobot precursor dendrimer. Biotin is also a suitable entity that forms a strong non-covalent bond with streptavidin protein, and this biotin–streptavidin conjugate is highly stable in physiological conditions. Now since our nanobot drug is less than 10 nm, therefore, the streptavidin protein at the surface will spherically wrap up the drug from all sides (Fig. 1). Thus, the stability of the nanobot, as well as travel through physiological fluid, would be enhanced. Such a protein shield would stop exposure of the drug to the enzymes of the physiological environment for a longer time. In Fig. 1b, we can see two variations of nanobots have different molecular machine rotors (MM1 and MM2). Figure 1c shows the stepwise synthesis of the new biotinylated nanobot drugs; here, we start from the fourth-generation PAMAM dendrimer (P) as starting material. In the first step of the synthesis, we physically encapsulate small Nile red molecules in the dendrimer cavity to create the central control unit and thus resulted in the Nile red encapsulated structure PC. Next, we use the PC as starting material and conjugated a small number (approximately 4 molecules per PC) of biotin molecules on the amine terminal of the PC and we

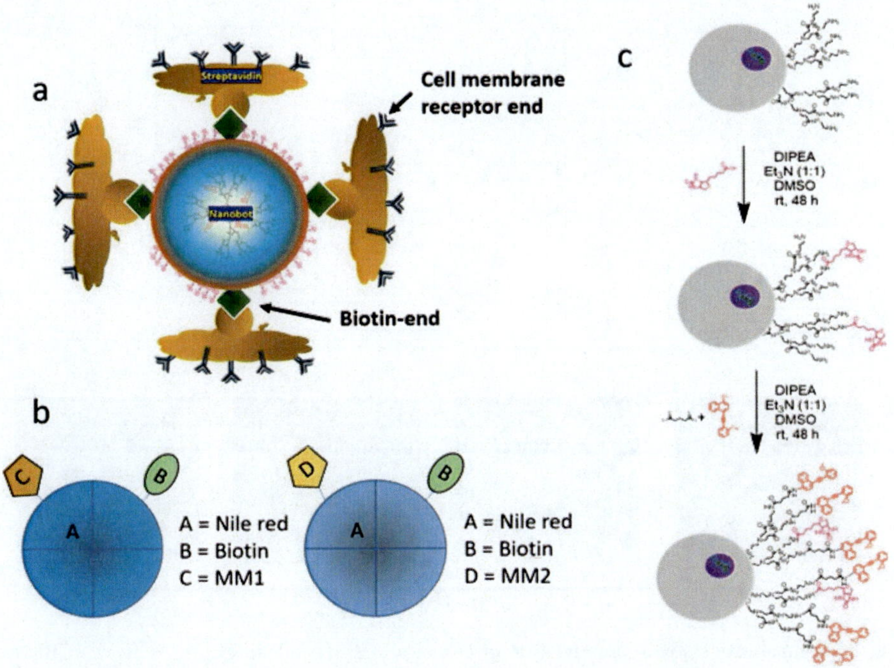

Fig. 1 **a** Representative structure of crypto-nanobot resonance drugs and the drug architecture is centralized and embedded in streptavidin capsule; **b** abstract presentation of two varieties of nanobots composed of PAMAM G4 dendrimer, Nile red, A, biotin, B, and two different molecular machines (MM1, C & MM2, D); **c** step-by-step synthesis of nanobot is presented, and in the first step, biotin is attached by chemical reaction then MMs are connected side by side on the same dendrimer surfaces

resulted in the molecular structure PCB. In the last step, we have functionalized the PCB with excess number of molecular machine rotors MM (either MM1 or MM2) and could furnish the final resonance nanobot PCBMs. The drug potential of the resulted in PCBMs are to be tested in different biological systems. We have designed the biotinylated nanobot drugs to observe the resonance energy transfer to the target object; However, we are also concerned about the unwanted interference of the protein shell of the streptavidin protein, which should not obstruct the resonance energy transfer pathways.

In Fig. 2, we see the proton NMR spectra of the step-by-step synthesis of biotinylated nanobots. The spectra (iv) and (v) confirm the formation of the two biotinylated nanobots PCBMs (PCBM1 and PCBM2), which are further confirmed by 2D-correlation NMR spectra. In the figure, we could see that there are certain new patterns of peaks in the specific chemical shift (δ ppm) position which attribute to the corresponding elementary molecular parts that are entangled in the systems in the stepwise chemical synthesis.

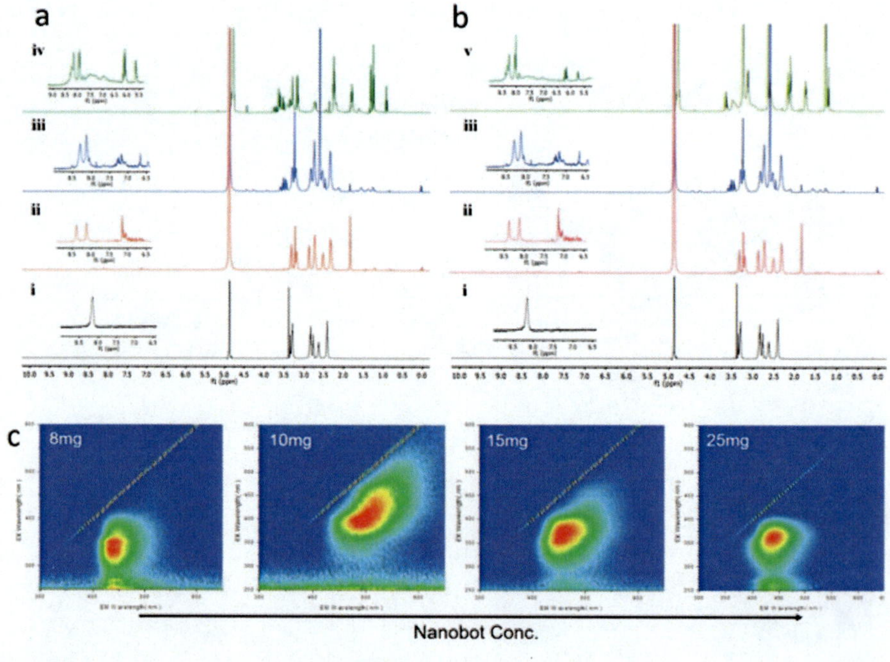

Fig. 2 Comparative study of proton NMR spectra of **a** PCBM1: i. P, ii. PC, iii. PCB, iv. PCBM1, **b** PCBM2: i. P, ii. PC, iii. PCB, iv. PCBM2. **c** concentration variation CEES spectrum of PCBM1 nanobot

Thus, two new and separate nanobot resonance drug architectures are prepared. However, due to structural complexity, assignments of the exact chemical shift position for individual elements are not done and work is under progress in the same direction to establish stipulated spectral characters. But in the characteristic pattern, we observe that the expected protons related to the aliphatic and aromatic environments can be identified. Some NMR peaks related to some component molecules are very weak because of their very low numbers of protons available in terms of ratio compared to the bigger number of protons of the large-sized dendrimer matrix molecule. Again, some peaks are not well segregated due to overlapping with other protons in the complex molecular structure. So, we have further elucidated the synthesized nanobot structures by concentration-dependent CEES spectroscopy and confirmation of structure is presented.

In Fig. 3, we have presented the energy-level diagram of the nanobot resonance drug molecule along with the different unit molecules that is used to compose the nanobot structure. From the energy diagram, we could easily see that compared to

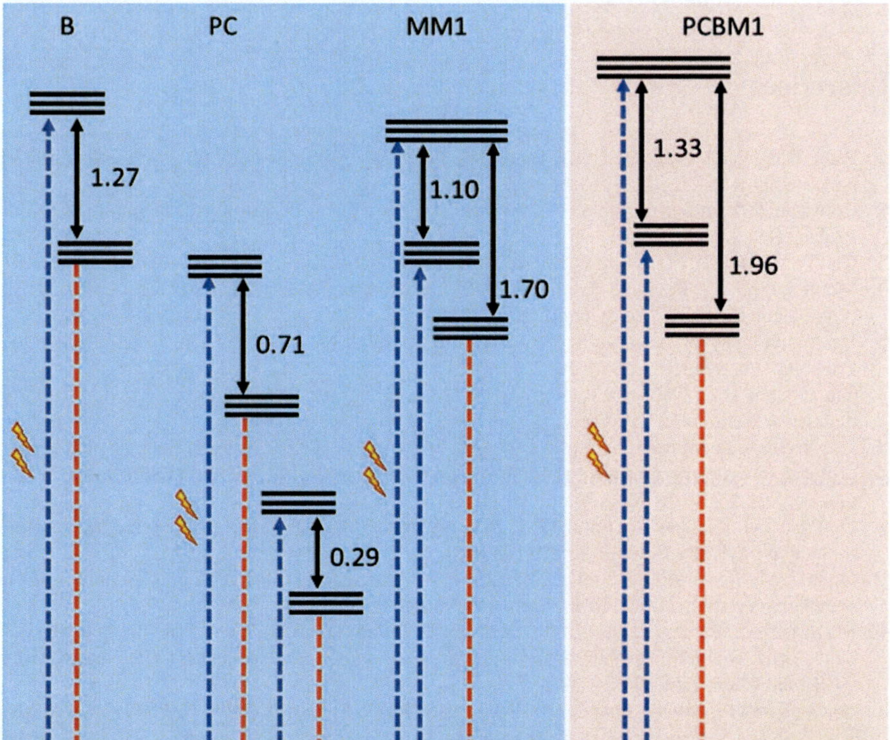

Fig. 3 Energy-level diagram calculated from CEES for B, PC, MM, and PCBM molecular structures

the component units, which has to build up the nanobot architecture, the nanobot has a greater capacity to hold energy (1.33–1.96 eV approx.). Therefore, this would have a property of more extended vibrational dynamics. This could play a major role in the potential resonance energy transfer process between the nanobot to the target objects located nearby around the nanobot.

3 Conclusion

In this study, we have found an advanced nanobot architecture, which is the potential to play the role of a future resonance drug with more precision toward the specified target. Crypto-drug would be a new advancement toward the novel resonance drugs. A further test of the synthesized nanobot is undergoing in vitro and in vivo levels. We will report further the advancement of biochemical interaction of our synthesized crypto-nanobots.

References

1. Patel GM, Patel GC, Patel RB, Patel JK, Patel M (2006) Nanorobot: a versatile tool in nanomedicine. J Drug Target 14:63–67
2. Couvreur P, Vauthier C (2006) Nanotechnology: intelligent design to treat complex disease. Pharma Res 23:1417–1450
3. Gao W, Wang J (2014) Synthetic micro/nanomotors in drug delivery. Nanoscale 6:10486–10494
4. Montemagno C, Bachand G (1999) Constructing nanomechanical devices powered by biomolecular motors. Nanotechnol 10:225–231
5. Bandyopadhyay A, Acharya S (2008) A 16-bit parallel processing in a molecular assembly. Proc Natl Acad Sci 105:3668–3672
6. Yin P, Choi HM, Calvert CR, Pierce NA (2008) Programming biomolecular self-assembly pathways. Nature 451:318–322
7. Ghosh S, Dutta M, Sahu S, Fujita D, Bandyopadhyay A (2013) Nano molecular-platform: a protocol to write energy transmission program inside a molecule for bio-inspired supramolecular engineering. Adv Funct Mater 24:1364–1371
8. Douglas SM, Bachelet I, Church GM (2012) A logic-gated nanorobot for targeted transport of molecular payloads. Science 335:831–834
9. Gulati NM, Stewart PL, Steinmetz NF (2018) Bioinspired shielding strategies for nanoparticle drug delivery applications. Mol Pharm 15:2900–2909
10. Magarkar A, Schnapp G, Apel AK, Seeliger D, Tautermann CS (2019) Enhancing drug residence time by shielding of intra-protein hydrogen bonds: a case study on CCR2 antagonists. ACS Med Chem Lett 10:324–328
11. Oh JY, Kim HS, Palanikumar L, Go EM, Jana B, Park SA, Kim HY, Kim K, Seo JK, Kwak SK, Kim C, Kang S, Ryu JH (2018) Cloaking nanoparticles with protein corona shield for targeted drug delivery. Nat Commun 9:4548
12. Carnemolla B, Borsi L, Balza E, Castellani P, Meazza R, Berndt A, Ferrini S, Kosmehl H, Neri D, Zardi L (2002) Enhancement of the antitumor properties of interleukin-2 by its targeted delivery to the tumor blood vessel extracellular matrix. Blood 99:1659–1665
13. Simon HA (1962) The architecture of complexity. Proc Am Philos Soc 106:467–482
14. Orrit M (2002) Molecular entanglements. Science 298:369–370

15. Sarovar M et al (2010) Quantum entanglement in photosynthetic light-harvesting complexes. Nat Phys 6:462–467
16. Sahoo P, Dastidar P (2012) Secondary ammonium dicarboxylate (SAD)—a supramolecular synthon in designing low molecular weight gelators derived from azo-dicarboxylates. Cryst Growth Des 12:5917–5924
17. Ghosh S, Chatterjee S, Roy A, Ray K, Swarnakar S, Fujita D, Bandyopadhyay A (2015) Resonant oscillation language of a futuristic nano-machine-module: eliminating cancer cells and Alzheimer Aβ plaques. Curr Topic Med Chem 15:534–541
18. Ghosh S, Roy A, Singhania A, Chatterjeec S, Swarnakar S, Fujita D, Bandyopadhyay A (2018) In-vivo & in-vitro toxicity test of molecularly engineered PCMS: a potential drug for wireless remote-controlled treatment. Toxicology Rep 5:1044–1052
19. Singhania A, Dutta M, Saha S, Sahoo P, Bora B, Ghosh S, Bandyopadhyay A (2020) Speedy one-pot electrochemical synthesis of giant octahedrons from in situ generated pyrrolidinyl PAMAM dendrimer. Soft Matter [accepted]

Performance Evaluation of Fuzzy-Based Hybrid MIMO Architecture for 5G-IoT Communications

Fariha Tabassum, A. K. M. Nazrul Islam, and M. Shamim Kaiser

Abstract Pervasive sensors connect many applications of our day to day lives and allow us to collect and understand ubiquitous data ranging from environment monitoring to healthcare. The proliferation of these sensor nodes creates Internet of things which demands massive connectivity, energy efficiency, high throughput, etc. Massive MIMO in the extended frequency band of mmWave can be a rising solution in this regard. This work presents a fuzzy-based switching algorithm for subconnected hybrid structure of mmWave massive MIMO system. The switching algorithm has also been proposed to improve the signal-to-noise ratio of the mmWave links. The main aim is to save energy and improve system capacity through a less complex structure in contrast to the existing structures. The performance evaluation of the proposed fuzzy-based algorithm achieves better energy efficiency at a moderate data rate compared to other methods.

Keywords Hybrid beamforming · Millimeter wave · IoT · Massive MIMO · Energy efficiency · Subconnected

1 Introduction

The future of the Internet of Things (IoT) will face two main challenges, i.e., the bulk connectivity and low latency [1]. Prediction says the future IoT is going to accumulate almost 50 billion IoT devices by the year 2020 [1, 2]. The fifth-generation (5G) wireless networks have been introducing some promising technologies to cope up with this demand for massive connectivity. Among these technologies, millimeter wave (mmWave) has been accepted to be a strong candidate to increase the network

F. Tabassum · A. K. M. Nazrul Islam
Military Institute of Science and Technology, Dhaka, Bangladesh
e-mail: abonty.mist@gmail.com

M. S. Kaiser (✉)
Jahangirnagar University, Dhaka 1342, Bangladesh
e-mail: mskaiser@juniv.edu

K. Ray et al. (eds.), *Proceedings of International Conference on Data Science and Applications*, Lecture Notes in Networks and Systems 148,
https://doi.org/10.1007/978-981-15-7561-7_24

capacity by extending the existing bandwidth (sub 6 GHz) spectrum [3]. The large unutilized spectrum at mmWave band (30–300 GHz) will improve the throughput gains significantly [4]. Moreover, the shorter wavelengths allow a large array of antenna elements at both the base station (BS) as well as the users. This induces the concept of multi-user multiple-input multiple-output (MU-MIMO) to massive MIMO [4]. Massive MIMO delivers reliable communication by enhancing capacity and offering high data rates. The integration of mmWave and massive MIMO will act as capacity boosters because of the extended bandwidth and higher spectral efficiency (SE) [5]. In this system, beamforming with a large antenna array is used to diminish path losses caused by mmWave communication with directional transmissions [6].

On the other hand, massive MIMO engenders excessive transceiver complexity and energy consumption because each antenna in MIMO requires a dedicated radio-frequency (RF) chain [7, 8]. Thus, huge RF chains will be used due to the use of an equally huge number of antennas in mmWave massive MIMO systems. The approximation says that RF components may absorb up to 70% of the total transceiver energy consumption [9]. Thus, massive MIMO becomes unrealistic in practice due to the high hardware cost and energy absorption caused by many RF chains. Thus, the conventional beamforming system (digital beamforming), though offers better SE, becomes unacceptable because of using a large number of RF chains as the number of antennas [10]. Analog beamforming, however, transmits single data streams only using one RF chain and several phase shifters.

However, the combination of analog beamformers with digital beamforming brings about the concept of hybrid beamforming structure (HBS) which gives a promising solution to these problems. HBS can transmit multiple data by saving the cost and reducing the complexity as well as energy consumption by lowering the number of RF chain in a massive MIMO system [11]. The HBS can be divided into two categories: fully connected (FC) and sub connected (SC) architecture. The FC structure provides better performance at the cost of hardware complexity whereas SC structure is less complex which causes lower beamforming gain [12]. In an SC-HBS structure, the RF chains are connected to a subarray of BS antennas only. Traditional FC-HBS structure with phase shifter (PS) requires many PSs to realize analog beamforming, where all the RF chains are connected to all the BS antennas via all PSs. The use of large-scale PS will increase hardware complexity and overall power consumption [13] which could be a hurdle for the resource constrain of IoT.

To overcome these challenges various type of suboptimal energy/hardware efficient HBS with PS has been proposed. In this work, a fuzzy-based energy-efficient HBS (FeE-HBS) architecture with PS has been proposed. The fuzzy-based switching enables less PSs when the requirement is less. This makes the system more dynamic and lessens the power consumption of the network. Thus, the proposed system reduces the hardware complexity of the network to a great extent, improves the energy efficiency and delivers desirable data rates. The main aim is to improve the overall system performance of the hybrid-mmWave-massive MIMO systems.

The rest of the paper is arranged as follows. The system model of the proposed HBS mmWave-MIMO architecture is initiated in Sect. 2. Section 3 proposes a fuzzy-based switching algorithm. Simulation results are shown in Sect. 4. Finally, the paper draws a conclusion in Sect. 5.

2 FeE-HBS with PS

The proposed FeE-HBS system model has one BS equipped with N_t antennas and N_t^{RF} number of RF chains such that, $N_t >> N_t^{\mathrm{RF}}$. The N_S data streams are being transmitted from the antenna array of the BS to users each with N_r antennas ($N_r = N_t$) and N_r^{RF} RF chains. Proposed FeE-HBS model is illustrated in Fig. 1. In the transmitter section, there are main three subsystems: digital precoder, RF chain and fuzzy-based analog precoder. The base band signal is processed by $N_t^{\mathrm{RF}} \times N_S$ dimensional digital precoder, then converted into $N_S \times N_t^{\mathrm{RF}}$ dimensional RF signal using N_t^{RF} RF chain. The RF chains are coded with fuzzy-based analog precoder. The output of the RF chain is fed into PSs for generating N parallel signals with same amplitude $1/\sqrt{N}$ and different phases. Finally, the signal with required phase will be fed into fuzzy-based switching matrix for transmitting with MIMO antenna array.

The signal transmitted by the BS can be expressed as

$$x = \mathbf{F}s \tag{1}$$

where the hybrid precoding matrix $\mathbf{F} = \mathbf{F}_{\mathrm{AB}}\mathbf{F}_{\mathrm{DB}}$ is applied to transmitted symbol vector $\mathrm{s} = \left(s_1\ s_2\ \ldots\ s_{N_S}\right)^T \in \mathcal{C}^{N_s \times 1}$ in the baseband with $\mathbb{E}\left[ss^H\right] = I_{N_s}.\mathbf{F}_{\mathrm{DB}}$ be an $N_t^{\mathrm{RF}} \times N_t$ baseband digital precoder matrix. For FC structure where the data streams

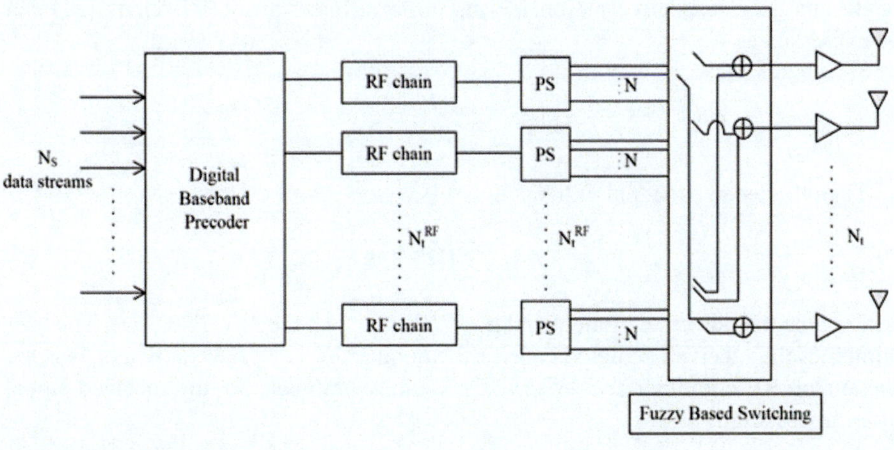

Fig. 1 FeE-HBS system model

are connected to all RF chains, $\mathbf{F}_{\text{DB}}$ is expressed as

$$\mathbf{F}_{\text{DB}} = \begin{pmatrix} d_{11} & \cdots & d_{1N_s} \\ \vdots & \ddots & \vdots \\ d_{N_t^{\text{RF}}1} & \cdots & d_{N_t^{\text{RF}}N_s} \end{pmatrix} \tag{2}$$

But we consider the SC structure where single RF chain is connected to each data stream. This reduces $\mathbf{F}_{\text{DB}}$ to a diagonal matrix $\mathbf{F}_{\text{DB}} = \text{diag}\big(d_{11}, d_{22}, \ldots, d_{N_t^{\text{RF}}N_s}\big)$. The digital precoded data is directly fed into the analog beamformer of the corresponding subarray. $\mathbf{F}_{\text{AB}}$ be the $N_S \times N_t^{\text{RF}}$ analog beamforming matrix which can be expressed as

$$\mathbf{F}_{\text{AB}} = \begin{pmatrix} a_1 & 0 & \cdots & 0 \\ 0 & a_2 & \cdots & 0 \\ \vdots & \vdots & \ddots & \vdots \\ 0 & 0 & \cdots & a_{N_t^{\text{RF}}} \end{pmatrix} \tag{3}$$

We consider the frequency-flat channel for narrowband frequency. Assuming the widely accepted Saleh-Valenzuela channel model for L number of paths, the channel matrix $\mathbf{H} \in \mathcal{C}^{N_r \times N_t}$ can be expressed as

$$\mathbf{H} = \gamma \sum_{l=1}^{L} \alpha_l \boldsymbol{a}_r(\phi_l^r) \boldsymbol{a}_t^H(\phi_l^t) \tag{4}$$

where α_l is the complex gain for lth path with $\mathbb{E}\big[|\alpha_l|^2\big] = 1$ and γ is a normalization factor. $\phi_l^r \in [0, 2\pi]$ and $\phi_l^t \in [0, 2\pi]$ are the lth paths angle of departure (AOD) and angle of arrival (AOA), respectively. $\boldsymbol{a}_r(\phi_l^r)$ and $\boldsymbol{a}_t^H(\phi_l^t)$ be the array response vectors of BS and user, respectively. Considering uniform linear array (ULA), $\boldsymbol{a}_t^H(\phi_l^t)$ can be expressed as

$$\boldsymbol{a}_t(\phi_l^t) = \frac{1}{\sqrt{N_t}}\Big[1, e^{j\frac{2\pi}{\lambda}d\sin\phi_l^t}, \ldots, e^{j(N_t-1)\frac{2\pi}{\lambda}d\sin\phi_l^t}\Big] \tag{5}$$

Then the received signal will be

$$\mathbf{r} = \sqrt{\beta}\mathbf{HFs} + \mathbf{n} \tag{6}$$

where β be the power coefficient and $\mathbf{n} \in \mathcal{CN}\big(0, \sigma^2\big)$ be the AWGN vector. The user combines the received signal via a hybrid combiner $\mathbf{W} = \mathbf{W}_{\text{AB}}\mathbf{W}_{\text{DB}}$ where $\mathbf{W}_{\text{AB}}$ be the analog RF combiner and $\mathbf{W}_{\text{DB}}$ is the digital combiner. So, the received signal after processing is

$$\mathbf{y} = \sqrt{\beta}\mathbf{W}^*\ \mathbf{HFs} + \mathbf{W}^*\ \mathbf{n} \tag{7}$$

3 Switching Algorithm

In our FeE-HBS system, the signal with required phase will be fed into fuzzy-based switching matrix for transmitting with MIMO antenna array. This switching is designed which aims to maximize the sum rate. The performance of precoder and switching matrix system depends on selection of appropriate PSs. However, optimal search algorithm increases the complexity of the HBS. The fuzzy-based low complexity design requires to compromise with the performance of the HBS.

Chafaa et al. [14] and Peng et al. [15] proposed estimation of mmWave channel parameters such as the path gain, the angle of arrival (AoA) and the angle of departure (AOD). In this work, we have considered that the CSI information is available at the receiver as per these estimation techniques. In addition, AoD of a group of uses takes values from [0, 2π] whereas the mean AoA of a group follows uniform distribution over π/3 sector.

Figure 2 shows fuzzy logic controller (FLC) for the switching matrix operation. Based on the fuzzy inference rules, FLC may select PS to be switched on considering differentiation of AoA, AoD, data rate demand. Each variable contains three linguistic

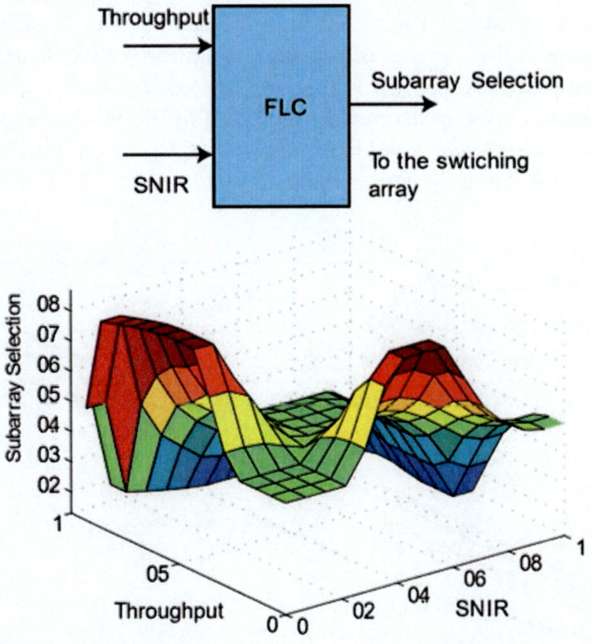

Fig. 2 Fuzzy logic controller selects subarray of the mmWave antenna (top) whereas subarray selection is a function of throughput and signal-to-noise plus interference ratio. The bottom figure shows the surf plot of the input and output membership functions

terms low (L), medium (M) and High (H) and they follow Gaussian membership function. Figure 2 shows the relation between the input and output membership functions.

4 Simulation Results

In this section, the performance of the proposed network is evaluated and compared to the conventional FC/SC system. To be more specific, we showed the change in energy efficiency (EE) performance with the change in a signal-to-noise-plus-interference ratio (SNIR) and in terms of the number of RF chain. Also, the behavior of data rate is monitored while changing the SNR value. The channel paths are assumed $L = 5$. The carrier frequency is predefined at 45 GHz. We considered $N_t = N_R = 64$, i.e., a large number of transmit and receive antennas are assumed to implement the concept of massive MIMO. A frequency-flat channel with perfect channel state information (CSI) is considered.

Figure 3 shows the EE behavior with the change of SNIR considering $N_t^{\mathrm{RF}} = N_r^{\mathrm{RF}} = N_s = 6$ and $N_t = N_R = 64$. We observe that the EE is reduced with the increase of PSs per subarray. Figure 3 shows the depletion of the EE for $N = 2$, 4 and 8. It is because of the fact that increasing the number of PSs lead to an increase in energy consumption.

The data rates of all the structures are analyzed in Fig. 4. The above-mentioned network parameters are considered here, i.e., $N_t = N_R = 64$ and $N_t^{\mathrm{RF}} = N_r^{\mathrm{RF}} = N_s = 6$. Both the BS and users are assumed to own perfect CSI. Figure 4 shows the

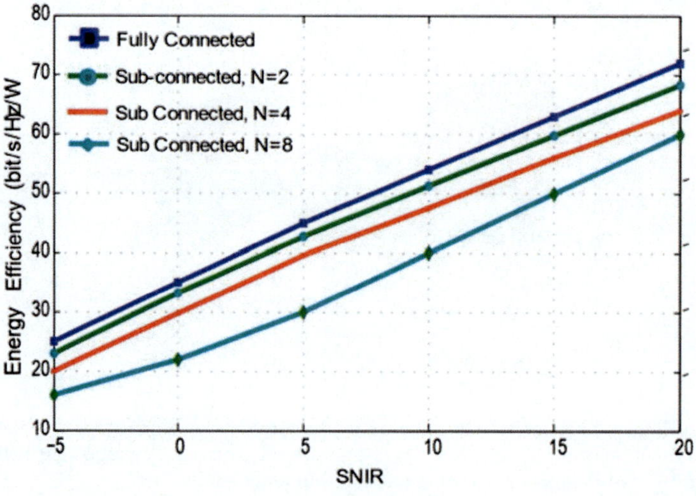

Fig. 3 Comparison of EE over SNIR of the network with $N_t = N_R = 64$ and $N_t^{\mathrm{RF}} = N_r^{\mathrm{RF}} = N_s = 6$

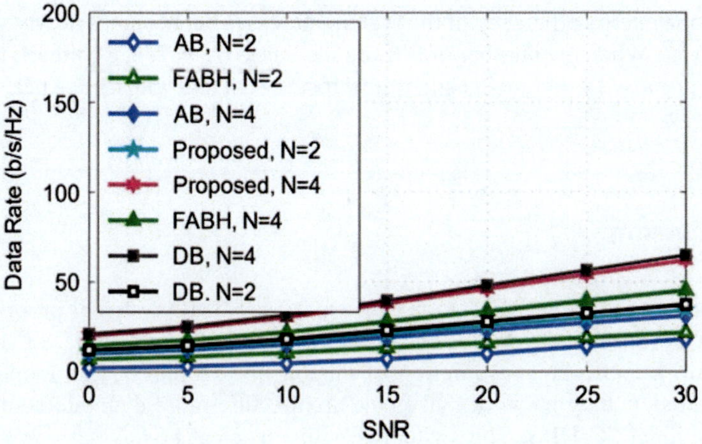

Fig. 4 The effect of SNR in dB on average data rate in b/s/Hz for $N = 2$ and $N = 4$

change in data rates in terms of SNR. As we can see, the analog beamforming (AB) scheme obtains the lowest rate for the same SNR requirement among all connections. Both the digital beamformer (DB) and our proposed scheme achieves higher data rates than all other structures. The increase in the number of PSs ($N = 4$) surely allows the same structure to achieve a higher data rate at the cost of lowering the EE.

In Fig. 5, the change in EE is observed with the change in RF chains. We observe that the EE of FC-HBS lags much behind than that of the proposed FeE-HBS system. Both the conventional SC-HBS and the proposed system outperform the FC structure in terms of EE especially when N_t^{RF} is low. Nonetheless, as the number of N_t^{RF} is

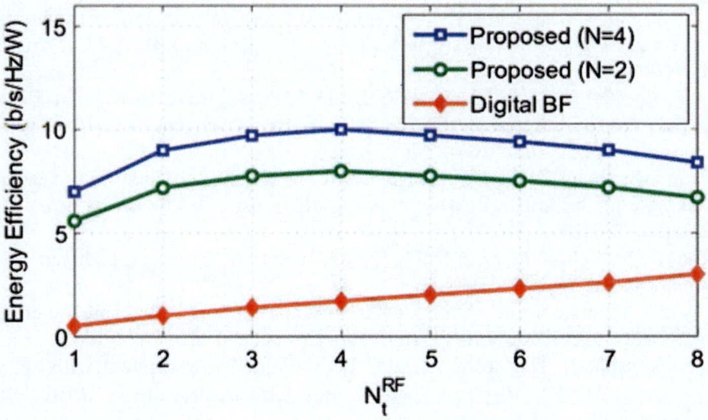

Fig. 5 The effect of N_t^{RF} on energy efficiency for $N = 2$ and $N = 4$

increased, the energy efficiency of the system increases but becomes nearly saturated at higher N_t^{RF}. When the number of PSs are increased ($N = 2, 4, 8$), the EE becomes less in our proposed structure but still outperforms FC-HBS with a huge performance gap.

5 Conclusion

In this paper, a novel FeE-HBS for mmWave massive MIMO system is proposed. The number of PS per subarray has been made lower than the traditional FC structure to deal with the energy consumption of analog PS. Moreover, the adoption of a fuzzy-based switching network makes the architecture more dynamic compared to the conventional SC-HBS. The switching helps to select PSs as per one's criteria. This enables the propose FeE-HBS system to achieve higher energy efficiency at desirable spectral efficiency than the conventional hybrid structures.

References

1. Lv T, Ma Y, Zeng J, Mathiopoulos PT (2018) Millimeter-wave NOMA transmission in cellular M2M communications for Internet of Things. IEEE Internet of Things J 5(3):1989–2000
2. Liu X, Ansari N (2016) Green relay assisted D2D communications with dual battery for IoT. In: Proceeding IEEE global communications conference (GLOBECOM), pp 1–6
3. Alkhateeb A, Leus G, Heath RW (2015) Limited feedback hybrid precoding for multi-user millimeter wave systems. IEEE Trans Wireless Commun 14(11):6481–6494
4. Almasi MA, Vaezi M, Mehrpouyan H (2019) Impact of beam misalignment on hybrid beamforming NOMA for mmWave communications. In: IEEE transactions on communications
5. Gao X, Dai L, Han S, Chih-Lin I, Heath RW (2016) Energy-efficient hybrid analog and digital precoding for mmWave MIMO systems with large antenna arrays. IEEE J Sel Areas Commun 34(4):998–1009
6. Ahmed I, Khammari H, Shahid A, Musa A, Kim KS, Poorter ED, Moerman I (2018) A survey on hybrid beamforming techniques in 5G: architecture and system model perspectives. IEEE Commun Surveys & Tutorials 20(4):3060–3097
7. Wang B, Dai L, Wang Z, Ge N, Zhou S (2017) Spectrum and energy-efficient beamspace MIMO-NOMA for millimeter-wave communications using lens antenna array. IEEE J Sel Areas Commun 35(10):2370–2382
8. Rusek F et al (2013) Scaling up MIMO: opportunities and challenges with very large arrays. IEEE Signal Process Mag 30(1):40–60
9. Salh A, Audah L, Shah NS, Hamzah SA (2019) Trade-off energy and spectral efficiency in a downlink massive MIMO system. Wireless Pers Commun 106(2):897–910
10. Vlachos E, Kaushik A, Thompson J (2018) Energy efficient transmitter with low resolution DACs for massive MIMO with partially connected hybrid architecture. In: IEEE 87th vehicular technology conference (VTC Spring), pp 1–5
11. Xu C, Ye R, Huang Y, He S, Zhang C (2018) Hybrid precoding for broadband millimeter-wave communication systems with partial CSI. In: IEEE Access, vol 6, pp 50891–50900
12. Zhao P, Wang Z (2018) Hybrid precoding for millimeter wave communications with fully connected subarrays. IEEE Commun Lett 22(10):2160–2163

13. Molisch AF, Ratnam VV, Han S, Li Z, Nguyen SLH, Li L, Haneda K (2017) Hybrid beamforming for massive MIMO: a survey. IEEE Commun Mag 55(9):134–141
14. Chafaa I, Djeddou M (2017) Improved channel estimation in mmWave communication system. In: Seminar on detection systems architectures and technologies (DAT), pp 1–5
15. Peng Y, Li Y, Wang P (2015) An enhanced channel estimation method for millimeter wave systems with massive antenna arrays. IEEE Commun Lett 19(9):1592–1595

Reducing Frequency Deviation of Two Area System Using Full State Feedback Controller Design

Shubham, Sourabh Prakash Roy, and R. K. Mehta

Abstract More than one generating power stations are interconnected, to meet the increasing energy demand and to maintain the energy equation. It is desired to have minimum deviation frequency in the system with the variation in dynamic load. There are various control techniques to achieve the desired dynamic system response. This paper models the two similar and dissimilar units. The generating station models used for analysis are thermal non-reheat, reheat and hydro turbine. The frequency deviation response for the unit is obtained with I-controller and compared with its full state feedback controller. The whole system is simulated in MATLAB Simulink.

Keywords Two area · I-FSB · Frequency deviation · Non-reheat · Reheat · Hydro

1 Introduction

Conventional and non-conventional generating plants are connected among themselves as grid via tie line to meet the increasing energy demand of the world. The former are those stations whose energy-producing sources are limited for example thermal, hydro, nuclear whereas, non-conventional energy sources are unlimited. The interconnection between the generating unit helps to meet the energy demand from the ever-changing load side. The excess power generated by any unit can be transferred to the unit not able to meet its load demand. However, change in the dynamic load of one unit will affect the other units connected to it. Thus, the performance of the entire unit gets influenced [1].

Shubham (✉) · S. Prakash Roy · R. K. Mehta
Electrical Engineering Department, NERIST, Nirjuli, Arunachal Pradesh, India
e-mail: shubham.roy3@gmail.com

S. Prakash Roy
e-mail: sourabh.roy2050@gmail.com

R. K. Mehta
e-mail: rkm@nerist.ac.in

K. Ray et al. (eds.), *Proceedings of International Conference on Data Science and Applications*, Lecture Notes in Networks and Systems 148,
https://doi.org/10.1007/978-981-15-7561-7_25

The desired performance of the areas can be achieved using control strategies. The control strategies are broadly classified as classical, optimal and modern intelligent approaches. Classical controllers are basically integral, proportional, derivative or any combination of them like PI, ID, and PID, etc. Modern control strategies include artificial intelligent networks (ANN), fuzzy, combination modern controller with classical ones. [2] Two are similar and dissimilar units are modeled with the combination of non-reheat turbine, reheat turbine and hydro turbine [1]. To make the modeling process simpler and easier to design the non-linearity like generation rate constraint, generation dead band and time delay in the interconnection are neglected. Also, the transfer function considered is of first-order system. The state feedback control strategy is used for the stabilizing the two area model. In state feedback controller, usually, one state variable output from the respective unit is feedback and whose value is compared with a reference value to get the desired output known as single state feedback. But in full state feedback (FSB) is a multivariable feedback control method in which change in each state variable is feedback and compared with its reference value to get the desired output. Thus, improving the performance of the system [3, 4].

The state variables are feedback to the controller with the gain defined by the gain matrix. There is various method of finding the gain matrix of FSB controller namely pole placement, linear quadratic method (LQR), eigenvalue technique [4–8]. This paper uses the technique of LQR to find the feedback gain matrix to design the full state feedback controller which is described in the later sections.

2 Subsystem Configurations

This section presents the generalized representation of two area system as shown in Fig. 1. It consists up of controller connected to the governor and turbine which is driving the system static and dynamic load. There are two closed-loops namely primary and secondary closed loop. The primary loop connects system output with the controller via speed regulator (R_1, R_2) which regulates the speed of the governor with the change in the system dynamic load (d_1, d_2). The secondary loop connected the system output with the controller via area control error (B_1, B_2) senses the change in frequency with the variation in d_1 and d_2.

The subsystem consists of the governor and turbine model connected in series. The first-order transfer function for the subsystem considered for non-reheat and reheat as well as hydro turbine and governor are as follows [1].

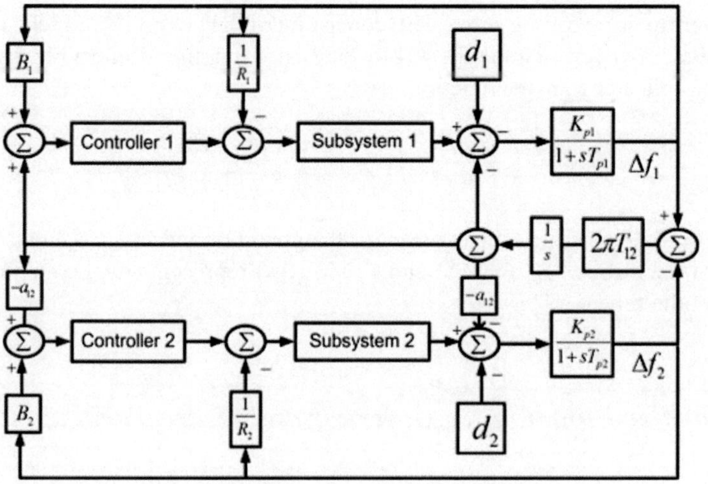

Fig. 1 Block diagram for two area system

2.1 Transfer Function for Governor and Non-reheat Turbine

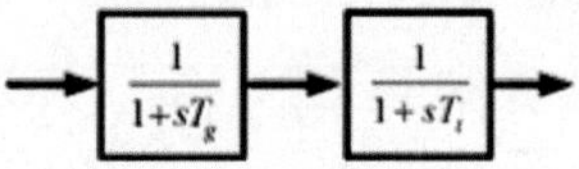

The transfer function for the combined governor and non-reheat turbine indicating the subsystem is the product of the individual transfer function of the governor and non-reheat turbine

$$G_{\text{TN}} = \left(\frac{1}{1 + sT_g}\right)\left(\frac{1}{1 + sT_t}\right) \tag{1}$$

where T_g (= 0.08 s) is the time constant of the governor and T_t (= 0.3 s) is the time constant of the non-reheat turbine [1].

2.2 Transfer Function for Governor and Reheat Turbine

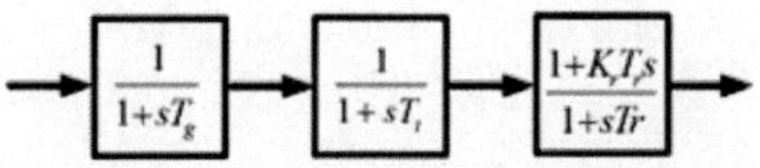

The transfer function for the subsystem consisting of the governor and reheat turbine is the product of governor and reheats block transfer function. Reheat block contains turbine and reheater transfer function.

$$G_{\mathrm{TR}} = \left(\frac{1}{1+sT_g}\right)\left(\frac{1}{1+sT_t}\right)\left(\frac{1+K_rT_rs}{1+sT_r}\right) \tag{2}$$

where T_g (= 0.08 s) is the time constant of the governor and T_t (= 0.3 s) is the time constant of the turbine. K_r (= 0.5) and T_r (= 10 s) represents the dc gain and time constant of the reheater [1].

2.3 Transfer Function for Governor and Hydro Turbine

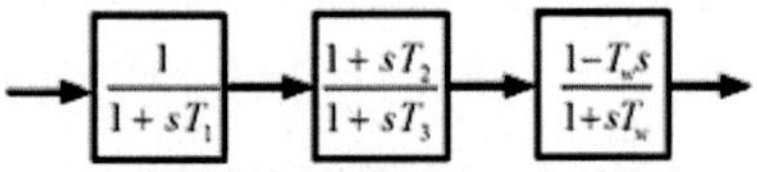

The transfer function for the subsystem consisting if the governor and hydro turbine is their transfer function product and given by

$$G_{\mathrm{TR}} = \left(\frac{1}{1+sT_1}\right)\left(\frac{1+sT_2}{1+sT_3}\right)\left(\frac{1-sT_w}{1+sT_w}\right) \tag{3}$$

where T_1 (= 48.709 s) is the time constant of governor and T_2 (= 5 s), T_3 (= 0.513 s) and T_w (= 1 s) is the time constant of the turbine [1].

The transfer function for the static power system load is

$$G_P = \left(\frac{K_P}{1+sT_P}\right) \tag{4}$$

where K_P (= 120) and T_P (= 20 s) in case of thermal turbine area and K_P (= 80) and T_P (= 13 s) for hydro turbine area [1].

Both the areas are interconnected to each other such that any disturbances in one affect the response of the other. The interconnection between them is via tie line system. The controller is designed to minimize the shoots in the frequency response of the areas with the change in the dynamic load.

3 Controller Design Methodology

From the transfer function of each block, overall transfer function of each unit is calculated and then converted into its state-space representation. In Eq (4), $\dot{x}$ is a matrix which represents state variable of the system, Y is output state matrix, A is the state matrix, B is the control matrix, T is disturbance matrix whereas X represents state vector, U is the controlling vector, and d is the disturbance vector. Since we are taking changes in load as step response so $Td = 0$. Also, for feedback controller $D = 0$ [3].

$$\begin{aligned} \dot{x} &= AX + BU + Td \\ Y &= CX + DU \end{aligned} \tag{5}$$

In full state feedback, the control vector (U) is defined as a sum of the scalar quantity multiplied with the feedback from each state variable $U = -KX$. The scalar quantity is the respective gain of each state forming feedback state matrix. As in two area model, there is two control input u_1 and u_2 can be written in the form of expression in Eqs. (6) and (7) gives the feedback state matrix [9].

$$\begin{aligned} U_1 &= K_{11}X_1 + K_{12}X_2 + K_{13}X_3 \ldots K_{1n}X_n \\ U_2 &= K_{21}X_1 + K_{22}X_2 + K_{23}X_3 \ldots K_{2n}X_n \end{aligned} \tag{6}$$

$$K = \begin{bmatrix} K_{11}K_{13}K_{13} \ldots K_{1n} \\ K_{21}K_{22}K_{23} \ldots K_{1n} \end{bmatrix} \tag{7}$$

The feedback matrix is designed in such a way that minimum deviation in frequency which the variation in the load. In expression (8) R is identity square matrix which gives weight to the control vector, B^T is the transpose of control matrix and S is the solution of the Riccati equation for the considered system. Following the above-mentioned step feedback gain matrix is obtained for each case [4].

$$K = \text{inv}(R) \times B^T \times S \tag{8}$$

4 Proposed Model for Case Study

4.1 Case Study 1

This case assumes both areas are having a thermal non-reheat system. The output frequency (Δf_1) of the area$_1$ is connected B_1 which is added with tie line and fed to

Table 1 Gain values for the case study 1

Gain for area 1	Gain for area 2
$k_{11} = 4.2661$	$k_{21} = 0.1628$
$k_{12} = 3.9634$	$k_{22} = 0.5928$
$k_{13} = 0.7646$	$k_{23} = 0.0090$
$k_{14} = 0.1628$	$k_{24} = 4.2661$
$k_{15} = 0.0598$	$k_{25} = 3.9634$
$k_{16} = 0.0090$	$k_{26} = 0.7646$
$k_{17} = -4.3506$	$k_{27} = 4.3506$
$k_{18} = 1.0000$	$k_{28} = -0.0000$
$k_{19} = -0.0000$	$k_{29} = 1.0000$

the controller. The difference of the output of the controller and the R_1 is given to subsystem1. The transfer function subsystem$_1$ of the area is given by Eq. (1). The difference of the subsystem$_1$, d$_1$ and tie line is given to static load. This gives the frequency response of area$_1$ of the assumed system. Area$_2$ can be connected in the same manner being a similar system as area$_1$. The output frequency of both areas is connected with each other via tie line. The area$_1$ is connected with the tie line in the same phase whereas area$_2$ is connected with tie line using a factor of $-a_{12}$ giving as 180° phase shift.

For the considered case, the no. of the state variable is nine and control vector is two so, K is of the order (2×09) matrix and its value is tabulated in Table 1.

4.2 *Case Study 2*

This case assumes only thermal reheat system for both the units. All the connection is the same as described in case study 1. The only difference lies in the subsystem connected for the case. The transfer function for the subsystem used is given by Eq. (2).

For the considered case, the no. of state variable is eleven and control vector is two so, K is of the order (2×11) matrix and its value is tabulated in Table 2.

4.3 *Case Study 3*

This case assumes two dissimilar areas of the thermal non-reheat system and hydro system. The transfer function of the subsystem considered in area$_1$ and area$_2$ is given by the expression in Eqs. (1) and (3), respectively, rest is same as case study 1.

Table 2 Gain values for the case study 2

Gain for area 1	Gain for area 2
$k_{11} = 4.4384$	$k_{21} = 0.4807$
$k_{12} = 5.3064$	$k_{22} = 0.2998$
$k_{13} = 3.8133$	$k_{23} = 0.3201$
$k_{14} = -0.8618$	$k_{24} = -0.0865$
$k_{15} = 0.4807$	$k_{25} = 4.4384$
$k_{16} = 0.2998$	$k_{26} = 5.3064$
$k_{17} = 0.3201$	$k_{27} = 3.8133$
$k_{18} = -0.0865$	$k_{28} = -0.8618$
$k_{19} = -7.0892$	$k_{29} = 7.0892$
$k_{110} = 1.0000$	$k_{210} = -0.0000$
$k_{111} = -0.0000$	$k_{211} = 1.0000$

Table 3 Gain values for the case study 3

Gain for area 1	Gain for area 2
$k_{11} = 0.6585$	$k_{21} = -0.0201$
$k_{12} = 1.1557$	$k_{22} = -0.0211$
$k_{13} = 0.2092$	$k_{23} = -0.0033$
$k_{14} = 0.1894$	$k_{24} = 0.1933$
$k_{15} = 0.9355$	$k_{25} = 0.6050$
$k_{16} = 2.3149$	$k_{26} = 12.6538$
$k_{17} = -0.2429$	$k_{27} = 10.3472$
$k_{18} = 0.9775$	$k_{28} = 0.4701$
$k_{19} = 0.7709$	$k_{29} = -0.6370$
$k_{110} = 0.6370$	$k_{210} = 0.7709$

For the considered case, the no. of the state variable is two and control vector is two so, K is of the order (2×10) matrix and its value is tabulated in Table 3.

5 Result and Discussions

This section analyses response of the areas obtained from MATLAB Simulation. The frequency in pu/Hz is plotted in the vertical axis and time in second is taken on the horizontal axis. The value for the governor and turbine is already given in the Sect. 2 and rest of parameters such as area control error speed regulator and integral controller (I) is shown in Table 4. All the values are taken in per unit. The dynamic

Table 4 Parameter values [1]

Parameter Name	Symbols	Values (pu)
Area control error	B_1 and B_2	0.425
Speed regulator (Hz/pu Mw)	R_1 and R_2	2.4
I-controller	k_1 and k_2	0.3

load (d_1, d_2) is considered as step input. The initial value for d_1 is taken as 0 and final value as 0.01 pu in the step time 0 s whereas d_2 is taken as 0.

5.1 Two Area with Thermal Non-reheat System

For the case1, frequency response (Δf_1 and Δf_2) using I-controller and I-SFB controller is shown in Figs. 2 and 3 respectively. The dashed line indicates the response obtained from I-controller and solid line indicates the response obtained from I-SFB. On apply d_1, the system response fluctuates giving rise to maximum understood in $\Delta f_1 = 0.022$ pu and $\Delta f_2 = 0.014$ pu with I-controller. Using I-SFB the understood in Δf_1 reduces from 0.022 pu to 0.012 pu and Δf_2 reduces from

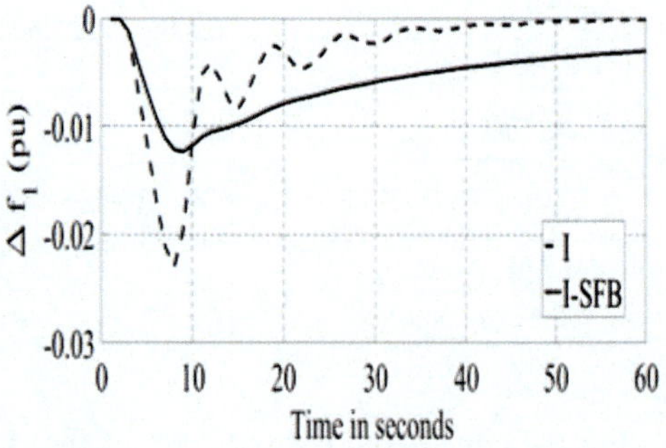

Fig. 2 Frequency response of area 1 with I and I-SFB controller

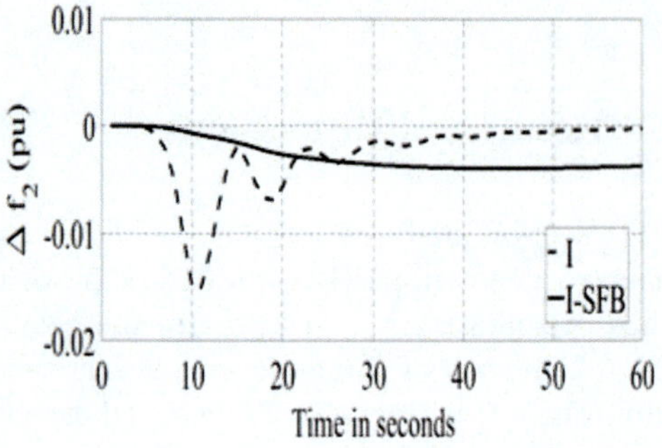

Fig. 3 Frequency response of area 2 with I and I-SFB controller

Table 5 Comparing the result obtained from I and I-SFB controller

Sl. No	Δf (pu/Hz)		I	I-SFB	Reduction
1	Thermal non-reheat system	Area 1	0.022	0.012	0.45
		Area 2	0.014	0.003	0.78
2	Thermal reheat system	Area 1	0.028	0.020	0.28
		Area 2	0.027	0.01	0.62
3	Thermal hydro system	Area 1	0.25	0.02	0.92
		Area 2	0.32	0.012	0.96

0.014 pu to 0.003 pu. The reduction in frequency deviation is 0.45 and 0.78% in $unit_1$ and in $unit_2$, respectively as given in Table 5.

5.2 *Two Area with Thermal Reheat System*

For case 2, frequency response (Δf_1 and Δf_2) using I-controller and I-SFB controller is shown in Figs. 4 and 5 respectively. With the change in d_1, Δf_1 have the maximum undershoot of 0.028 pu which reduces to 0.020 pu using I-SFB and Δf_2 gives the

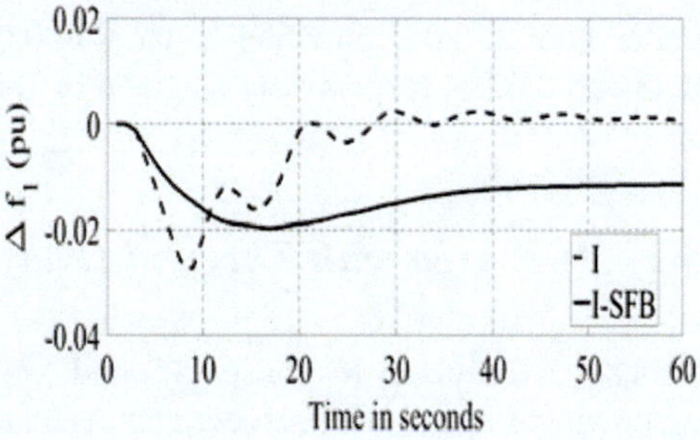

Fig. 4 Frequency response of area 1 with I and I-SFB controller

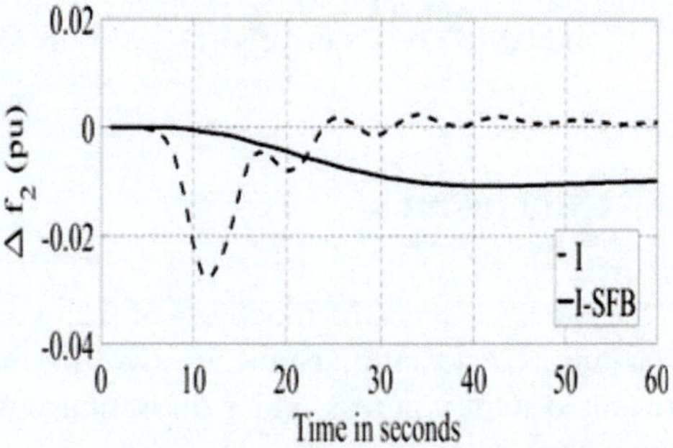

Fig. 5 Frequency response of area 2 with I and I-SFB controller

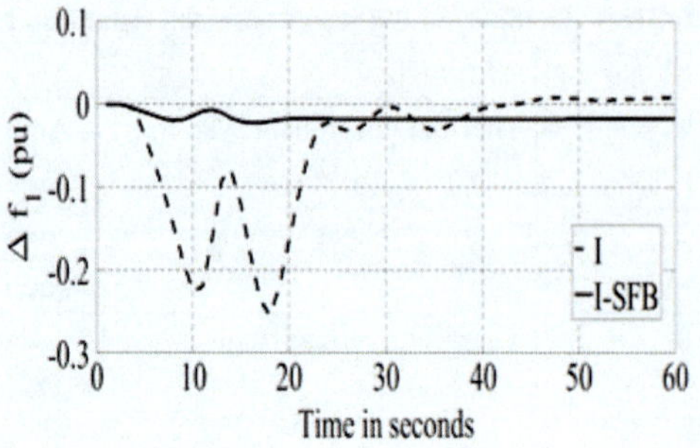

Fig. 6 Frequency response of area 1 with I and I-SFB controller

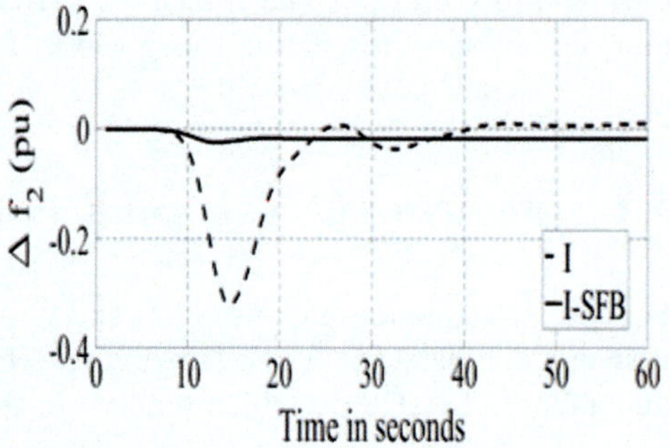

Fig. 7 Frequency response of area 2 with I and I-SFB controller

undershoot of 0.027 pu which reduces to 0.01 pu. The reduction in unit_1 and unit_2 is 0.28 and 0.62%, respectively as given in Table 5.

5.3 *Two Area with Thermal Hydro System*

For case 3, frequency response (Δf_1 and Δf_2) using I- controller and I-SFB controller is shown in Figs. 6 and 7 respectively. With the change in d_1, Δf_1 have the maximum undershoot of 0.25 pu which reduces to 0.02 pu using I-SFB and Δf_2 gives the undershoot of 0.32 pu which reduces to 0.012 pu. The reduction in unit_1 and unit_2 is 0.92 and 0.96%, respectively (as given in Table 5).

6 Conclusions

This paper presents the modeling of the two area system with the similar and dissimilar unit. Under similar unit are two thermal non-reheat systems and two thermal reheat system whereas under dissimilar unit thermal non-reheat and hydro turbine have been taken.

From the comparison of the frequency response of both the unit for different cases, it has been observed that I-FSB given lesser undershoot in comparison to I-controller. However, there is a presence of some steady-state error but the system becomes stable after settling time period.

References

1. Saadat H (2010) Power system analysis, 3rd edn. PSA Publishing
2. Shankar R et al (2017) A comprehensive state of the art literature survey on LFC mechanism for power system. Renew Sustain Energy Rev 76:1185–1207
3. Elgerd OL (1982) Electric energy systems theory; an introduction, 2nd ed, McGraw-Hill Inc
4. Subbaram ND (2002) Optimal control systems. CRC press
5. Barsaiyan P, Purwar S (2010) Comparison of state feedback controller design methods for MIMO systems. 2010 International conference on power, control and embedded systems. IEEE
6. Mariano SJPS et al (2012) A procedure to specify the weighting matrices for an optimal load-frequency controller. Turkish J Electr Eng Comput Sci 20(3):367–379
7. Vinodh KE, Jerome J, Srikanth, K (2014) Algebraic approach for selecting the weighting matrices of linear quadratic regulator. In: 2014 international conference on green computing communication and electrical engineering (ICGCCEE). IEEE
8. Fujinaka T, Omatu S (2001) Pole placement using optimal regulators. IEEJ Trans Electron Inform Syst 121(1):240–245
9. Cheok KC (2002) Simultaneous linear quadratic pole placement (LQPP) control design. IFAC Proc 35(1):289–293
10. Johnson MA, Grimble MJ (1987) Recent trends in linear optimal quadratic multivariable control system design. IEE proceedings D (control theory and applications), vol 134. no 1. IET Digital Library

Author Index

A
Ahmad, B. H., 123
Ahmad, M. R., 123
Anand, B., 59
Archana, 73, 249
Ashour, Amira S., 59

B
Bandyopadhyay, A., 123
Bandyopadhyay, Anirban, 1, 281
Bernardin, Delphine, 33, 133, 197
Bhadani, Rakesh, 85

C
Cerecedo-Núñez, H. H., 263
Chowdhury, Bristy Roy, 111
Chowdhury, Linkon, 111

D
Dey, Nilanjan, 59, 111, 209

E
Eduardo Lugo, J., 33, 133, 197
Esha, Naznin Hossain, 47

F
Faubert, Jocelyn, 33, 133, 197
Fujita, Daisuke, 1

G
Ghosh, Subrata, 1, 281

H
Hassanien, Aboul Ella, 209
Huq, Silvia, 47

I
Islam, Md. Maynul, 99

J
Jagatheesan, K., 59
Jain, Jinendra Kumar, 241
Jain, Sonu, 145
Joshi, Pooja, 221

K
Kaiser, M. Shamim, 47, 99, 289
Kapoor, Gaurav, 153
Khosravy, Mahdi, 59
Kumar, Parveen, 241
Kumar, Rajesh, 59

L
Lugo-Arce, J. E., 263

M
Mahmud, Mufti, 47, 99
Mehta, R. K., 299

K. Ray et al. (eds.), *Proceedings of International Conference on Data Science and Applications*, Lecture Notes in Networks and Systems 148,
https://doi.org/10.1007/978-981-15-7561-7

Mejia-Romero, Sergio, 33, 133, 197
Mithal, Amit, 273
Mittal, Rohit, 273
Monika, 187

N
Nazrul Islam, A. K. M., 289

P
Padilla-Sosa, P., 263
Pathak, Vibhakar, 273
Prakash Roy, Sourabh, 299

R
Rajinikanth, V., 111, 209
Raji Saikot, Ali Mual, 99
Rawat, S., 123
Ray, K., 123
Ray, Kanad, 1, 281
Rodríguez-Méndez, Rosa Ma, 263
Roy, K. C., 85, 221
Roy, Krishna Chandra, 165
Rumman, Israt Jahan, 99

S
Sachin, 73, 249
Sahoo, Pathik, 1, 281
Sahu, Satyajit, 1
Salim Reza, S. M., 99
Saxena, Komal, 1
Shamma, Zarin Subah, 99
Sharma, M. D., 145
Shrivastava, Bhasker, 177
Shrivastava, Vishal, 187
Shubham, 299
Singhania, Anup, 281
Singh, Neha, 165
Singh, P., 123
Singh, Phool, 73, 249
Singh, Pushpendra, 1
Singh, Taniya, 145

T
Tabassum, Fariha, 289
Tasmim, Mst. Rubayat, 47

V
Vijay, S. K., 123

Y
Yadav, Ajay, 145
Yadav, Anil Kumar, 177